WALKING THE BONES OF BRITAIN

www.penguin.co.uk

WALKING THE BONES OF BRITAIN

A 3 Billion Year Journey from the Outer Hebrides to the Thames Estuary

CHRISTOPHER SOMERVILLE

doubleday

TRANSWORLD PUBLISHERS
Penguin Random House, One Embassy Gardens,
8 Viaduct Gardens, London SW11 7BW
www.penguin.co.uk

Transworld is part of the Penguin Random House group of companies
whose addresses can be found at global.penguinrandomhouse.com

First published in Great Britain in 2023 by Doubleday
an imprint of Transworld Publishers

A CIP catalogue record for this book
is available from the British Library.

ISBN 9780857527110

Typeset in 11.5/15.5pt Minion Pro by Jouve (UK), Milton Keynes
Printed and bound in Great Britain by Clays Ltd, Elcograf S.p.A.

The authorized representative in the EEA is Penguin Random House Ireland,
Morrison Chambers, 32 Nassau Street, Dublin D02 YH68.

Penguin Random House is committed to a sustainable future
for our business, our readers and our planet. This book is made
from Forest Stewardship Council® certified paper.

2

For Matilda, Arthur, Sam, Mina, Vincent and Rufus,
hoping that they will enjoy hunting for rocks and
fossils, and trusting that they will look after this
extraordinary, irreplaceable world.

Contents

Timeline: Precambrian to the Present

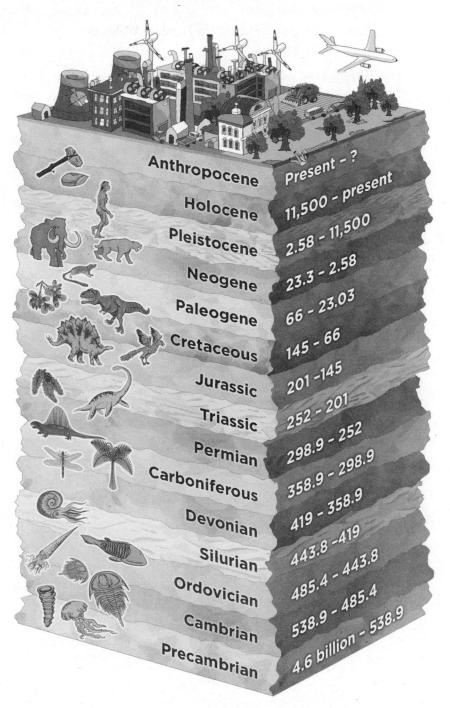

Period	Time (millions of years ago)
Anthropocene	Present – ?
Holocene	11,500 – present
Pleistocene	2.58 – 11,500
Neogene	23.3 – 2.58
Paleogene	66 – 23.03
Cretaceous	145 – 66
Jurassic	201 –145
Triassic	252 – 201
Permian	298.9 – 252
Carboniferous	358.9 – 298.9
Devonian	419 – 358.9
Silurian	443.8 –419
Ordovician	485.4 – 443.8
Cambrian	538.9 – 485.4
Precambrian	4.6 billion – 538.9

The Route, Chapter by Chapter

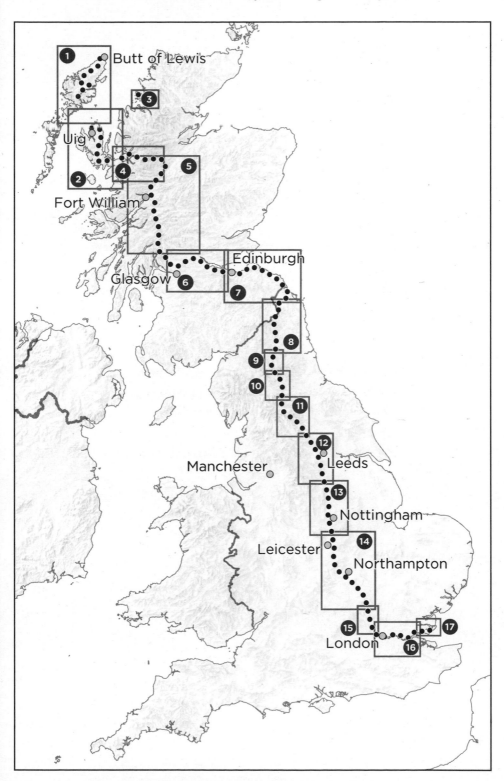

Today, out walking, I considered stones.
It used to be said that I must know each one
on the road by its first name, I was such a dawdler,
such a head-down starer.
I picked up
a chunk of milk-seamed quartz, thumbed off the clay,
let the dry light pervade it and collect,
eliciting shifting gleams, revealing how
the specific strength of a stone fits utterly
into its form and yet reflects the grain
and tendency of the mother-lode, the mass
of a vanished rock-sill tipping one small stone
slightly askew as it weighs upon your palm,
and then I threw it back towards the sun
to thump down on a knoll
where it may move a foot in a thousand years.

Extract from 'Evening Alone at Bunyah' by Les Murray

Introduction

At school I was bored by geology – so, so bored. The drama and colour of the story passed me by completely. I would droop half-asleep at my desk, watching the sun illuminate the hairs in Mr Watson's ears as he stood at the blackboard and droned on and on. Volcanoes held a brief excitement, like a firework. But deposition of sediment in a river system? Andesitic sheets of laminated rhyolites and tuffaceous breccias? Yawn. How boring it must be to be a teacher. Whatever I do when I leave school, it's not going to be that.

I became a teacher. And one lucky day, flicking through a tattered copy of the *Philip's Modern School Atlas*, my glance snagged on the geological map of Britain. The sheer beauty of that little illustration, its swirls of colour that bent in parallel like the ribs of a skeleton, beguiled me. I stole the map. Ripped it out and took it home. *Mea culpa, mea maxima culpa.* There I studied it with guilty pleasure. Scotland's bright pink face was spotted with scarlet like a teenager's, its neck scarfed in a random spatter of colours. The icy-blue shoulders of England gave way to an orderly simplicity further south, where gracefully sinuating stripes of chocolate, lemon, blueberry mauve and mint green all converged on the Dorset coast a dozen miles from where I was standing at that moment.

This crude map for schoolchildren told a tale, the remarkable story of the rocks that lay unseen beneath these islands, the bones of Britain from their first formation 3,000 million years ago all the way through to the present day. That was evident at first glance. And something else struck me as I studied the torn-out map. The oldest rocks were at top left, the youngest at bottom right; northwest to southeast, Outer Hebrides to Thames Estuary. Couldn't one simply take a walk through this story from beginning to end, starting in the Scottish Isles among the hot red volcanics of the ancient rocks, and finishing among the grey clays and moody marshes still being created by the sea along the Essex coast? The story was there, first to be understood, then to be told, if one could unwrap it from its coat of many colours and prise it away from the intimidating technical vocabulary which scared off most sentient beings if they tentatively lifted the stone under which geology lurked.

Couldn't *I* do that? Not now, not with my young family and my mortgage and my increasingly unremarkable teaching career. But one day, maybe, when I had written my way out of the classroom. Yes, and that day pigs would become airborne, I thought as I put the map into a folder marked 'Ideas' and stuck it in a drawer.

There it stayed while teaching gave way to writing. I walked over every corner of Britain, trying to put into words my curiosity and excitement about its hills and valleys, birds and animals, its vivid and violent history, how East Anglian farmers grow their wheat, why Yorkshire lead mines lie up on the moors and Lancashire cotton mills stand down in the dales. I found that everywhere I walked, the bones of Britain kept calling from under the ground. A hand-flaked flint in a Lincolnshire field, bent-up cliffs in Dorset, 1,500 feet of Chiltern chalk, volcanic ash baked hard in Glencoe, Hadrian's Wall surfing the wave of the Whin Sill. This country enjoys, for its size, the most varied geology in the world, and that impinges on the variety of wildlife in any one place, on soils, on human activity from farming to steel-making, on buildings and what they're made of. I came to see that our lives and ways of going about things are profoundly affected by what happened ten thousand, or a million, or a thousand million years ago under the ground we walk on.

I didn't grasp the scale of geology's influence over our daily lives till I had walked and written about these islands for forty years or so. I got the drama, the fiery upheavals below and above ground, the epic floods and mountain-building, the incredible timescales, the fabulous colours

of rocks, a child's excitement at unearthing a fossil. Though most geological phenomena lay tucked away underground and out of sight, I saw
for myself the layers of rock exposed in cliffs and quarry faces, horizontal from bottom to top in chronological order, or mixed and contorted
beyond immediate understanding. But it was like picking at a plate of
starters in the geological restaurant. There was a whole gourmet meal
waiting for me if I could learn how to savour it, if I could get over myself
and look geology and its technicalities squarely in the face.

Various things helped to rob geology of its terrors. The internet, for
one. No need to travel to Berwick-upon-Tweed public library to locate,
unearth and burrow around in the *Proceedings of the Northumberland
Geological Society*, Volume XXXVII, Issue 7 (October 1889), when it
was available (if really wanted) at the click of a keyboard. Likewise, the
technical terminology, so daunting to the uninitiate. 'The Whin Sill's
petrographical affinities are dominantly tholeiitic, but geochemically
transitional between alkaline and tholeiitic' is a sentence that would
have sent me screaming for cover before I found the excellent website
of the BGS (British Geological Survey), which enabled me to interpret
it as follows: 'Thin segments of this rock studied under a specialist
microscope reveal that its volcanic magma was rich in iron and picked
up plenty of magnesium as it formed into the hard grey basalt called
dolerite.'

Another advantage was that I was seeing this stuff every time I went
for a walk. I'd met the Whin Sill in great rock waves and cliffs along
Hadrian's Wall, out at the Farne Islands off the Northumbrian coast,
and down in Upper Teesdale. To know that it had been a vast hot tongue
of viscous magma that rose up from the depths 290 million years ago
and pushed its way sideways between layers of limestone and sandstone, cooling into a wedge of hard dark dolerite rock that still lies there
beneath nearly 2,000 square miles of northern Britain, was a feast for
the imagination. Geology seemed to sneak its way into every conversation I had during my wanderings on foot. I crouched on a narrow ledge
above the River Tees in Upper Teesdale while a patient and friendly

National Nature Reserve manager showed me how the dolerite magma had baked the rocks, among which it so rudely intruded, into crunchy crystals called sugar limestone that met the nutritional needs of the valley's beautiful and rare spring gentians and bird's-eye primroses. That gave geology an immediacy and relevance, and an ecological value.

I bought the BGS's big two-part geological map of Britain and stuck it on the wall. I stood in front of the great multicoloured chart and compared it with the ripped-out page from the *Philip's Modern School Atlas* that I'd retrieved from its folder. Oh Lord, the real thing is more complicated than this little schoolroom map led me to believe. But even more fascinating. On the BGS map I marked all the walks I'd done, on the gneiss and the basalt, the sandstone and limestone, the dolerite and gritstone and greensand. Britain's greatest national treasure, in my view, is the country's network of public footpaths. Without the benefit of these rights of way, open to all and ramifying out, root and branch, for 140,000 miles, I'd hardly have been able to set foot to earth anywhere in these islands. But here like a gift across the geological map were hundreds of walks I'd done and hundreds more I planned to do, travelling over dozens of rock types, all carved and ground up by the Ice Age glaciers and smeared across the landscape as clay and rubble when those great ice behemoths rumbled away into history. What a hell of a story. I had to try to tell it.

Checking with the admirable website of the Long Distance Walkers Association, I puzzled out a succession of long-distance paths running for 1,000 miles with scarcely a break from the Isle of Lewis to the River Thames. For anyone inspired to follow my Bones of Britain route, they go like this:

Scotland. In the Isle of Lewis, from the Butt of Lewis to Callanish along the clifftops, then to Stornoway along the Pentland Road. The Skye Trail from top to bottom of the Isle of Skye. The 'Famous Highland Drove' route from Skye going east to the Great Glen. The Great Glen Way to Fort William, followed by the West Highland Way to the outskirts of Glasgow. East from here along the Forth & Clyde Canal, then the John Muir Way and the North Sea Trail round the coast of southeast Scotland as far as the English border.

England: St Cuthbert's Way from Lindisfarne west to the Cheviots, then the great wild Pennine Way running south along the backbone of Britain as far as Edale in the Peak District. The Limestone Way through the White Peak, the National Forest Way through the regenerated land-scapes of the industrial Midlands. The Midshires Way running south across the Leicestershire Wolds, and from the outskirts of Northamp-ton the Grand Union Canal towpath to the River Thames at Kew. From there a boat ride east through London, shadowed by the Thames Path; a muddle of short paths along the built-up Essex shore of the lower Thames; and finally the Roach Valley Way from Rochford out to the moody shores of Wallasea Island.

I threaded the walks along this route like beads on a string, and I found that with a few sidesteps and detours they crossed a sequence of terrain that followed the geological story, more or less. Most were on public footpaths, but there were some roads to follow, too, some sea passages to cross at the start, and a couple of inland stretches by boat that looked like fun. I wasn't going to try to take in every star in the geological firmament of Britain: that would result in just another add-ition to the bulging library of general books on the 'Geology of Britain'. It might be possible to devise a swerving, snaking course hither and yon across the UK map that included every variant and location of terrain and geology, but how exhausting, long-winded and ultimately boring that would be. I'd be in my grave before I completed it, too. No – I wanted to pick a route it would be possible for ordinary mortals to explore, a journey on which they could see for themselves what shaped the landscape and the lives lived there.

I got out my old notebooks and refreshed my memory and my appe-tite. I waited as impatiently as a dog shut indoors for Covid restrictions to ease. And then, at last, I set off for the Butt of Lewis and the oldest rocks in Britain.

*

Over the course of a couple of years I revisited the places I'd already walked along my Bones of Britain route, picking the walks that best showed up the geology. I found the gaps and filled them in, travelling south and east across Britain, mostly on foot but also in places by boat and bus and car, piecing the story together, talking to geologists and drystone wallers, farmers and canal enthusiasts, coal miners, musicians, local historians and random folk I met along the way. I learned about the fragility of these islands, their unsettled foundations, the never-ending cycles of creation and destruction that are so bound up with their processes. Theirs was a story so dynamic, so immediate, that I knew I'd never see them in the same light again.

At Wallasea Island on the Essex shore the journey of 1,000 miles and 3,000 million years through the bones of Britain came to an end on a bitter winter's day. I'd brought with me the stolen geological map that had kicked the whole adventure off some forty years before. Standing on the flood wall where humans and nature were conspiring to build a new geology, I got out the tattered page and traced the path by which I'd come here across those layers of bright primary colours. It still looked as magical and mysterious as the day I tore it out and took it home with me.

On Garraidh beach, a rock stack of gneiss 3 billion years old.

1

Geology Bombshell: Isle of Lewis

You jump out of your car at the Butt of Lewis, most northerly point of the Outer Hebrides, and a geology bombshell goes off in your face. Not a tweedy lecture or an incomprehensible page of strange words and facts, but a proper gasp moment. You stand by the tall brick stalk of the lighthouse, and the seaward view is of a riot of strikingly coloured cliffs and sea stacks, as naked and contorted as when they first formed in the nascent Earth's crust 2,000 million years ago.

It looks as though a toddler giant has been at work with a box of plasticine, a toddler with a penchant for oranges and greys, whites and blacks. These rocks at the northernmost tip of the Isle of Lewis have been jammed together in layers and thoroughly squeezed out of the horizontal and out of kilter. Directly opposite the lighthouse, bands of orange, grey and brown bend in a tight arch among fellow layers all canted out of the sea at 45 degrees. The neighbouring stack appears to be similarly slanted until you look down near the roots and see how the toddler's fists, pushing from both ends, have caused the rock to crumple and sinuate, each roller-coaster wave outlined in white quartz. To the right a fat stubby stack as treacle brown as newly dug peat has been turned inside out and upside down, the edges bent vertical, the flanks swooping over and down in extravagant arcs. The violence of the contortions, the implication of gigantic heat and force and pressure, are mind-boggling.

This Lewisian gneiss was formed when the world was young. It's twice as old as the oldest forms of life ever discovered in Europe. Yet there are rocks not far away on Lewis, even more sensationally shaped and coloured, that came into being more than 1,000 million years before these sea stacks at the Butt of Lewis.

I sat on the rabbit-nibbled turf in the sunshine and wind, staring. Last time I'd been here it had been in a week of solid rain and mist, conditions that seem to love the Outer Hebrides island chain, and particularly the Isle of Lewis. They had clung to its dour moors, its wide flat bogs and thousands of tiny lochans and skimpy coastal settlements as though reluctant to part company. In those days the lighthouse had still had a keeper living on site, and he'd allowed me to climb the tower and gaze on many miles of rainswept moor and white-capped sea. For all that, back then I hadn't taken any account of the wild and ancient geology on display right under my nose.

It's not only easy but almost obligatory to fall into conversation with anyone you meet in the open air of the Outer Hebrides. Just along the road from the lighthouse I fell in with one of the Morrison clan. 'They're all Morrisons round here,' he said. 'I was born here, and I love the place.' I remarked on the number of elongated fields into which the land either side of the road was divided, narrow strips 6 or 7 strides wide and up to 400 metres long. 'Those are the crofts, and they go with the house, each one belonging to a family. People in the old days would farm sheep and cattle, corn and vegetables on their crofts. Potatoes in the ridge beds – you'll see those down on the cliffs. And peat from the moors. Everyone had their peat bank, and we'd cut them and dry them and use them for heat in the kitchen and in the fireplaces.' Morrison smiled diffidently. 'I have the electricity now, but I still like to have the peat fire. The smell of it, and the warmth and light it gives off.'

I mentioned the cliffs, the banded colours, the aeons of their infinitely slow formation. 'No, no, it is the Higher Power that has created them,' said Morrison. 'Don't you believe in a Higher Power? I used to be an evolutionist, but' – he spread his hands for emphasis – 'no way! Everything in this universe is so finely balanced, don't you think? There has to be a Higher Power that has ordered it.' There was total sincerity in his voice.

A screaming posse of Arctic terns saw me off on my journey, graceful red-beaked flyers with long streamer tails and fishwife voices, apt to divebomb and try to peck your brains out. A squadron of gannets flew south in strict line astern a mile out to sea, their sharp black wing-tips beating regularly. Far out beyond them fifty of their fellows splashed and plunged soundlessly after an invisible shoal of fish.

The fissuring of the ancient rock has given the cliffs a multitude of ledges, crevices and hollows. Guillemots had built tuffet nests out of seaweed and cliff plants. The tubby little seabirds were tucked away in the shadows, but their location was betrayed by a generous whitewash of guano immediately below each nest. A shag with an iridescent green neck squatted above them in a little rock garden of thrift, painstakingly pulling out plants, assessing and then rejecting them. One by one he dropped them down the cliff past the guillemot tenement, until at last he succeeded in pulling out a lump the size of his head with a long root attached. It must have exactly fitted his nest design, for he flew off clumsily, unbalanced by the root dangling from his beak. It swung like a pendulum, causing him to flap his wings frantically.

The cliff path led across short springy turf created by the nibbling of sheep and rabbits, a delight to walk on. It ran round the rim of a deep dark geo or narrow inlet where each successive wave sucked hungrily with an echoing thump and hiss at the converging black rock walls. This west coast of Lewis, open to the full force of Atlantic gales and stormy seas, has few seabirds nesting on its cliffs. The turf may be short and springy, but the peaty soil underneath is meagre, in most places only a couple of inches thick over the acid, impermeable gneiss. The underlying geology forbids flowers to flourish in quantity, trees to grow, food-rich scrub for birds and insects to take hold. But man, forever hunted by necessity, has tried for centuries to wrest a living here. The silent witnesses to the endeavour are the potato drills or ridges, known as lazybeds – what a misnomer! – laboriously scraped together and moulded out of peat, sooty thatch, sand, seaweed, and human and animal excrement. They still lie, grassy green in parallel ridges, sloping down gentle declivities towards the cliff edge where they end at a transverse ditch running the length of the lazybed patches. To lose a cow or a sheep over the cliff was a tragedy for a family's home economy, and anything that could be made by way of a barrier to prevent that mishap was worth a try.

Only the fulmars braved these Atlantic-facing cliffs today, a few pairs tucked in out of the weather, chubby birds with snow-white breasts and prominent 'nostrils' called naricorns that exude a saline dribble to help the bird keep desalinated after its fishing in the salty ocean. Fulmars can spit a foul orange fish oil at anything or anyone they deem a threat. If this vile substance gets on your clothes its 'fragrance' will cling there through many washes; if it gets on the plumage of a predator such as a great black-backed gull it glues the feathers together, crippling and sometimes killing the aggressor. Despite this, the fulmar is a very endearing bird, fond of indulging in neck-to-neck frottage with its partner, and apt to cruise past on stiff wings glancing sideways at you with a round black eye as though weighing you up.

A skylark overhead trilled out its seamless ecstatic song as I walked on round Cunndal Bay, a segment of untouched sand dipping through pale green shallows to clear steely blue water. In the shallows, rafts of black seaweed rippled with each wave, smoothly and heavily, as though oiled. Soft turf led on down to Eoropie Bay and a poignant cliff-top memorial to a fishing tragedy on 5 March 1885. Two boatloads of fishers from the tiny community of Eoropie were lost in a storm here; and the worst aspect of this catastrophe was that it took place close inshore under the eyes of the fishermen's families and their friends and neighbours who had gathered on the shore. The surf and undertow made it impossible to save anyone, even the last man who clung to his wrecked boat for over two hours, looking shoreward and hoping for rescue, before being washed away.

The *Highland News* commented:

> For the most part, our fisher-folk pursue their toilsome and danger-
> ous occupations unthought of beyond their own limited circle. When
> most of us are sound asleep these stalwart fellows are tossing in their
> open crafts on the broad Atlantic, cold and wet, with their lives in
> their hands, earning what is at best but poor fare for those dependent
> on them ... The Ness crews have left nine widows and twenty-two
> children. All these poor creatures are totally without means, a num-
> ber of the families being landless.

The addresses of the drowned men are given on the memorial, along with their names and ages. 'Angus Morrison, 36 Eoropie, age 28; John Murray, 34 Eoropie, age 29; Donald Macleod, 23 Eoropie, age 24; John Macleod, 23 Eoropie, age 29 ...'. Almost all young men in their prime, and all from a community so small and close-knit that no street names were needed. In fact there were no streets at all in the island villages back then; only footways that edged around the close-packed houses. The dark colour of their drystone walls and the smoke-blackened thatch of their roofs gave the dwellings the name of 'blackhouses'.

A few miles down the coast the rough cliff path reached the recon-structed blackhouse village of Garenin. The houses were built close together to block out the brutal island weather, their windows and doors small, their corners rounded to deflect the wind and rain. Each roof, thatched with barley straw or rushes from the stream or marram grass from the dunes, was held down by a chain of woven ropes, each rope kept taut and vertical by an anchoring stone of heavy gneiss. These roofs were not expected to last more than a year. In summer the smoke-pickled thatch would be stripped off, broken up and spread in due course on the crops as fertilizer and as a protective layer against frost. The house, meanwhile, was thatched afresh to combat another year of wind and weather.

The blackhouses of Garenin looked cosy and neighbourly in their tight community. The design did not really change for thousands of years – double walls with straw or turf in the gap for insulation, rounded corners, a communal door for people and animals, and a beaten earth floor with a central fire in the single long room inside. Animals occupied a separate compartment or byre at one end. No house in Garenin boasted a chimney; the smoke from the peat fire in the hearth was left to drift up through the thatch or out of the door.

With cattle and pigs, sheep and hens in the byre, up to twenty adults and children sharing the rest of the building, no privacy and masses of pungent peat smoke, winter must have been particularly 'snug'. You'd have needed a strong constitution to survive it. Coughs, lung complaints and itchy eyes were endemic. What a relief it must have been when the stormy weather finally relented and everyone could spend the day outside, in the fresh air and away from one another. Out above Poll Domhain Meadhonaich their lazybed strips still run down to the cliff edge. Looking around, there seems just about everything at hand for a subsistence life: fresh water from a stream that plunges over the pink and grey cliff, grazing for animals, the outlines of old peat banks further up the hill, and the trace of a hazardous path leading down to the beach and the fishing.

A little further up the coast at Dail Mòr beach, twin cemeteries occupy opposing grassy headlands. Here Macdonalds, Mackeys, Macleans and Macleods lie peaceably in each other's company. Late on this spring evening with a northwest wind pounding the waves onshore under a blue sky, each headstone cast a long eastward shadow. Rimming the dunes at the back of the beach, a litter of striped and faded boulders of gneiss lay around the stream-gully. I walked the soft coarse sand towards the eastern cliffs, and was brought to a standstill by the extraordinary spectacle of the mother rock. This Lewisian gneiss, 3,000 million years old, was formed when the Earth's crust was still fighting itself and being torn by the violent turmoil below. Magma under huge pressure from below pushed up and out, freeing itself to burst through the still solidifying crust of the Earth at a temperature of a couple of hundred degrees centigrade. The seabed stretched, broke and fragmented in epic slow-motion collisions, and the magma twisted and wriggled as it cooled, the minerals from deep in the mantle staining it in delicate colours.

All this showed itself in vivid exhibition at Dail Mòr. A great dyke or finger of black volcanic rock, long since liberated from the gneiss it originally penetrated, gave the impression of having only just shot out of the sea and across the beach. It diminished to a pipe in the sand, the size and shape of a conger eel, then rose up the cliffs as a narrow rock-walled channel quite distinct from the surrounding rock.

Beyond the dyke a litter of boulders lay at the foot of the cliff – boulders strikingly banded, much more vividly than those at the back of the beach, the stripes canted this way or that as chance dictated when they fell from the cliff. As for the cliff, that gave the appearance of a living organism convulsing in a sheet of fire, its normally spaced bands squeezed vertical, others snaking off horizontally towards the sea in pale green tendrils before turning back on themselves – this movement only one of a hundred opposing and colliding streams of solidified magma. Nearer to the sea a detached stack was shot with veins of quartz, of deep orange matrix and of black gneiss that might have been polished marble.

If you wanted a visual introduction to the epic violence of ancient geology, this was it.

The Morrison, Macleod and Murray fishers and crofters of bygone Lewis may have pondered on the origins of these native rocks of theirs, and they may have appreciated their dramatic beauty, but they didn't put such thoughts on paper for posterity. As their successors do today, as did their distant ancestors, they recognized the usefulness of what lay to hand. Heavy, rounded stones of gneiss held down their thatch, ground their corn, anchored their boats and built their blackhouses. Two thousand years ago the inhabitants of Carloway, a few miles along the west coast from Garenin, felt so much oppressed by some unrecorded threat that they built themselves a broch of gneiss, a hollow defensive tower with a winding staircase embedded in the double walls. I'd somehow expected the broch to be no more than a circle of boulders, but arriving there on a wet evening I found it rising from its rock outcrop like a great stony tree, splintered off 10 metres above the ground. The now familiar stripy patterns in the individual stones, green, pink and grey, split along their lines of banding, showed up as though varnished under the polishing effect of the rain. The doorway was hip-high; any attacker would have had to crawl inside, giving the defenders a golden opportunity to remove head from neck.

According to local tales, however, Donal Cam MacAuley of Uig, a one-eyed and hot-tempered warrior, did manage to breach the defences. During one of many episodes of clan fighting and feuding early in the seventeenth century, the Morrisons of Ness made away with the MacAuleys' cattle. Pursued by Donal Cam, his sidekick Big Smith and a posse of MacAuleys, the Morrison rustlers took refuge in Carloway broch. On arrival Donal Cam blocked up the doorway. Then this formidable man scaled the outer wall of the broch, hauling himself up by means of a pair of dirks which he dug in between the stones, and helped by the slight inward batter or slope of the wall. Upon summiting he threw clumps of burning heather into the open top of the broch. The Morrisons inside were either smothered by the smoke or burned to death. Donal Cam ordered the broch to be slighted. The MacAuleys got their cattle back. On they went.

Lewisian weather was reverting to type. It was still raining next day when I came to Callanish. The crowd of tall slim figures that occupied the ridge, coming and going through the murk, took some time to resolve itself into an assemblage of standing stones, and longer still to assume an orderly resolution into a stone circle and associated avenues of stones. In every direction the field walls, sheds and old houses of Callanish were built of the convenient gneiss in boulders, blocks or roughly shaped cobbles. And the stones on the ridge, though erected long before any of these mundane buildings, were carved from the same material, all that was at hand for the job.

The central circle has thirteen stones standing erect around a central monolith about two and a half times the height of a man. It was constructed some time between 2,900 and 2,600 BC, nearly a thousand years before the use of metal was known in the islands. The builders had tools of stone and of animal bone; they could utilize heat and gravity. But what hard repetitive labour it must have taken to shape these monoliths, some with sloping 'shoulders' and protuberant 'heads' that give them the appearance of ancient grey figures, motionless watchers. The masons split them almost exclusively along the lines of the gneiss's bedding planes so that the grey and white bands stand vertical, and that must have helped the work, though a couple are shaped against the grain with angled stripes that contrast bravely with the adjacent stones. The northerly edge of the king stone features streaks and surface coils, as though held still between one movement and the next.

When you consider the perfection of most prehistoric stone alignments the world over, the meticulous calculation of the dimension of the circle, the straightness of the flanking avenues, the orientation with solstice sunrises, the Callanish stones seem endearingly wonky. The circle, centrepiece of the whole creation, is rather too small to show off the height and bulk of its standing stones to best advantage, and it's squashed flat along its east side. The central monolith isn't quite in the centre. It shares the space with a chambered tomb which looks a bit too close to the nearest stones, as if it was squeezed in by a builder's mate who didn't really understand his instructions. The stone rows that march towards the circle aren't quite aligned with the cardinal points of the compass, and they all aim a little off the exact centre of the monument.

These eccentricities notwithstanding, there's something truly compelling about the Callanish stones, especially when cloaked in drizzle on an icy wind. They dominate their ridge, soldierly figures that have endured five millennia of this sort of weather. When they were erected the climate of Lewis was calmer and drier, and the peat had spread no further than the damp hollows of the gneiss. A farming community could make the most of the land in the interior of the island to grow grain and graze their herds. But slowly things got windier and wetter, and the peat and heather moorland crept on upwards and outwards to smother all the interior and drive the people out to the coasts, to hazardous fishing and poorer farming.

Next day I set out along the Pentland Road, a narrow ribbon that straggles 16 miles across the waist of Lewis from Carloway to Stornoway. If there is a more desolate, more barren stretch of land in these islands than the interior moors of Lewis, I've yet to see it. The name of Lewis may have Norse origins, but it's tempting to connect it to the Gaelic *leogach*, which means 'boggy'. This is by far the largest island in the Outer Hebrides, with the highest population. Yet it has very little arable land. Away from the coast, it's the blanket bog known as the Black Moor that takes up the space. It overwhelms the interior, feeding on rainwater and moisture in the air. From the Pentland Road it is blanket bog as far as you can see. And it's all due to the geology.

During the last peak of the Ice Age some 22,000 years ago, extensive sheets of ice covered all but the highest summits of the Outer Hebrides. As they moved out into the Atlantic Ocean they scoured away loose rocks and projections and dug out shallow basins in the gneiss. Over these the ice dragged sheets of glacial till or rubble, glued together with sticky deposits of boulder clay. This was the landscape that emerged around 11,500 years ago as the glaciers receded and the climate began to turn warmer and wetter. Heather, juniper and grass arrived to colonize the thin stony soil, followed by trees – birch and hazel, pine, then oak woodland. By 7,000 BC human settlers were well established, hunting and gathering their food in this forested land. But the climate continued to warm and to moisten. More rain fell, flushing away minerals and salts from the soil. The underlying gneiss is hard acid rock. It doesn't erode, and doesn't drain well. Fallen trees, dying heather and juniper couldn't rot down into good productive soil. The bacterial activity needed to promote such activity could not get going. So the hollows scraped out by the ice filled with peat, organic material that never decayed, but became thicker and more compressed. The peat lined the hollows where, unable to drain away through the impermeable gneiss, rainwater soon formed lochans or stagnant lakelets.

Around five thousand years ago, when the inhabitants of Callanish built their stone monument, there was a period, a brief flash of geological time, when the climate dried off, trees began to return, and agriculture became more viable. The boulder clay exposed by digging through the peat could be teased into fertile soil if mixed with shell sand and seaweed. But the climate soon turned warmer and wetter again. The peat spread in an ever-widening and thickening blanket across the low hills and ridges shaped by the glaciers. The classic Outer Hebridean landscape of 'knock and lochan' came into being – small, rounded hillocks interspersed with lakelets and smothered with heather moorland rooted in a blanket of peat. As for the trees of Lewis: those that escaped the axes of the early farmers couldn't find the nutrients they needed, and their decline continued. What the Mesolithic inhabitants started, Viking settlers finished after their arrival late in the eighth century AD. Within three hundred years the island had been all but denuded of trees. But the peat prospered. These days the average thickness of the peat across the Black Moor is over 2 metres. At its maximum you can dig 6 metres down and still not hit the gneiss that lies beneath.

The houses of Carloway are widely scattered, and once the few trees associated with them had fallen behind me, it was knock and lochan all the way. The Carloway River, a shallow rush of peat-stained water, soon curled away from the road. The Black Moor stretched away on all sides, not flat, but undulating in long drab-coloured crumples, swelling like a heavy brown sea. Spits of rain came across on a cold north wind. The Pentland Road ribboned eastward, rising and falling in front of me between outcrops of gneiss striped in orange and black. Big mountains marched on the southern skyline where Lewis and Harris met on their mysterious, undefined border. But they seemed a world away from the low and sodden moor I had entered.

The saga of building the Pentland Road stands for all the problems and failures encountered down the years in Lewis by 'improvers'. By the 1890s a decent road already existed a little way south. What was really wanted was a railway to bring the newly established Loch Roag fisheries' catch from nearby Carloway, with its natural harbour and good fishing grounds, across the island to Stornoway. But no one in the island was prepared to subscribe towards the cost of a railway. A new road was the next best thing, and would shave 7 miles off the journey – an important saving of time when dealing with fish, considering their rapid rate of decomposition.

Work started in 1894, but the local engineer had underestimated the amount of work needed. He and two contractors were dismissed. 'This work has been unsatisfactory from the very beginning,' scolded the 1898 Report of the Congested Districts Board for Scotland. The original grant of £15,000 from the Scottish Office was nowhere near enough; £13,000 of it had been spent for no real results by the time of the report. And it took another sixteen years for the road to be fully ready. Opened on 6 September 1912, it was named after John Sinclair, Lord Pentland, Secretary of State for Scotland. 'A gold-lettered day,' announced the *Highland News*. But Neil M. Macleod, writing in the same paper only three weeks later, grumbled, 'For the development of fishing in the west of Lewis the Pentland Road, when all is said and done, is little better than a cart without a horse.' And as things have turned out, rather than boosting the fishing industry of Lewis, the main benefit of the Pentland Road has been a more modest one: the transport of cut and dried peats from the moor to domestic hearths around Carloway.

The Black Moor of Lewis is a sullen landscape, for sure, but a compelling one. You cannot imagine ever getting on terms with such a monotony of blanket bog, of strolling or picnicking here in the same way as on chalk downland or the limestone dales. There's nothing pretty or comforting about the Black Moor. Impressions of this landscape are directly attuned to the weather, generally wet and often challenging. But it is massively impressive all the same. Its peat can be dug for fuel, its lochans fished. But it can't be tamed for agriculture. Its sheer extent, its sogginess and sourness, are indefeasible. It possesses what could be described as strength of character. Like the island itself, the Black Moor has rejected the best endeavours of a succession of well-meaning outsiders to turn it to account.

In 1844 Sir James Matheson, an East India merchant who'd dabbled in the Chinese opium trade and grown rich, bought the Isle of Lewis from the Mackenzie family for £190,000. He built the opulent Lews castle at Stornoway, cleared the land for his park by paying for about five hundred residents to emigrate to Canada, and imported soil for growing trees and shrubs. Those were desperate years, with a disastrous potato blight hitting Scotland in 1846 and lasting on and off for the next decade. The potato was the chief source of nourishment in western Scotland, and over half the population of Lewis became reliant on the Destitution Fund, largely made up of contributions from Sir James and Lady Matheson. In the first winter of the potato famine the Mathesons spent £40,000 on flour for their poverty-stricken tenants. There were no sheep evictions in Lewis, unlike most of the other isles. Sir James spent altogether about a quarter of a million pounds on improvement schemes in the island, including establishing a lobster fishery, building new harbours and roads, draining wetlands and reclaiming peat moorland for agriculture. He seems on the whole to have done his best for the island, including paying the passages of 1,800 islanders to Canada when times were at their toughest. But the islanders, independent in mind and inclination if not financially, weren't over-enthusiastic, and the death of Sir James in 1878 was not mourned with universal wailing in Lewis.

Another essentially benevolent despot and 'improver' was Lord Leverhulme, the 'Sunlight Soap Baron', who bought Lewis in 1918. He, too, had grand designs for the island, including turning Stornoway into a major fishing port with fish-processing and canning factories and an ice plant. New harbours were to be founded at the end of new roads. And Lewis-caught fish was to be sold all over the UK through the nationwide chain of Mac Fisheries fish shops – the one solid gain, as it turned out, from all this planning. Leverhulme aimed to establish dairy farms, encourage the weaving industry and connect up remote parts of the island with a road-building scheme. What he didn't want his tenants doing, and did his damnedest to discourage, was crofting, the traditional part-time husbandry of a small piece of land. As a tremendously successful industrialist, Leverhulme saw crofting as an inefficient industry, distracting people from the work they should be concentrating on. He was right about the viability of crofting as a full-time occupation, especially on land as poor as that in Lewis. But experience told the islanders that the croft was a necessary plank in their domestic economy, something to fall back on when times were hard.

Lord Leverhulme badly misread the mood among many of the islanders. He failed to appreciate the cultural and emotional hold of crofting on the community of Lewis. The islanders were cautious about adopting new trades and new ways, and they proved reluctant to dance to a rich man's tune. And it was the new laird's bad luck to be riding his hobby horses at the end of the First World War. The social upheavals brought about by ordinary folks' service in the war, during which Jock felt he had proved himself as good as his master, reached as far as the Isle of Lewis. Returning servicemen didn't want to work for and be beholden to Lord Leverhulme. Instead, they believed that the government had promised to allot them land for crofting. When they found this was not forthcoming, some of the ex-servicemen began a series of 'land raids', seizing farms that Leverhulme had created by amalgamating crofts, and redistributing the land to former crofters.

The tide of history, and the mood of the Scottish Office and of Parliament, were against Lord Leverhulme. In 1922, after a four-year stand-off, the Board of Agriculture stepped in and compulsorily purchased enough tenanted farmland to provide a hundred crofts. The following year Leverhulme, stymied in all his schemes, decided to quit Lewis. In a final act of magnanimity he offered the islanders as a gift the freehold of their crofts. Some individuals accepted, but Lewis District Council, excessively cautious and thinking the sums would never add up, rejected the opportunity. Lord Leverhulme was left with no alternative but to sell to whoever would buy. No one wanted to possess a large remote island of peat bog and poor land with sitting crofter tenants, so Lewis was sold off piecemeal as small estates to mostly absentee landlords. No one really benefited, in the end, from all those good intentions.

Thirty years had passed since I last trod the Pentland Road. Back then there was no tarmac underfoot. It was a rough track that crossed the Black Moor, stony and hard on the feet. Rain had lashed at me all day, soaking through every stitch I wore. It was the Sabbath in this stern stronghold of the Free Presbyterian Church, a day for furtive boozing in the corrugated huts of peat cutters along the way. I heard the clink of bottles and the phlegmy chuckling of smokers as I squelched by. The pages of my notebook for that day are crisp and yellow now, smeared with peat and blurred with rain, but they still set the scene before me:

A day of continuous drizzle, mist and wind. A landscape scarred and riddled with peat workings. Banks, ramparts, pools. The plastic bags of the peat cutters – white, blue, green – the only colour in the dreich rainy landscape. Empty carry-out cans line the roadside ditches.

An exposed place, the low-lying shape of the land allowing free play to wind and rain. The blue and red tin shacks of the peat cutters, each with its single black chimney and whiff of peat reek, have iron rings let in above the windows, through which cables are passed to anchor the hut to its rock foundation. A stiff windstorm can bowl an unsecured hut clean over.

I step aside on to the moor and the peat quivers beneath me. Sphagnum loose underfoot, gurgle of water settling as my weight transfers to next tuffet. Back to the Pentland Road. An old bridge of rusty box girders, only 5 yards long, with neatly shaped and curved abutments, allows a little brown burn to go under the road. A lorry full of stones trundles across and the whole structure trembles, the bog on either side shivering from the shock waves like jelly in an earthquake.

Four shepherds drive their flock across the moor, the sheep seething like a pack of white maggots, while the black-and-white dots of their seven collies whirl round the flock. All the shepherds in working blue overalls, carrying big crooks, as they corral the sheep towards a pen for shearing. Each dog crouches, quivering with tension, tongue out, head held low, then at a whistle from its master races away, belly to the heather, bounding over pools and tussocks, sending up a splatter of peat particles as it heads off escapers. Energy on tap. Skylarks and golden plover the only other animated beings in this utterly drab and downbeat landscape.

Halfway along the Pentland Road large mountains began to loom again to the south through the moisture-laden air. On the opposite side of the road a long, gently sloping bank of peat rose away. I trudged up to where the view opened northward to a skyline barred with dark cloud that trailed a ragged hem of rainfall. Out there stood a black dome that the map named 'Stacashal', a low hummock of a hill, but at 700 feet the highest point on the moor for many miles. I took aim and set off among soft tuffets of sphagnum and deep channels of sloppy peat. From the cotton grass larks ascended singing over Loch nan Geadh, little slivers of joy in a bleak place.

The back of Stacashal was a maze of peat hags 7 and 8 feet tall. I stumbled and slithered my way up between them. The northward view from the summit was hidden till the last moment. Then the badlands of Lewis lay suddenly open, a glimpse of sandy machair on the west coast perhaps 10 miles off, ruined shielings or summer bothies of ancient stone dotted across the vast plain, a thousand watery eyes of peat pools winking. A prospect of the Black Moor stripped to its basic elements of bog and lochan, bog and lochan, all the way out to the Butt of Lewis 20 miles away.

Ten miles north of Stornoway on the east coast of Lewis, the circle of the forgotten improvers' roads is closed. Here from Tolsta Bay the coast road, already narrow and wriggly, crosses a handsome concrete bridge and winds on north for a mile or so before petering out, as if in recognition of its own futility. It was one of Lord Leverhulme's pet projects, intended to connect the northern port of Ness with Stornoway for the rapid transport of fish to the processing plants at the capital town. But it fell by the wayside, along with Leverhulme's other schemes of improvement, when he sold the island in 1923 and removed to Harris to lick his wounds. Now a rough coast path meanders the 10 miles north to Ness, a ghost road that sketches out what might have been.

Tomorrow I would be leaving Lewis, crossing the Minch on a ferry bound for the Isle of Skye. Today was for a final immersion in the ancient geology of the island, on Garraidh Beach at the foot of the road that never was. As I clicked the gate by the little loch above the beach, a delegation of sheep came to see if I had sandwiches – one subdued-looking ewe, and four curly-horned rams intent on pushing the fence over by leaning on it with their muscular thighs.

Once past the welcoming committee, I found a perfect semicircle of untrodden sand between dark cliffs, with the sharp profiles of stacks as tall as churches standing out on the sands. That was one's first impression of the rocks. Down in the stream at the back of the beach lay a tumble of sea-sculpted boulders of gneiss, some squared, others rounded, all shot through with those telltale bands of light and dark. Not just two shades of grey, though. Here were salmon pink, white and a veneer of green so intense and milky it looked like gloss paint.

The sand was table-smooth, though the individual grains quarried by the sea from the gneiss were large and glassy, streams of them hastening seaward, blown by the wind, others settling with dark peat particles in ribs along the stream bed, an object lesson in the fluid dynamics of deposition.

As for the cliffs and sea stacks – their true colours and shapes only revealed themselves to close inspection. At the base they were black from sea-washing and thickly furnished with barnacles, but above head height they wore coats of many colours. These rocks emerged from the Earth's crust as it was still cooling and hardening 3,000 million years ago, metamorphosed by heat and pressure, squeezed against other nascent rocks, crystallized and hardened. There were bands as black and shiny as freshly cut coal, shot with molten-looking streaks of fiery orange and enamel green. The face of one blank stack had cracked and a big fragment fallen away, like a square window opening to expose a block of multifaceted quartz, a creamy orange-grey.

The sea stacks were tremendously jagged, looking as though they had been jammed together as shards in white heat by some impatient smith. Looking at these colours and fluid shapes, cooking words came to mind – boiled, baked, roasted, melted. They felt as cold as glass under my trailing fingers. No fossils, no sediments, just the mother rock, now teeming with sea life – brilliant emerald-green leaf worms, bunches of mussels, mats of algae, millions of barnacles, seaweed and anemones. I stood and stared my fill as fulmars planed across the sky and the sea washed round my feet.

Incredibly, though, rocks older still were lying not far away. 'I can tell you where the oldest dated sample on the Outer Hebrides was collected,' Dr Kathryn Goodenough, Principal Geologist at the British Geological Survey, had advised me, 'and that is in North Harris, near Hushinish. The sample was collected at roughly NA990124, just by the jetty at Hushinish . . . Some of the zircon crystals in the rock had cores (the oldest parts) that were dated at 3125 mya [million years ago].'

Before catching the Skye ferry at Tarbert, down in North Harris, I drove a very long and winding road to reach the tiny landing place of Hushinish. From the end of the road a sandy track led on across the machair and down to a concrete jetty that sloped into the calm waters of the Strait of Scarp. Four men were unpacking a rubber rib at the seaward end. Their collie dog pattered between them and the stranger, uncertain which to herd.

It seemed perfectly obvious which were the oldest rocks. They lay just to the east of the jetty in heaps, spectacularly striped and coloured, glittering with crystals, eye-catching and beautiful. I scrambled down and filled my camera. But something nagged at the corner of my mind. Kathryn Goodenough had mentioned cores being taken, and here there was an absence of boreholes. Rugged Scarp island, boats and the sparkling evening sea made an alluring picture, but still I had the feeling I was settling for the spectacular rather than the true.

It wasn't till I had turned for home that I glanced over the rail on the west side of the jetty and spotted the hole in a rock down there, a modest cylinder a couple of inches across. I slithered down and saw a dozen or so neat holes in a cluster of rocks, humped and smooth, their surfaces eroded by the aeons, but still distinctive with their bands and swirls careering seaward. I patted and stroked their rough surfaces. Here they were, the oldest rocks in all the isles, survivors through 3,125 million years of this planet's existence.

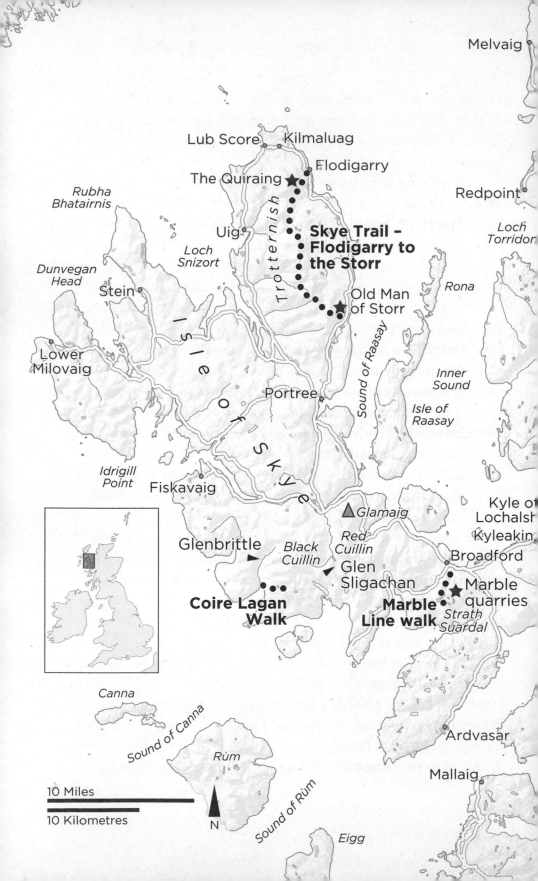

2

Purple, Red and Black: Isle of Skye

THE FERRY TROD A SMOOTH if not a speedy path across the sea to Skye. This is the way to approach, crossing the Minch from the Outer Isles with the harbour of Uig on Skye's craggy north coast gradually growing nearer. There's something fundamentally disappointing about arriving in Skye the other way by driving over the bridge from the mainland at Kyle of Lochalsh. The sea loch is so narrow that Skye hardly looks like a separate island. Once on the bridge, you're there in half a minute. And there's that pesky earworm about a bonny boat speeding over the sea to Skye that diminishes the impact of the real island by reinforcing the notion of a remote isle far out in the ocean.

The Minch, the sea channel between the Outer and Inner Hebrides, is not always as benign as it was today. 'White the Minch is,' goes the Mingulay Boat Song, and when I last crossed it in a summer gale it was white with driven spume and extremely lumpy, tossing the ferry around like a bully. This spring morning it deposited me as gently as a kindly hand on the jetty at Uig. I leaned on the rail and pondered what the British Geological Survey had to say about the Isle of Skye. Its rocks were younger by some 2,700 million years than those I'd left on Lewis earlier that day. It's a hell of a jump to make through deep time from those to these. But age was not the only criterion on this journey through the bones of Britain. On previous visits to Skye I'd been more interested in sampling malt whiskies and trying to play the bagpipes.

Now I was going to be immersed in the story of the younger rocks that built this island's remarkable landscape, at first with trillions of light-as-a-feather depositions over millions of years, then in sudden fire and brimstone.

During the Permian (298 mya–252 mya; mya = million years ago) and Triassic (252 mya–201 mya) eras, what is now Scotland sat in the centre of a large continental mass, hot and dry. The Minch was a broad shallow basin into which rivers flowed from higher ground, depositing masses of sand and pebbles from older eroding mountains round its rim. At the start of the Jurassic (201 mya–145 mya), the Minch basin was flooded by shallow seas. Thick layers of sediment – mudstone, sandstone and limestone – built up on its bed. Corals grew there; oysters embedded themselves; ammonites with spiral shells and bullet-shaped belemnites multiplied. Plesiosaurs as long as minke whales swam and hunted in the warm tropical waters.

The oldest rocks on Skye's Trotternish peninsula, the most northerly projection of the island, date from this time of slow submarine accretion. Driving from Uig towards the tip of the peninsula, I passed the rocky shore of Lub Score and crossed the level plain of Kilmaluag, underpinned by these sedimentary layers. The proper drama, however, was waiting just beyond the turn of the road at Flodigarry. The first full sight of the 700-foot curtain of purple-black basalt that hangs in folds from the spine of the peninsula jump-started my imagination forward to a fiery time, the last of Britain's major volcanic cataclysms.

Sixty million years ago the tectonic dance entered a new phase, rotating and fracturing old partnerships. The lands that had been jammed together for nearly 300 million years to form the supercontinent of Pangaea were in the final phase of their break-up. Was this convulsive movement caused by an upward gush of hot matter from deep below the Mid-Atlantic Ridge, the boundary between the Eurasian and North American tectonic plates? Or were the plates moving apart anyway, allowing shallow 'melt' to push its way between rocks already stretched thin and fragile? Geologists, as ever, agree to disagree. In any case, the land masses that would become our current continents of Africa, the Americas and Europe were easing apart, and between the latter two a new ocean, the North Atlantic, began to open and expand. Sea levels fell, and land appeared that was roughly the shape of the British Isles we know today.

The crust of Scotland stretched and thinned. Magma from the depths forced its way up and out of the weakest spots in blasts of volcanic activity. But much of this melted rock never reached the surface. Instead, it squeezed its way between layers of Jurassic rock, shrinking as it cooled to form thick sills of the igneous rock called basalt, some of it crystallizing into tall hexagonal columns. It is this basalt that forms the escarpment or cliff, fractured and fallen in many places, that runs 20 miles south along the length of Trotternish. The older sedimentary rock that once sandwiched the basalt has gone, eroded clean away, leaving one of geology's most celebrated *tours de force* naked to the eye.

I parked the car and found the squelching track that led past Loch Langaig with its peaty shore. On up to pass Loch Hasco, cradled by the stone buttresses of Leac nan Fionn – the Tomb of Fingal, a.k.a. Fionn mac Cumhaill, a suitably high and mighty resting place for the mythical Celtic warrior and strongman. Beyond loomed the Quiraing, an extraordinary jumble of pinnacles, free-standing stacks and boulder slides, its interior hollows dark with shadows and wraiths of mist. Landslips have detached these big, jagged outliers, 50 feet tall or more, from their parent cliffs, some to go tumbling in fragments, others to remain erect. The flat-topped Table, a chunk slipped away from the summit plateau; the twisty spire of the Needle; the turrets and ramparts of the bulky Prison. The curious springy name of Quiraing derives from the Norse *kvi rang* or 'round fold'. The Vikings knew the place well; it was here that the islanders would drive their cattle as soon as they sighted the longships, corralling them among the rocks in the hope that the invaders would pass them by. The Quiraing is the only place along the escarpment still occasionally in motion – witness the frequent repairs to the road below as miniature earth movements twist it out of place.

I sidled with the mist among the stacks and spires, pressing my palms to the basalt. The last rocks I had caressed in this way, only yesterday, had been those on Garraidh Beach, smooth, varicoloured sea stacks of ice-cold Lewisian gneiss 3,000 million years old. This purple basalt of Trotternish, as rough as sharkskin and one-fiftieth of their age, felt curiously warm to the touch, as though a flicker of its fiery origins remained.

For drivers on the coast road, the Trotternish escarpment forms a breathtaking backdrop just inland, close by below the Quiraing, then swinging 3 or 4 miles away before curving back to dominate the road near the free-standing pinnacle of the Old Man of Storr. Walkers, however, can scramble up a rocky path behind Leac nan Fionn to reach the top of the cliffs. From here a high ridge path, the Skye Trail, follows the basalt escarpment all the way south to Skye's capital town of Portree at the foot of the peninsula. Sometimes just below the cliffs, sometimes on top of them, it's a tricky walk of 18 miles along the roof of Trotternish, with views unrolling over the backward-sloping cliffs of the escarpment to a spatter of lochs and the sea coast.

Scrambling up towards the Old Man of Storr I raised my eyes from these captivating nearer prospects to see, 20 miles off in the south, the dark jagged profile of the Black Cuillin mountains. They look like, and in fact they are, the roots of a volcano that erupted around 65 million years ago, part of the upheavals that attended the final break-up of Pangaea and the opening of the North Atlantic Ocean. They are built of layer upon layer of gabbro, a coarse crystalline cousin of basalt that cooled deep in the Earth's crust. Under the jagged surface effects of weathering and erosion, the Black Cuillin stand humped like a pile of saucers stacked upside down. They are shot across with hundreds of close-packed dykes, rapidly cooled tongues of basalt. On the geological map of the southern half of Skye these dykes show up as a dense crowd of short purple streaks like tracer bullets, all shooting southeast. There's something a little disturbing in the sight, their sheer number and uniformity of direction across the map. They give the appearance of being in rapid motion, as intent and purposeful as a swarm of locusts. And that's the collective noun by which geologists know them – a swarm of dykes. Around the Black Cuillin the swarm loses its directional discipline and goes crazy, the individual dykes elongating, bowing and bending, fouling each other's passage and turning back on themselves as though trying to burst outwards like a Catherine wheel. The gabbro, younger than the surrounding rocks, was relatively cool and hard, more resistant to intrusion, and the basalt had to fight and twist to lodge itself in such cracks and weaknesses as it could find.

Another range of mountains, the Red Cuillin, is separated from the Black Cuillin by the flat-bottomed valley of Glen Sligachan. Younger and less grim in aspect than their neighbours, the Red Cuillin are formed of granite, yet another coarse and crystalline rock. They are lighter in colour than the Black Cuillin because their granite contains more quartz and pink alkali feldspar crystals than the gabbro, and they are less formidable-looking, more rounded, thanks to the greater propensity of granite to yield to the scraping and shaping of the glaciers when the succession of big freezes began around 2.5 million years ago. Only the peaks of the Black Cuillin stood above the ice dome that smothered the hills. Only the gabbro resisted the smoothing and polishing of the immense rivers of ice.

Everything about the Black Cuillin sets them apart – not just the notorious difficulty of climbing their long saw-backed ridge, but their geographical isolation in their own ragged segment of southern Skye, the harsh volcanic gabbro rock which makes them so spiky and unyielding, and their lordly power and presence on the skyline. 'Lordly' might be an extravagant label for a mountain range, but there is something demanding of respect about these black teeth that for 65 million years have been bared against ice, earthquake, storm and landslip, and still defy the little climbing figures of men and women.

*

I had to be realistic about my chances of traversing the Black Cuillin ridge without breaking my neck. They were nil. All the same, I had a walk in mind, a path up from sea level into the very heart of the range. But I woke to a washout day. The Mountain Weather Information Service had it in with a vengeance for the Black Cuillin: westerly gales, persistent rain mixed with hail, cloud shrouding the mountains all day. Tomorrow I was due to leave Skye for the mainland, and tomorrow in the Black Cuillin, according to the MWIS, looked even worse. Those mountains were tall, though, with most of the central massif over 3,000 feet. They might catch the worst of the weather before it could cross them. There could be a nice, sheltered rain shadow beyond them, around Broadford on the east coast, couldn't there? 'Might' and 'could' – better than 'can't', anyway. And hadn't I read something about Broadford marble, too?

The uses of Skye's varied rocks have been many. Some of them are crunched up small and spread on the roads, like the durable 200-million-year-old sandstone from Sconser, between Portree and Broadford. Diatomite, a kind of china clay, was dug out of the cliffs at Inver Tote on the east coast of Trotternish early last century for brick-making and pottery glazes. Sand and gravel from Kyleakin, a tiny ferry port opposite the mainland near Kyle of Lochalsh, go to the building and road-construction industries. But marble? It brought visions of Michelangelo's David in ice-smooth Carrara marble, of languid Victorian sylphs recumbent in purest white. What an exotic thought.

About 450 million years ago, thick layers of carbonate sediments began to settle on the bed of the shallow tropical sea that covered these regions. These sediments compressed and solidified into dolostone, a harder form of limestone containing magnesium, which today outcrops along the hillsides of Strath Suardal behind Broadford. When the major volcanic upheaval that formed the Cuillin mountains took place 65 million years ago, magma chambers in the Earth's crust radiated such intense heat that the Red Cuillin granite, in contact with the dolostone, metamorphosed it to form the bands of marble that lie along the eastern flank of Strath Suardal.

Specific details of the uses of Skye marble are hard to come by, but it certainly built a fabulous grand staircase at Armadale Castle, seat of the Macdonalds of Sleat. Blocks of it were shipped abroad, and there is a belief that it was used to decorate the Vatican. Small-scale marble quarrying went on for centuries in Strath Suardal before a concerted effort was made early in the twentieth century to extract it on an industrial scale. A light railway was built from Broadford pier to the quarries, along with workshops for cutting and polishing the marble. But Strath Suardal marble didn't add up as a commercial venture. The infrastructure was too limited, the quarries were soon worked out, and there was a grittiness in the marble's consistency that rendered it suitable for the prosaic manufacture of millstones, but also made it hard to cut with precision for more delicate work. By 1913 the quarries were finished and the marble railway closed. All that's left of it these days is the Marble Line footpath, the trackbed of the light railway, an easy route that snakes on a gentle incline up the strath to the old quarries in the hillside.

Just south of Broadford I joined the Marble Line. Part stony and part grassy, it swung along above the curlicues of the Broadford River. West across the strath rose Beinn na Caillich, an outlier of the Red Cuillin, a smooth conical shape, its rampart curving to the south around the deep hanging valley of Coire Gorm. I scanned the skyline beyond for a sight of a jagged ridge with the notches, pinnacles and angles of the Black Cuillin, but they stood invisible under a thick grey cap of raincloud which the southwesterly gale had drawn over the gabbro. It was a lazy wind; it couldn't be bothered to go round the mountains, but came straight through both the Black and Red Cuillin to smack me in the face.

Native woodland of birch, oak, willow and hawthorn had recently been planted along the lower slopes of Strath Suardal, fifty thousand trees to reclothe the valley. Big, rounded lumps of dolostone outcropped near the line, a grey abrasive karst with deep hollows and cracks like the limestone pavements of Yorkshire and the Burren in County Clare. The young trees were growing well on the dolostone, with plenty of lime in the rock to counteract the eternal peat and heather that swathe the upper areas of the glen.

The scattered white houses of Suardal fell behind as the old railway gradually gained height to reach the lower of the two marble quarries. Spoil heaps lay beside the track, along with the square foundations of the former workshops. Otherwise, it was hard to distinguish what had happened here not much more than a hundred years ago. Looking more closely at the rock I saw shock ripples crowded together, twisting into arches and flows, bunching like paper plates that had been crumpled together: a graphic illustration of pressure and power as the molten granitic magma baked the limestone into marble. Not that you would guess the presence of that handsome rock under the blackened weathering of the modest quarry walls. However, despite a notice begging amateur geologists not to hammer the rocks, someone had recently done exactly that. The flake they chipped off had exposed a pure white surface, finely lined with dark grey veins, lying immediately beneath the thin skin of oxidation. A beautiful clean stone. No wonder they went to all that trouble to get at it.

Now the track steepened, turning rougher and passing through more young woodland, its surface covered in chunks of marble. Up at the top of the incline a circle of stone and concrete with three rusted bolts in the centre showed the location of the winch that let the heavy-laden wooden trucks down from the upper diggings to where the quarry's little saddle-tank locomotive *Skylark* waited to take them to the pier at Broadford. Here at the upper level squared-off banks of spoil surrounded a much more extensive quarry-delving, the blackened rock faces hiding their white treasure.

On a lump of dolostone on the hillside above I sat and ate an orange. My picnic rock stood at the edge of a sea of wild flowers – alpine lady's mantle, wood anemone, primroses and celandines, early purple orchids, violets and milkwort, windflowers and creamy-white mountain avens. From this eminence I stared west as the clouds shredded away to reveal the toothed spire of Blà Bheinn, the 'Black Mountain', outlier and herald of the Black Cuillin ridge. Over there tomorrow was going to be a better day. The horsetail streaks in the sky said so, even if the Mountain Weather Information Service wasn't here to read them.

Down in the bottom of the strath a side path was signposted 'Coire Chatachan'. Looking across the river I could see the white farmhouse of that name, but it was impossible to pick out the ruin of the dwelling where Dr Samuel Johnson and James Boswell arrived, very late at night on Saturday, 25 September 1773. The two men of letters, the caustic lexicographer and moralist Johnson and his perma-cheerful companion Boswell, were embarked on their great Hebridean Tour. I had brought their combined accounts of that jolting journey, Johnson's *A Journey to the Western Islands of Scotland* and Boswell's *The Journal of a Tour to the Hebrides*, to Skye with me, and was relishing Johnson's know-all brusqueness and Boswell's endless flattery and appeasement of his crusty mentor.

At Coire Chatachan the two were guests of Mr McKinnon, better known as the Laird of Corrichatachin, or 'Old Corrie', and his wife. The travellers found themselves weather-bound for three days in this very cheerful, not to say boozy, house, where life during their stay became one continuous ceilidh. By the time they left, the following Tuesday, Boswell had suffered a succession of cruel hangovers as the result of drinking an enormous amount of whisky punch at the instigation of Old Corrie and friends. As for the staid and portentous Dr Johnson, he had mellowed to the point of flirting with one of the ladies of the house. Boswell delighted in reporting how this . . .

. . . lively pretty little woman good-humouredly sat down upon Dr Johnson's knee, and, being encouraged by some of the company, put her hand round his neck, and kissed him. 'Do it again,' said he, 'and let us see who will tire first.' He kept her on his knee some time, while he and she drank tea. He was now like a *buck* indeed.

Next day, of course, was absolutely fine, with fair weather over the Cuillin Hills and a fresh breeze zipping in from the sea to blow me up the mountains. I crossed the foot of Glen Sligachan and headed south down the length of Glenbrittle's narrow road. The saw-toothed gabbro ridge of the Black Cuillin rose ever higher and more forbidding on the east flank of the glen. Down at the wide beach I parked on a bumpy mud patch, the sole parking space left on this afternoon in early May, and started up the broad green hillside towards those mesmeric black castles in the air. A sunny afternoon was spread over the coast of Skye and the offshore islands with their distinctive shapes – low-backed Canna, mountainous Rùm, Eigg with its great sharp basalt prow. But as the sunlight ascended the slope towards the mountains, it seemed to lose energy and courage like an unfit hiker. Before it could touch the feet of the mountains it had dwindled, faded and disappeared. The Black Cuillin stood unilluminated, dark and forbidding in their own cloud shadow, the only entities not lit up by the sunshine.

The path rose steadily, a rocky track well pitched and drained, as befitted one of the best-known scenic walks of Skye. Heather and boggy turf stretched away on all sides, knobbled with outcrops of gabbro. Grass and heather gleamed with threads of water as the overloaded bog relieved itself before the next cloudburst should top it up again. I plodded on uphill. Violets sheltered with milkwort under rocks. A solitary butterwort flower of deep purple had emerged at the end of its slender 4-inch stem, but the lime-green leaves of dozens of the plants lay nearby, waiting to ingest unwary insects. Curlews bubbled and skylarks rose, twittering melodically.

Up in the cloud-shadow zone the temperature dropped noticeably. Blocks of gabbro the size of supermarket lorries lay where they had fallen from the peaks. The path climbed past Loch an Fhir-bhallaich, the 'Loch of the Wall Men' as my OS place-name glossary transliterated it. Could there be a more poetic name for rock climbers than Fhir-bhallaich, the Wall Men? A group of Wall Men and Women passed me with ease, toting their coloured ropes and clinking shackles as though weightless. Soon the path steepened, zigzagging up a wide and well-used track. A little scrambling in a gully up a staircase of gabbro blocks with the Coire Lagan burn hissing down its rock face nearby, and then the reward at the climb's end, the beautiful small loch hidden in Coire Lagan, and the spectacle all around of the great peaks and buttress walls at the heart of the Black Cuillin.

Clouds flew, fled, reassembled and flew once more from the peaks, so that there were only rare moments when the whole ridge could be seen. But their bulky bodies and steep slopes, all in glinting black, filled the view majestically. Sgùrr Mhic Choinnich, Mackenzie's Peak (948 m/3,110 ft), rose at the top of steep screes. Local guide John Morton Mackenzie (1856–1933), Britain's first professional mountain guide, was the first to climb it. An archetype of the Fhir-bhallaich, Mackenzie was famous for his courtesy and his quiet dignity. Poet George Donald Valentine, who often climbed with him, characterized this man of the mountains: 'Always alert, always cheerful, he was the perfect companion, but it was when the mist came swirling down on the wet rocks that his true worth was known.' From Sgùrr Mhic Choinnich the ridge dips and swoops up to the shark's fin of Sgùrr Thearlaich, and finally takes a remarkably steep upward line to the peak of Sgùrr Alasdair at 992 metres (3,255 feet). This is the highest peak in the Scottish islands, and summiting it is a tremendous feat by any walker who makes it up there, not least because the way from the loch is up a very scary-looking and steep slope of scree named by past Wall Men, with the heroic pragmatism typical of their tribe, the Great Stone Chute.

These black mountain walls and jagged ridges are utterly mesmeric, and it's hard to tear yourself away for the scramble down the gully and a moorland walk back to Glenbrittle by way of glittery Loch an Fhir-bhallaich. On the way down I stopped to look back at the Black Cuillin, shrouded once more in lowering mist, and saw a golden eagle lazily turning circles against the clouds, its wings outspread and uptilted into fingers, the familiar of these mountains.

Driving back up Glenbrittle I tried and failed to put a shape on the 'story so far'. It seemed impossible to get my impressions of the two islands, Lewis/Harris and Skye, into the same frame. These baby steps

in geology had me enthralled, but my understanding was still stumbling after the subject's enormous reach through time, set against its
tightly constricted space. The ancient gneiss of Garraidh Beach and
Dail Mòr on Lewis, not to mention the even older rocks by Hushinish
jetty on Harris, had felt like the ultimate prize. What could be as striking, or moving, as standing in the presence of the oldest rocks in these
islands? Yet at this moment and in this place the answer seemed to be:
Trotternish basalt and Black Cuillin gabbro, a forward leap in time of
3,000 million years, but a mere 70 miles sideways across the Minch.

I needed a reset of the dials, a walk to restore some perspective. The
Coigach peninsula, high up and far out on Scotland's mainland coast,
looked the best bet for that.

Basalt cliffs of the Quiraing dominate the mountain road to Flodigarry.

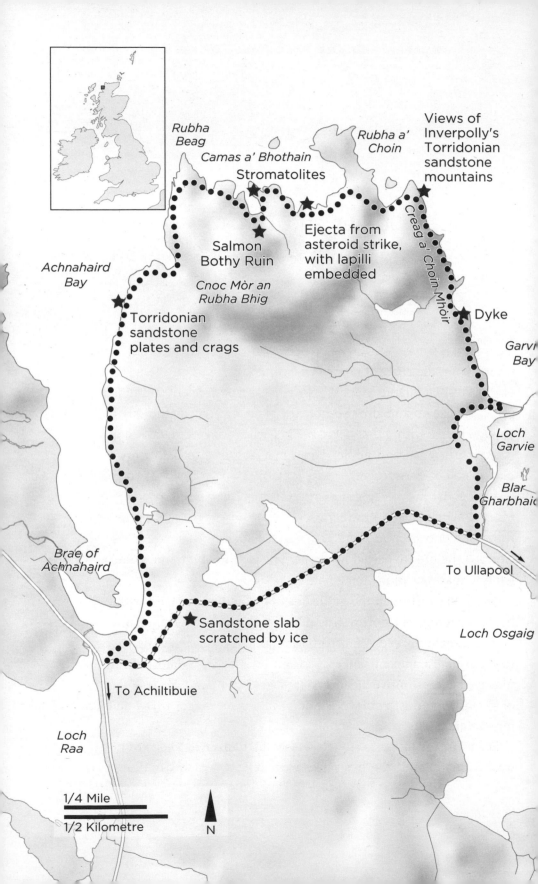

3

Strikes and Thrusts:
Coigach Geotrail and Knockan Crag

THE RAGGED PENINSULA OF COIGACH sticks up like a much-bitten thumb from the coast of northwest Scotland. It's part of the UNESCO North West Highlands Geopark, formed so that resident communities can capitalize on and also promote the amazing geology they are lucky enough to live with. I first heard of Coigach from Peter Drake, a quiet-spoken fisherman and fiddler I met at a weekend ceilidh nearby. We played a bit of music together and got talking about walking. Peter was modest about his role, but it turned out that he and his colleague Patrick Cossey had been prime movers in establishing a 45-mile geotrail through 2,000 million years of natural wonders round the coasts of Coigach. Before we parted he slipped me a copy of the trail guide and left it to marinate in my imagination.

As I looked through the guide, a stubby bullhead of land with twin horns, standing out northwards between the bays of Achnahaird and Garvie, caught my eye. A footpath ran round the edge. Lewisian gneiss, said the notes: ancient sandstones and mudstones, contorted rock, a massive fault, primitive life forms, Ice Age glacial scratches and the site of a catastrophic asteroid strike 1,000 million years ago. That would do for starters.

Peter Drake's interests stretch widely. Like most residents of northwest Scotland, and the lively if very remote Coigach community of

Achiltibuie in particular, he's tremendously proud of the place he belongs to. He's been an active fisherman in the coastal waters round Coigach for some fifty years – not just a rod-and-line man, but a salmon netter, a langoustine catcher and a trapper of lobsters in creels. He's a fine fiddle player, too, and a man who takes pleasure in construing the rocks of these beaches and inlets he knows so well. 'I'm very aware that geology's a complicated thing,' he told me as we set off up the beach of Achnahaird, 'but there's so much that's fascinating and unique about Coigach's particular story that we really wanted to make it accessible – understandable, you know – to anyone who came to walk here.'

Violets, primroses, milkwort, bluebells and silverweed spattered the low grass bank behind the shore on this early May morning. I picked green-brown pods of bog myrtle and sniffed them for their rich curry scent as I walked along, stumbling and slipping over small boulders. Oystercatchers with red pickaxe bills were strutting on the shore. Orange bladder wrack lay draped across plates of ancient Torridonian sandstone. Pale green bushy usnea lichen sprouted on the dusky red rocks. And cast up on these tilted red slabs of sandstone, themselves 1,000 million years old, was a big round knob of rock the size and shape of a seal, sea-smoothed and banded in pale and dark grey. Here was a solid lump of the Lewisian gneiss on which the sandstones rest, a basement rock three times as old as they were, formed some 3,000 million years ago. Hello, old friend. I bent and patted its curves, polished to a glossy sheen by the sea, and felt again a stab of wonder. This rock under my fingers was formed when the Earth's crust had not even properly solidified, when all was fire and fury, when there was no life and no oxygen anywhere in the molten world.

A swell of green sward, rising to low hills, rimmed the pristine sand beach of Achnahaird Bay. 'The Coigach Fault runs along the opposite side of the bay,' remarked Peter. 'The west side has dropped about four kilometres.' Four kilometres? Woah there! What? Say that again, please. The opposite side of this bay, a couple of hundred metres away, has fallen or slid down a matter of *two and a half miles* below its former location? When did that happen, Peter? 'Oh, well over a billion years ago, before these sandstones were deposited. Eventually it stabilized with its base five miles or so below the current level of the sea.'

How does the casual reader of the landscape process that? A cliff half the height of Everest? I heard the words, and I understood the distance, but I couldn't visualize it. What caused such an enormous shift in the Earth's crust? Put simply, according to Peter, it was the old tectonic shuffle. The uneasy seabeds were resettling themselves anew, the colliding plates forcing the Earth's crust ever upwards into gigantic mountain chains, to be spread and diminished and fanned ever outwards and downwards by erosion once more. I'd had a glimpse of these forces at play already in Lewis and in Skye, but I couldn't just carry on going 'Wow!' and 'Amazing!' all along this journey through the geological bones of Britain without trying to grasp the processes that lay behind, or rather underneath, those extremely ancient stripy gneisses and gritty red sandstones. I needed to get a better grip on the fundamental story, and here was a chance to row back and catch up with the essentials.

It's hard to be sure – and what a frequently used phrase in geology that is – exactly when the Earth's crust began to break up and move around as separate pieces. The science of plate tectonics only established itself in the 1960s, and there's almost no ancient evidence to rely on. But a summary of current ideas would say that our planet came together as a ball of gas and dust around 4.5 billion (thousand million) years ago, and over the next billion years solidified around a hard inner core of iron and nickel within a fluid outer core of the same minerals. Surrounding the inner and outer cores was a mantle of molten rock, as viscous as melted toffee, which extended in every direction up to the surface of the globe. When first formed, the Earth had no solid material around its rim. The surface was a boiling, shifting mass of semi-liquid rock. Gradually that settled and cooled into a solid state, initially forming a continuous crust or outer layer that sealed in the molten rock of the mantle. Water arrived on Earth at roughly the same, some 4 billion years ago, perhaps from strikes by comets bearing water or ice, helped by the hydrogen already present on the planet.

That theory was given a boost in 2021, when a fragment of meteorite fell on a driveway in the Gloucestershire town of Winchcombe. The householder, Rob Wilcock, had the presence of mind to put on a pair of Marigolds and collect the fragments into freezer bags. Analysed by scientists at the Natural History Museum, the Winchcombe Meteorite was identified as a carbonaceous chondrite meteorite, a tiny chunk sheared off from an asteroid with an icy surface that orbits between Jupiter and Mars. The fragment that landed on Rob Wilcock's driveway was found to contain hydrogen very similar to the hydrogen in our world's water. It also contained amino acids, ingredients for building proteins and potentially triggering life on Earth.

Probably the continuous crust of the early world was already covered by oceans when, some 3.8 billion years ago, it began to break apart. In response to the immense pressure and outward impulse of the boiling rock and gas immediately beneath it, the crust cracked into separate solid plates which began to move around the surface independently. These plates were tethered to the mantle below them by solid or semi-solid roots, and each reacted with individual movements as it was tugged or shoved up, down or sideways by pressures from the ever-shifting molten rock just below.

Inevitably these plates of rock, some of them thousands of miles wide and moving freely across a limited sphere, began to come into contact with one another. Where two plates propelled by such enormous forces collided, one had to give way. This 'loser' plate or crust would subduct, sliding as though on a conveyor belt beneath the leading edge of the opposing plate, to be remelted and reabsorbed into the boiling rock of the mantle. Meanwhile new crust was forming at spreading centres, points where two plates were separating rather than colliding. At these places the old crust thinned and cracked as it stretched apart, and the ever-waiting magma (molten rock from the mantle) seized its chance to rise and push through the gaps and weak points. Once out in the open the magma rapidly cooled into basalt, the building material of the ocean floor. This ongoing cycle, reabsorbing the old and forming the new crust at roughly the same rate, has maintained the balance ever since.

The early plates of solid rock, known as cratons, coalesced into giant supercontinents. Impelled by tectonic forces, they formed and broke up as the aeons rolled by, as Lewisian gneiss was formed and grew old in its turn. Trying to picture this early submarine world as I stood with Peter Drake on the tideline of Achnahaird Bay, I felt the depth of my ignorance. It seemed remarkable that I had never heard of these enormous ancient supercontinents with their evocative, fantasy-fiction names: Vaalbara and Ur, Kenorland and Rodinia.

The northern and southern halves of Britain only joined up about 400 million years ago. Before that, when these rocks of Achnahaird Bay were forming and for all the ages before that, northern and southern Britain were separate entities, drifting around the globe many thousands of miles apart. About 1,500 million years ago Scotland was part of the supercontinent of Columbia. This giant, comprising what is now the world's land masses from Australia to Amazonia and China to Greenland, was in its turn beginning to fragment. A new supercontinent, Rodinia, began to take shape, far south of the equator, out of the bits and pieces of Columbia. The tectonic plate carrying the emerging Rodinia (and what would one day be Scotland) collided with the edge of an ancient craton or mass of solid rock called Laurentia that comprised most of what's now the United States. As Rodinia and Laurentia pushed at one another with ever-increasing force a mass of mountains was thrust skywards, all along the eastern seaboard of North America and across to the northwest tip of Scotland. This 'Grenville orogeny' or mountain-building phase (named after the village of Grenville in Quebec) tore great holes in the Earth's crust, heaved portions several miles into the air, and shook up everything near and far. Slips like the Coigach Fault, displacements a couple of miles deep, were chicken feed compared to the size of the mountain upthrust, perhaps twice as high as the Himalayas.

Mountains are not eternal structures, though. As soon as they are formed they begin to erode and dissipate as sediment. A massive amount of red sandstone and silt from the Grenville mountains built up along the borders of the ancient craton of Laurentia and the new super-continent of Rodinia, layer upon layer, up to 15 miles thick – an extraordinary fact to get one's head around. Starting 1,000 million years ago and continuing over the next 200 million years, this sediment settled, hardened and formed the Torridonian sandstone that now underpins most of northwest Scotland. Geological time is so long and weighty that most of the sandstone has eroded away, chip by chip and grain by grain, over the 800 million years between then and now. Only a few stubs of the great wedge of Torridonian sedimentary rock still stand as mountains. Just round the corner from Coigach, the wild land of the Inverpolly Estate holds some of the tallest and most striking – Suilven, Canisp, Cùl Mòr and Stac Pollaidh, famous names and shapes in a walker's imagination, the enduring remnants of Rodinia.

No plants grew in Rodinia 1,000 million years ago. It was a hostile environment, a barren floodplain layered in red sediment. No birds disturbed the air, no fish swam the seas, no insects buzzed, no animals stirred. But Rodinia, with only 5 per cent of the oxygen that we enjoy, was not entirely devoid of life. Microscopic single-celled organisms grew and multiplied on the red rocks of the shore. They could photosynthesize and reproduce. And the calcium in their structures facilitated the production of oxygen, so vital to the development of more sophisticated forms of life. These Cyanobacteria, minuscule and primitive, were our ancestors, a thread that binds us to oystercatcher, lobster, seaweed and bluebells.

Out at the tideline, waves crept in from the green shallows to cream on the sands of Achnahaird. The path forsook the shore for the clifftops, a boggy sheep track across an acid heathland of coarse grass, gorse, heather and flea-sedge with small mustard-yellow beaky seeds.

A black-throated diver swam on the far side of rocky Camas a' Bhothain, its white breast flashing as it ducked neatly and disappeared. A flotilla of seals idled in the water, heads up, all watching us with that fixed curiosity that seals always seem to have about the inscrutable ways of human beings. One gets so used to being the source of fear and flight in interactions with the natural world, to seeing a succession of rapidly departing rumps, that there's an odd feeling at being the object of attention from animals with no intention of making off except in their own sweet time.

'It's not just seals that do that,' said Peter. 'We were fishing from a small boat one time, and a minke whale became fascinated by us. It swam alongside, dived under the boat, rolled on its side to view us, and came vertically up from the surface until its eye was clear of the water to get a better sight of us. I said we'd for sure never see anything like that again, but the very next day it was there too, rolling and diving and rising out of the water, just watching us with that big eye. We never did see it again after that.'

We rounded the corner of Rubha Beag's long rock promontory. All along the path there were black tarry smears full of fish parts – bones, scales and shrimp shells and claws. 'Otter spraint,' said Peter. 'I've never see one here, though. Oh, wait . . . what's that on the headland?'

With a flash of white chest and a wriggle of spiky brown back, a lithe otter was having a whale of a time up there on a little grassy saddle, rolling on its back and yawning in a spray of long whiskers. Every now and then it leaped up and bounded sinuously to and fro, its long rudder snaking behind it, full of energy, athleticism and what looked like pure joy in its own existence. We watched spellbound. The otter, if it even knew of our presence, couldn't have cared less. Later we saw it swimming on its back in the bay, grasping a fish between its front paws and crunching away with tremendous vigorous splashing, quite different in attitude and quality of charisma from the sober seals who floated there gleaming and bald like fat old men in a Turkish bath.

In Camas a' Bhothain, the 'bay of the huts', we paused at the ruin of a salmon bothy. 'My favourite place on this walk,' Peter said, 'a really special place.'

The bothy or fisherman's hut had been built with sandstone boulders and blocks roughly squared and shaped, and was probably in use well into the last century, said Peter. A battered aluminium teapot and an iron cooking pot on the griddle were the spoor of its last residents. Standing by the bothy Peter talked of his youth in the 1970s at the salmon fishery, using nets, getting 49½ pence an hour, a bonus when you had caught a thousand salmon, another if you doubled that. The season started in mid-May and was over by 26 August, but salmon were abundant, and those Coigach-caught salmon, Peter claimed with pride, were the best of the lot. Most fishers packed their salmon in straw for onward transportation to Billingsgate market in London, 'but we would lay them in foil, then on polythene. "Every scale lost is a penny dropped," was our skipper's maxim, and we really took tremendous care with our work in those days.'

We rounded the corner of the inlet beyond the bothy. 'Come this way if you want to see the stromatolites,' said Peter, 'but watch your step.' Good advice, for the rocks were treacherous underfoot. An ancient stream bed curved seaward, with a sequence of rocks I would never have figured out on my own, evidence of massive events that catapulted life on Earth up a gear.

On the basement rock of gneiss, paper-thin layers of a delicate red rock had been deposited, one on top of another, interspaced occasionally with a gritty layer where an ancient flood had laid down sand, then resuming, veneer upon veneer. In a landscape composed almost entirely of ancient metamorphic rock and sandy sediment, it is astonishing to realize that this finely leaved rock is limestone several feet thick. It is a stromatolite, an accretion of minuscule Cyanobacteria or single-celled organisms like algae. They exuded calcium carbonate which began to build up on the gneiss in the form of limestone some 1,000 million years ago. These tiny organisms, appearing all over the world, produced not only calcium but oxygen through photosynthesis – so much oxygen, globally, that the oceans became suffused with it, one of the building blocks of life on Earth. Here on Coigach these stromatolites represent the oldest form of life ever found in Europe.

Teetering round the base of the sandstone cliffs on slippery boulders, I heard Peter's urgent 'Don't miss this'. He indicated a lump of red rock. Looking closely, I saw that this particular sandstone had a molten texture, as though it had been boiled and then baked. And that's more or less what happened around 1,200 million years ago, before the main deposition of Torridonian sandstone sediments, when an asteroid half a mile wide slammed into the planet just about here at 25,000 miles an hour and plunged deep beneath the surface. In 2018 researchers from the University of Oxford's Department of Earth Sciences discovered the site of the impact at the bottom of the Minch, about 12 miles out from Coigach – although, geologists being the contentious debaters that they are, others claim an impact site near Lairg, 35 miles to the east.

Wherever the precise location of Ground Zero, the shock of the impact liquidized the Earth's crust in the vicinity and spattered it far and wide in a great viscous splash of ejecta or molten rock. Its eroded remains form a red-brown streak running inland, just round the corner from Camas a' Bhothain. Peter handed me a chunk, rough to the touch, with little green flecks that might be particles of gneiss bedrock, or could be fragments of the asteroid itself. Geologists don't agree on this, either. The petrified ejecta is pitted all over with tiny round nodules about 2 millimetres across, like acne on a teenage face – spherical lapilli, or droplets of molten rock, that cooled and hardened as they fell out of the sky from the massive volcanic cloud which billowed up above the site of the strike. In places the pitted rock is itself overlaid by a smoother layer of solidified volcanic ash, which drifted down for weeks or months after that mighty collision. The extraordinary fact is that the cataclysm itself can't be dated even to within 100 million years of its occurrence, but the effects which immediately followed it and lasted no more than an hour or two – the shower of red-hot lapilli out of the sky – are preserved in freeze-frame for evermore within the rocks. Looking at the ejecta splash, I felt again the power of geology to turn time upside down.

'What I find extraordinary,' Peter remarked as we contemplated neat round boreholes left in the cliffs by geology students, 'is the extent of these rocks' travelling. At one time they were twenty or thirty degrees above the equator, then the same below, and now they're fifty-five degrees north. That's twenty-five thousand miles of travel.'

That was a remarkable thought. Later I checked on it. A billion years ago, as part of the arid supercontinent Rodinia, the rocks of Coigach were well south of the equator. By 530 million years ago they were down at around 40 degrees south at the edge of Laurentia, on the west side of the Iapetus Ocean that had opened roughly where the North Atlantic lies today. The southern half of the British Isles, meanwhile, was across the other side of the Iapetus Ocean on a micro-continent named Avalonia. Seas were rising, and the first hard-shelled creatures were leaving a multiplication of fossils. Over the following 200 million years the Iapetus Ocean closed and Laurentia and Avalonia collided. This brought northern and southern Britain together as one body, drifting northwards to 20 degrees south, before scooting further north during the Carboniferous period over the next 50 million years to cross the equator into subtropical latitudes teeming with life underwater and on land. By the middle of the Triassic period, 230 million years ago, the British Isles were where the Sahara now lies at 20–30 degrees north, and 20 million years later, as Jurassic dinosaurs roamed the Earth, they came into temperate conditions at 35–40 degrees north. Nowadays they rest, and Coigach with them, at 50–61 degrees north, in cool climes that may not remain so for much longer if climate change continues to heat the globe.

Such thoughts brought to mind the singer Nic Jones and his heartfelt rendition of the old song 'Ten Thousand Miles':

> Ten thousand miles,
> My own true love,

Ten thousand miles or more;
The rocks may melt
And the seas may burn
If I should not return.

Although such love pledges hold true in human hearts, the rocks have indeed melted and the seas burned during these islands' 3-billion-year zigzag journey up and down the latitudes, as they may one day do again.

We rounded the top corner of the peninsula and the magnificent Torridonian sandstone profiles of the mountains of Inverpolly came into view – or they would have done had it not been for the dense white clouds hiding everything. At one moment the sharp crown of Stac Pollaidh broke through for a moment, with Suilven's twin peaks a tantalizing momentary glimpse beyond. Then the celestial scene-shifter moved the cloud shutter. Away went Stac Pollaidh and Suilven, and out for a second popped the high spine of Ben More Coigach, to be instantly snatched away again. On the rough path down the shores of Garvie Bay I kept glancing east, but the mountains continued to play hide and seek.

'Bit of a shame about the view,' said Peter as we parted by the sands of Achnahaird, 'but those clouds should clear by evening. You'll get a good view into Inverpolly around Knockan Crag.'

If you're looking for a classic example of the pig-headedness of geologists, their dedication and their genius too, Knockan Crag supplies it. A great curved cliff, it flanks the A835 on the way north from Ullapool. In a landscape of so much wild and mountainous drama there's nothing particularly compelling at first sight about Knockan Crag. But this grey cliff carries irrefutable evidence of massive subterranean upheavals so powerful that they could force ancient chunks of the Earth's crust upwards and sideways over younger neighbouring rock. The somewhat self-assured hierarchy of the Victorian geological establishment refused to believe such a thing was possible until they had had their noses well and truly rubbed in the facts.

Halfway up the track to Knockan Crag stands the Rock Room, a covered exhibition shelter. By the door I passed a bronze resin sculpture of two men in stout shoes and knickerbockers, one clean-shaven fellow sitting on a stone bench, his moustachioed companion leaning on a stick nearby. These are Ben Peach (1842–1926) and John Horne (1848–1928), geologists, friends and colleagues, whose painstaking and detailed work solved, or rather confirmed the solution of, a riddle that split the geological world.

I climbed on up the hillside to where, beyond the corner of the cliff, two distinct types of rock came together, one on top of the other. Below lay grey limestone in contorted bands. Over it was spread a coaly-black crystalline rock, apparently in the act of crawling forward. The lower layer was Durness Limestone (geologists now call it the Durness Group, but the former name seems to evoke better the actual rock itself), deposited during the Cambrian and Ordovician periods about 508 million to 465 million years ago along the shores of the Iapetus Ocean. The black stuff on top was Moine Schist, and it was metamorphosed from sandstone and siltstone around 800 million years ago. Geologists of the early Victorian era, those who recognized the rocks for what they were, scratched their heads. Everyone was agreed – weren't they? – that the Earth's rocks were laid down over time in regular layers, oldest at the bottom, youngest at the top. So how, they wondered, could the older rock have got on top of the younger?

During the mid-nineteenth century, geologists, and particularly those of intellectual Edinburgh society, revolutionized people's view of the origins and composition of the Earth. But leading figures as they were in their field, they themselves stood on the shoulders of a giant of the previous century: Scottish farmer, mineralogist, chemist and pioneer geologist James Hutton (1726–97). Hutton, a man of private means with plenty of physical and intellectual energy, followed his inquisitive nose wherever it led him, from the study of medicine and innovative agriculture to chemical experiments and, increasingly, towards geology. His friend, mathematician John Playfair, recorded Hutton telling him as a young man how 'he had become very fond of studying the surface of the Earth, and was looking with anxious curiosity into every pit or ditch or bed of a river that fell in his way'. It was evident to anyone with a pair of eyes that rocks lay in layers, with the occasional squiggle disturbing their steady upward growth from oldest at the bottom to youngest at the top. But Hutton didn't accept the generally held view that Earth was around six thousand years old, and that most of those layers of rock so visible in cliffs and quarries had been deposited during Noah's Flood. He observed that many rocks, even those high up in the layers of deposition, were made of the bits and pieces of creatures from the bottom of the sea, and the thickness of those layers must have required vast, unimaginable amounts of time to build up. He saw how the sedimentary layers had been disturbed and contorted by mineral veins and volcanic intrusions. Many layers of rock were bent into arcs or tilted up on end; many showed signs of having once been under the sea, then hoisted high, then submerged once more. As with the passage of time, the forces required to effect all these phenomena must have been enormous; only tremendous heat and pressure from inside the globe, allied to massive changes in sea level, could have produced them. And – more challenging to orthodoxy than anything else – Hutton concluded that these cyclical and long-drawn-out processes of upheaval, erosion and deposition were almost certainly still in operation. The world was not a fixed entity settled by God six thousand years ago; it was a dynamic, self-perpetuating machine for change and renewal, then, now and in the future. He presented his heftily titled paper *Theory of the Earth; or an Investigation of the Laws Observable in the Composition, Dissolution, and Restoration of Land upon the Globe* to a meeting of the Royal Society of Edinburgh in the spring of 1785, concluding, 'The result, therefore, of our present enquiry is, that we find no vestige of a beginning, – no prospect of an end.'

James Hutton struggled to get his theory widely promulgated during his lifetime. His writing style was jerky, his explanations dense. A book-length expansion of his paper in 1795 failed to sell. It wasn't until his friend and faithful supporter John Playfair produced a more concise and readable version, *Illustrations of the Huttonian Theory of the Earth*, five years after Hutton's death, that the wider world of science began to take his notions seriously.

By the mid-nineteenth century the world of geology had more or less accepted Hutton's theories. But the geological establishment of the day was still bestraddled by imposing figures, severe and authoritarian. It wasn't easy to argue against someone with the gravitas of Sir Roderick Impey Murchison, Bart., KCB, DCL, FRS, HonFRSE, MRIA (1792–1871), President of the Royal Geographical Society and Director General of the British Geological Survey. Murchison had enough clout to shut down any opinion he didn't care for. He had proposed the Silurian System of sedimentary layering with the oldest rocks at the bottom and the young-est at the top. He opposed Charles Darwin's theory of gradual evolution, believing that God had created new forms of plants and animals, fully formed, in sudden bursts of energy over long periods of time. He and others of the influential old guard remained convinced that older rocks could not be positioned above younger ones. Murchison had seen for himself on field trips that what he knew as 'Primitive' rocks could lie on top of younger rocks. In such cases, he insisted, the ones on top might *look* older, but as that didn't make sense, was against logic, and was essentially impossible, they must actually be younger. And that was that.

Once other geologists who were younger, nimbler in the field and less closed of mind had taken a look at the geology of the northwest Highlands, battle lines were drawn. Charles Lapworth (1842–1920) sug-gested that Murchison's belief was wrong, and went looking for corroboration. Up near Cape Wrath at Loch Eriboll he found older rocks on top of younger ones, and between them a thin layer of the old rock that had been ground up into splinters, evidence that it had been scraped over the younger rock in a sideways, up and over movement.

Some massive force had displaced the old rock, pushing it miles westward, up and over the younger.

In fact, Lapworth had found two thrust faults, weak places in the Earth's crust where such forces could displace rocks over many miles. One such event was the Ben Arnaboll thrust which had pushed Lewisian gneiss over younger sedimentary rocks; the other was the Moine Thrust, an unstoppable force emanating from a huge fault running from Loch Eriboll, just east of Cape Wrath, southwest for 120 miles through the Sleat peninsula on Skye and on towards Ireland. Enormous movement along this thrust zone around 430 million years ago had shoved older schists (flaky metamorphic rocks) diagonally upwards and westwards, depositing them on top of younger ones in a movement that carried them at least 10 miles. Lapworth wrote:

> Conceive a vast rolling and crushing mill of irresistible power, but of locally varying intensity, acting not parallel with the bedding but obliquely thereto; and you can follow the several stages in imagination yourself ... Shale, limestone, quartzite granite and the most intractable gneisses crumple up like putty in the terrible grip of this earth machine – and all are finally flattened into thin sheets of uniform lamination and texture.

These thin sheets are what is now called mylonite, formed between two strata when the upper one is moving against the lower, grinding and compressing it to form a new layer of crushed metamorphic rock.

The mechanics of thrusting, once the principle is accepted, are not too hard to grasp. The horizontal force of the thrust pushes along a weaker layer of rock until it fractures and forces itself up diagonally through the stronger layer above. This forms a sloping ramp up which the stronger rock, up to many hundreds of metres thick, is pushed in its turn, to level out and ride along, more or less horizontally, until it comes to a stop, in some cases dozens of miles from its starting position. If the thrust continues, more layers can be pushed up, each riding piggyback on the previous one, each one folding down over the leading edge of the layer below as it comes to rest, forming a stack of rock sheets that are the building blocks of mountains.

Lapworth worked tirelessly in the field, but couldn't get his insight accepted. It was just too outrageous a suggestion for the geological establishment to swallow. He eventually suffered a nervous breakdown, recovering in time to publish his findings in 1885. But his work had already earned the imprimatur of the trusted and respected geologists Ben Peach and John Horne. The previous year Sir Archibald Geikie, OM, KCB, PRS, FRSE, FRS (1835–1924), Director of the British Geological Survey, had commissioned Peach and Horne to lead an expedition into the northwest Highlands to look into the claims of Lapworth and others. Geikie, a dedicated advocate of Murchison's views, expected his men to find against Lapworth. Instead, Peach and Horne considered the evidence at Knockan Crag without prejudice and concluded that, yes, the thrust zone to the east, what we now call the Moine Thrust, had indeed pushed the older rocks on top of the younger.

This was epoch-making. It showcased the violence of Earth's formative processes; it demonstrated how pliable and malleable rocks could be, and how geological formations anywhere in the world needed to be examined and understood in this new light. Mountains could be turned upside down and rocks squeezed upwards and outwards for many miles, either over tremendously long periods of time, or suddenly and cataclysmically as long-closed wounds in the Earth's crust were ripped open once more.

The dark 800-million-year-old schists of Knockan Crag lie like melting marzipan on top of the pale limestone 300 million years their junior. Staring up at the crag in the evening sunlight, the process seemed so self-evident. An amateur like me, of course, had to rely on handy information boards and a bit of reading to tell the schist from the limestone. But that geologists of the first water could ever have doubted the evidence of Knockan Crag was hard to credit. That, however, was judgement with the benefit of hindsight. Once a theory has clicked into place, it can be as hard to shift as to push a mountain over on its back and turn it upside down.

To get a proper view of the mountains of Inverpolly, I went a long way around an endlessly unrolling ribbon of roads. A scramble up the path from the Achiltibuie road beside Stac Pollaidh was rewarded with a viewpoint into Inverpolly National Nature Reserve. Here golden eagles, peregrines and merlins, otters, red deer and wildcats share nearly 30,000 acres of truly wild country of rivers, lochs, woodland, wetlands and coasts. The handful of Torridonian sandstone mountains that remain here are footed in a low lumpy landscape of 'knock and lochan', very reminiscent of that in the Isle of Lewis. It is Lewisian gneiss the mountains stand on, the bedrock down to which the Ice Age glaciers whittled all the Torridonian sandstone save these few more resistant survivors.

Suilven ('Pillar Mountain', 731 metres) was the most eye-catching this evening, dramatically sunlit, rising abruptly from the surrounding country, striped horizontally in sandstone layers. Glacial ice flow sculpted it, a slim-hipped animal shape seen from the east, a sphinx-like double hump from north or south. Just to the east stood Canisp ('White Mountain', 846 metres), its shark-fin profile capped by gleaming grey-white quartzite, deposited on top of the sandstone around 540 million years ago. South of these twins curled the long spine of Cùl Mòr ('Big Back', 849 metres). The sharp sandstone tooth of Stac Pollaidh ('Peak of the Pools', 613 metres) guarded the southern approaches to Inverpolly like a butte in Death Valley, and south again rose the buttresses of Ben More Coigach ('Coigach's Big Mountain', 743 metres), now clear of cloud.

Since starting this journey at the Butt of Lewis a few truths had already made their way home to me, chief of which was that nothing in geology was quite as obvious as it looked. Nevertheless, laid out in front of me here as plainly as a picture book was the story of how the erosion of 800 million years, hastened and exacerbated towards its end by the juggernaut scouring mechanism of the ice sheets, had chopped, smoothed and dismantled a massive wall of sedimentary sandstone down to its very roots, the Lewisian gneiss, the mother rock on whose back all the others stand.

Now I felt properly launched on my way, the time had come to head south and east again, travelling on down the chronological tunnel to a younger era. This next leg of the journey would take me through the tumbled mountains and deep winding glens of the western Highlands, from Kyle of Lochalsh to the Great Glen by way of Kintail, Glen Affric, Glen Quoich and Glen Garry, regions with a fine heroic ring to their names. For companions in spirit, if not in actuality, I'd be travelling with a herd of Highland cattle and a bunch of drovers on their way from the Isle of Skye to the Great Glen. The path I had chosen was the Road to the Isles, and what could be more romantic than that?

But the next phase of the geology – I couldn't make head or tail of it. Everything seemed to be upside down or back to front, or changed from what it had been to something else entirely. I turned for help to my godson Andy Harrison, a professional geologist with a happy knack for clear and easy explanation. 'Oh, the western Highlands? To use a technical term, that's all FUBAR.' Fubar, Andy? Some kind of metamorphic rock that wasn't in the literature? 'No – it means Fucked Up Beyond All Recognition.'

Stac Pollaidh rises like a sandstone butte in Death Valley on the edge of Inverpolly National Nature Reserve.

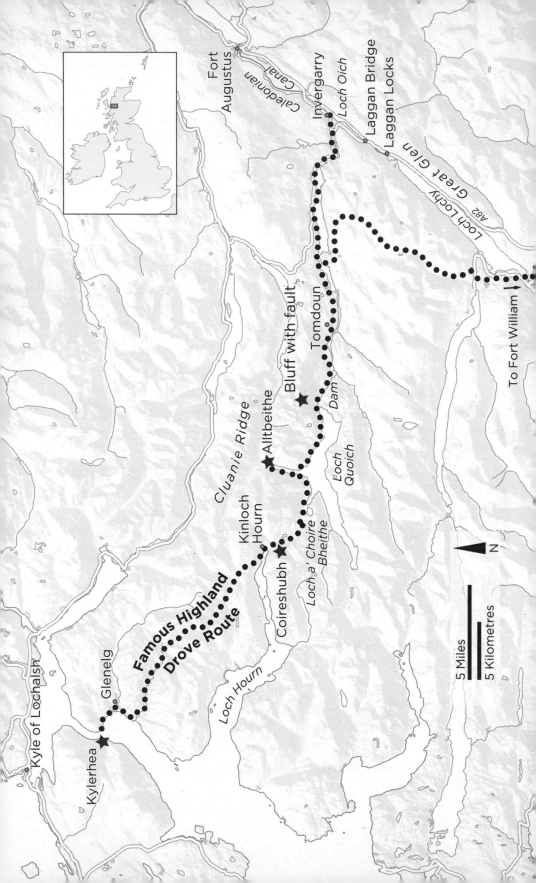

4

The Famous Highland Drove:
Isle of Skye to the Great Glen

IF YOU'RE GOING TO DRIVE a herd of thirty cattle 200 miles on foot across some of Scotland's toughest country, it doesn't hurt to be backed by a whisky distiller. Not that the Famous Highland Drove of 1981, sponsored by the makers of the golden nectar Famous Grouse, was a tipsy lark or a pub jape. It was the brainchild of Himalayan explorer and writer John Keay and his wife Julia, who wanted to see whether the old mountain route along which herds of black cattle used to be walked from the Isle of Skye to market in Crieff, 200 miles to the east, was still passable on foot for man and beast. They set out in October of that year from Glenbrittle in the shadow of the Black Cuillin with five friends, three garrons or hill ponies to carry their loads, and thirty Highland cattle with long fringes and horns – twenty-nine bullocks and one cow, Matilda, leader of the herd. Hillwalker and writer Irvine Butterfield joined them at various stages along the route, and afterwards he wrote a book about the adventure, *The Famous Highland Drove Walk* (Grey Stone Books, 1996), part dry-witted account, part guidebook for walkers to follow.

Irvine Butterfield (1936–2009) was a big name among Scottish walkers and mountaineers. In 1985 he published a best-selling guide, *The High Mountains of Britain and Ireland*, which did for the Munros or mountains over 3,000 feet what Alfred Wainwright's series *A Pictorial*

Guide to the Lakeland Fells did for the Lake District – it encouraged ordinary walkers, as opposed to mountaineers, to explore high country. Though he climbed all the Munros, the Yorkshire-born Butterfield called himself a hillwalker rather than a mountaineer. He was very active in organizations such as the Munro Society, the Mountain Bothies Association and the Mountaineering Council of Scotland. But while he wanted everyday walkers to feel that they, too, could enter and enjoy the mountains, where his heart truly lay was in conserving the remote and wild corners of Scotland. He gave his energy, his skill with pen and camera and many of his book royalties in the service of this cause. When it came to recording the wild adventure of the Famous Highland Drove, Irvine Butterfield was the man to do it.

As a guide to follow eastwards through the mountains to the Great Glen, Butterfield's account of the Famous Highland Drove was perfect for me. The route that the drove took is not waymarked, nor in any way prepared for public consumption. I ended up having to dodge in and out of it as weather and time dictated. But it showed me a way through the heart of the mountains, and lent me a sense of direction and purpose. Butterfield eschewed geological description, and I was left to make the best I could of a geology that my godson had labelled FUBAR – 'Fucked Up Beyond All Recognition'. But even among such research books and websites as dealt with this region, weltering in a word-soup of metamorphic events, of orogenies, of psammite and pelite and microgranodiorite, I found that things at their most basic boiled down to a cycle I'd already met – sand, mud and silt eroding off those Grenvillian mountains of 1,000 million years ago, settling and turning to stone in mighty layers, then transformed and cast up at odd angles by massive movements in the Earth's crust. There was more to it, for sure, but those seemed to be the essentials still.

In *A Journey to the Western Isles of Scotland*, Dr Samuel Johnson opined that 'the wealth of the mountains is cattle'. It had been cattle for centuries before the doctor travelled through western Scotland in the autumn of 1773. There was no other object of commerce in those remote glens and islands whose inhabitants never moved outside their clan borders. Cattle were wealth on the hoof; they were objects of pride and envy, measures of social and political status, often stolen by other clans, often reclaimed with bloody consequences. Where they lived, life revolved around them. But if they were to be realized as capital, they had to be walked to where the markets and fattening pastures were, away south and east in larger towns and easier country than the western Highlands and Islands possessed.

From the fourteenth century onwards, droves of cattle took place annually, generally in autumn when the animals, after a summer of quiet feeding, were in their best condition to face the arduous 200-mile walk over rough mountain tracks with patchy poor grazing, to the trysts or markets at Crieff and Falkirk. These beasts were kyloes, small black cattle, ubiquitous in the west, semi-wild and hard to govern. Against the stony condition of the tracks their hooves were shod with removable shoes of iron, no mean feat in itself as they had to be tipped over and laid on their backs for the shoes to be fitted, something they didn't take kindly to. They could manage up to 20 miles a day at best, often less, and they needed plenty of water and good grazing on hand at the stances where they stopped each night. Fords and sea lochs were often flooded by autumnal rains and could baffle the drove, and early snows might block the mountain passes. Cattle, men or dogs could fall prey to sickness or accident, slowing or even halting progress. Thieves might be waiting to ambush the drove in one of the remote glens the trail passed through, or they could sneak up at night to spirit the beasts away. And even if the drove reached market safely, that wasn't the end of it for the cattle. They would be taken on by the purchaser's drovers to better grass further south, often across the border in England, where they would be rested and fattened for slaughter and subsequent sale as meat. That could take place in any of the centres of population hungry for beef – Manchester, say, or even down at Smithfield Market in London, a cumulative foot journey for some cattle of the better part of 1,000 miles.

As for the drovers, popular image has them as a rough bunch, massively bearded, hatted and booted, a law unto themselves, all too ready for riot and randy. Like the canal and railway navvies who were their counterparts, public prejudice of the eighteenth and nineteenth centuries had them down as dangerous customers on the loose, wild men from the mountainous badlands on the lookout for something to steal or someone to outrage. There was some truth in that caricature, but it was far from the whole story. They were practical men, competent to deal with any emergency they might encounter on the trail. They dressed for any and all weathers in broad-brimmed hats, stout boots and homespun socks whose soles they coated with soap to guard against blisters. Unless they pitched up at a stance or wayside bothy for the night, they slept out of doors. The all-enveloping plaids worn by the Highland drovers did duty as both cloak and sleeping bag. When hard frost sent the night temperature plummeting, drovers had been known to soak their plaid in a burn before wrapping themselves in it to sleep, well insulated by the layer of ice that formed on the outside of the garment. They carried home-made whisky, and oatmeal which they would moisten with blood drawn with a knife cut from one of the cattle. With them they had pack ponies for the heavy lifting, and a pack of dogs running alongside that needed to be clever and biddable when it came to marshalling the beasts, but capable of turning nasty to deter cattle thieves.

They were men apart, hard customers to look at, and there were certainly some rogues and chancers in the business. The way they treated the cattle, thrashing and bullying them past obstacles, wouldn't be tolerated today. But a good drover had to do much more than force the cattle along, goad them through floods and doctor their ailments on the fly. He needed to possess the character to inspire confidence in the poor and remote-living clansmen who entrusted their few cattle to him each autumn, and in the laird or clan chief whose property the cattle ultimately were, and he needed to be tough and canny with the market dealers he had to bargain with at the other end of the long haul. He carried money and promissory notes to and fro, and was often entrusted with messages to deliver along the way. He had to keep accurate accounts and present them on his return to those who raised and those who owned the cattle. And although he could and would pocket any profit on the side from the sale, he also had to make up any shortfall from his own purse. No rough-arsed ignoramus would last long as a professional drover, and the best of them were trusted colleagues and lynchpins in the commerce of the Highlands and Islands.

To get from the Isle of Skye to the mainland the drovers of yesteryear would swim the kyloes across the strait of Kyle Rhea, the narrowest crossing point. It is only 500 metres from shore to shore, but the strong currents and the beasts' natural reluctance to venture out of their depths made it a Styx-like crossing. Roped nose to tail in strings of six at a time, they were forced across with cries and blows from the drovers who accompanied them by boat. Many cattle panicked; some had their tails bitten or pulled off; some were said to have deliberately rolled on to their backs and drowned themselves rather than continue with what to them was inexplicable torture.

In 1981 there was no bridge connecting Skye to the mainland. The Famous Highland Drove adventurers could not employ their predecessors' harsh methods with the cattle, and a previous experiment with swimming the beasts across a narrow sea loch ended in failure. They simply turned round and headed back to shore. So Matilda and her consorts sailed across the strait of Kyle Rhea in style aboard a ferry.

It's the best part of 20 miles through the mountains by the old drove route from Glenelg on the mainland shore to Kinloch Hourn, an oasis of flat grazing land and abundant fresh water at the head of Loch Hourn. By the time they got there, two tough days later after fording rivers and struggling up rough paths in mist and rain, the cattle were already suffering from sore hooves. Matilda had helped alleviate the problem by leading the bullocks into a cooling bog along the way where they loitered up to their knees in the soothing black muck, impervious to the drovers' entreaties. But it was already clear that these animals, accustomed from birth to gourmet grazing, kind words and gentle handling, were very different beasts from the hardy black kyloes that walked this route before them. One of the Famous Drovers was a qualified vet, and there were regular visits from a uniformed officer of the Scottish Society for the Prevention of Cruelty to Animals, known to one and all as the Cruelty Man.

The tiny settlement of Kinloch Hourn at the head of its long sea loch is one of the remotest in Scotland. The nearest main road runs down the Great Glen, more than 20 miles eastward across tough country. 'Shut in by high mountains,' wrote Irvine Butterfield, 'the few inhabitants of Kinloch Hourn go unpretentiously about their business without fear of disturbance. Their only link with the outside world is a twisty, narrow road which in the space of a mile struggles 400 feet from the level of the sea.' Setting out on a spring morning to follow that long and lonely road, I found that little had changed since the Famous Highland Drove passed through some forty years before. The diminutive harbour lay still and silent. A solitary lobster pot stood on the jetty. The shore at low tide was strewn with bright orange seaweed, a dab of brilliant colour in a landscape growing misty with the threat of the same dreich weather that had plagued the drove of 1981. No one stirred in Kinloch Hourn. In a plantation beside the road a red deer with velvet thick on its horns chewed the bark off the young trees without fear of disturbance.

Immediately the route began to climb, twisting up narrow gorges with water crashing down the hillsides and under humpback bridges after days of heavy rain. The road had the air of an alpine switchback, and looking back I saw Kinloch Hourn already far below, scattered idyllically like a Swiss mountain village on a green sward edged with tall conifers. At the top of the climb the ruin of a cottage stood by the road on a patch of flat grass. This was the old stance of Coireshubh, a wayside resting place where drovers could shelter and the cattle would get some grazing. The Famous Highland Drove halted here on a night of torrential rain. The cattle, indifferent to the weather, grazed and slept contentedly, but Irvine Butterfield and the drovers spent a miserable night at Coireshubh.

The building is but a hollow shell roofed in parts by wooden cladding and rusty corrugated iron sheets. Some of the discarded metals provided a rough mattress on a damp earthen floor reeking of sheep droppings. The more fortunate were able to secure a space under the remaining roof, free from the constant drip drip off the ragged edges.

Today, roofless and sodden, Coireshubh wouldn't have sheltered a mouse. It rains an awful lot in the western Highlands, and today was a classic. In the brief intervals when the rain eased off, the air seethed with moisture blown on a sharp east wind. The rivers roared, the wind whistled and the wayside lochans slapped their shores, a wash of sound like the sea in a dream, from which the occasional dash of rain in the face awoke me.

Beyond the stance the road levelled out into a long pass at the watershed between two rivers, the Lochourn River tumbling westward down to Kinloch Hourn and the sea, the Caolie Water running east into the big bow-shaped expanse of Loch Quoich. The rainy air had leached the colour from the mountains on either side, leaving the milky greys and insubstantial rippling shapes of a Chinese landscape painted on silk. They were anything but insubstantial, though, these east–west barriers interspersed with extenuated lochs lying parallel to the peaked ridges of the mountains. Ice Age glaciers deepened the lochs and moulded the slopes, but beneath the surface covering of peat and heather, here was the same stuff that had formed the striking shapes of the Inverpolly mountains, the underlay for most of Scotland – sandstone, mudstone and silt.

By the corner of Loch a' Choire Bheithe, one of a string of watery beads along the pass between Coireshubh and Loch Quoich, a broken tooth of rock stuck up from a knoll beside the road. Its diagonal stripes would not have disgraced a cad's blazer. Seen in sunlight they would contrast spectacularly, and even in today's rainy air they showed up boldly in dark and light grey, purple and a muddy yellow the colour of French mustard. Adhering to this smooth rock face was a thick wedge of crystalline quartzite, itself stuck to a block of shiny grey rock as dense

as baked mud. I wiped the rain off my notebook and took a stab at identification. Here was a chunk of psammite, a sheet of quartzite and a
wedge of pelite. Quartzite – I sort of knew that already, a shiny whitish
metamorphic rock, incredibly hard, formed when huge pressures
bonded sand grains and silica together. Psammite (from the Greek
psammos, meaning sand), that was a rock metamorphosed from coarse
sandstone. And pelite was metamorphosed mudstone and siltstone,
with a grain finer than psammite. I ran my fingers across the three tex-
tures. There they were, all three clinging together intimately, familiar
old sand and mud, metamorphosed, transformed and tilted. Just how
tilted I only appreciated when I took a proper look at the geological
map and saw how between Glenelg and the Great Glen the old drove
road crossed from psammite into pelite, from pelite into psammite and
back to pelite again, then psammite once more, and so on and on. It
took an effort of imagination, even so, to conjure up the horizontal
layers as they were deposited, fine silt and mud on to coarse sand, more
sand and more mud in flat bands one above the other as rivers snaked
and sea shallows waxed and waned; to metamorphose them into psammite and pelite, then tilt them up at 45 degrees as the tooth of rock by
Loch a' Choire Bheithe was tilted; to smooth them off flat across the
angles by hundreds of millions of years of erosion; and to run the old
drove road across the surface between the alternating rock types, as a
drover might run his herd between patches of sunlight and shadow.

And what caused this metamorphosis, this change from sandstone to
psammite, from mudstone and siltstone to pelite? It was part of a long-
drawn-out series of severe geological jolts to the settled order of things
between 800 million and 500 million years ago. The theory is that as the
emerging supercontinent of Rodinia and the ancient craton of Laurentia went head to head (a tectonic shoving match I'd learned about on
Peter Drake's geotrail in Coigach), enormous amounts of matter were
scraped off the leading edge of Laurentia and transferred to the rim of
Rodinia. As ever with plate tectonics, action produced reaction. The
pressure of this build-up of alien crust on the Rodinian shore caused

the closure of a big rock basin and a consequent clash along the line of the Moine Thrust. The strain of sections of crust colliding and grinding resulted in a major upheaval, generating enough heat and pressure to metamorphose the fine-grained siltstone and mudstone into pelite, and the interleaving sandstone into the coarse-grained psammite.

The main movements of the Moine Thrust took place around 430 million years ago, during the period of mountain-building activity known as the Caledonian orogeny (490–390 mya) when the Iapetus Ocean was closing. These 100 million years of tectonic jostling pushed together the building blocks of most of Scotland's mountains. There were two main phases, first when the ancient craton of Laurentia came into collision with a chain of volcanoes around 470 mya, and secondly when Laurentia butted heads with the supercontinent of Baltica around 440 mya. As a skyscraping range twice as high as the modern Himalayas formed, rose and began to erode, life on Earth was developing in superabundance. Most branches of life as we see it today had already developed by the start of the Caledonian orogeny, and during the ensuing hundred million years the seas saw the arrival of masses of plankton, reef systems with corals and sponges, sea scorpions 2 metres long and huge predatory fish with jaws. On land, insects made their appearance as plants began to creep out of the water on to the shore, and pioneer amphibians were soon to follow them. But of all these living wonders there's no trace in the metamorphosed, squashed and massively displaced rocks underpinning the country that the old drove road was leading me through.

There's a fine geological word, terrane, which describes large pieces of the Earth's crust separated one from another by faults or displacements of the rocks. Five terranes make up Scotland. Conveniently for the amateur geologist who likes a simple layout, they lie side by side, fanning out slightly, their separating boundaries running more or less northeast to southwest. From top to bottom of Scotland they are the Hebridean Terrane (the Outer and Inner Hebrides, Skye and the northwest coast, including Coigach and Inverpolly); the Northern Highlands Terrane (everything else that's west of the Great Glen, including the mountains crossed by the Famous Highland Drove); the Grampian Terrane (basically the mighty Grampian Mountains chain of the central Highlands, including Ben Nevis and the Cairngorms); the Midland Valley Terrane (everything south of the Grampians to the northern edge of the Scottish Border country, including Glasgow and Edinburgh); and the Southern Upland Terrane (the Scottish Borders).

Scotland's five terranes are separated by four pretty straight fault lines, tectonic fractures or seams that also run roughly southwest to northeast, right across the country from coast to coast. The Hebridean and Northern Highlands terranes have their border along the Moine Thrust (southernmost Skye to the north coast at Tongue). The Northern Highlands and Grampian terranes meet along the Great Glen Fault (Fort William along Loch Ness to Inverness). The Grampian and Midland Valley terranes are divided by the Highland Boundary Fault (Firth of Clyde to Stonehaven); and the Southern Upland Fault (Rhins of Galloway to Firth of Forth) separates the Midland Valley and Southern Upland terranes.

My journey across the bones of Britain would eventually take me through all five of Scotland's terranes. I'd already crossed the Hebridean Terrane from Lewis and Skye to the Coigach peninsula, and now the route of the Famous Highland Drove had taken me deep into the Northern Highlands Terrane. And here it was the influence of the Moine Thrust that I saw all around me, the great tectonic push north and west that fractured and folded sedimentary layers, flipping them and tilting them, lifting and transporting them as much as 60 miles from where they were laid down, squeezing them not only out of shape but into new existence as metamorphic rocks.

For the Famous Highland Drove of 1981 the road to the Great Glen was a purgatorial procession, a long road through uncompromising country in miserable weather. Constant contact with the hard road surface had cracked and infected the cattle's hooves. The beasts limped along at one mile an hour, a rate of progress that frustrated the drovers. 'Such a pace is tedious on the finest of days,' wrote Irvine Butterfield. 'On a grey day of rain it tempts even the most well-balanced of drovers to thoughts of ending it all in the convenient waters of Loch Quoich.' Coils of mist above the tumbled stones of ruined houses reminded the writer of the peat-smoke that once marked the dwellings of hundreds of Mac-Donnell clansmen, while the ribbon of road put him in mind of the evicted folk trudging to the emigrant ships. As for Loch Quoich: 'The loch is one which seldom sparkles in the sun, the grey depths of its waters imparting a deadness to an already wild and inhospitable landscape.' This is pure 'school of Alfred Wainwright' (the famously gruff chronicler of the Lake District fells), and to dispel my own wet weather glooms I imagined an encounter between those two bluff Yorkshiremen, Butterfield and Wainwright.

'Morning.'

'Is it?'

'Happen.'

'What's it to thee?'

'Nowt.'

'Daft bugger.'

'Back on thee.'

'Think on.'

'I'm off.'

'Get on, then.'

As I neared Loch Quoich the river at my right hand began to hasten, sliding fast down its chutes to meet the battleship grey of the loch. The water of this remote mountain loch is used for generating hydro-electricity – hence the string of pylons that march in close companionship with the old road all the way from Skye to the Great Glen. Loch Quoich is shaped like an eastward-leaping shark, rather snub-nosed, with two inlets like a pair of fins sticking out of its curved back. The road passed the first inlet and hurdled the neck of the second over a concrete bridge.

Here a good firm stone track swung away from the road and shadowed the shoreline of the inlet, heading north up the slim arm of water into the hidden cleft of Glen Quoich.

Modern road surfaces are hard on the feet of both man and beast. Irvine Butterfield detested the long and tedious miles of tarmac with which much of the old drovers' route had been spread since the black cattle came this way. For the greater ease and enjoyment of those who might follow in the footsteps of the drovers, he details in *The Famous Highland Drove Walk* a number of alternatives for sidestepping the more monotonous sections by detouring through higher and wilder country. His alternative route to the north of the Kinloch Hourn road via Coire Sgoireadail and Glen Loyne sounded good, and although I couldn't follow both detour and main route at the same time, here was a chance to catch a glimpse of what lay beyond the immediate surroundings of the old road. The shore track up the northern arm of Loch Quoich enters the hidden glen at the point where it divides into Wester and Easter Glen Quoich. Here, beside a rain-swollen burn, stood the old two-storey farmhouse of Alltbeithe. What a quiet and lonely existence it must have been for the inhabitants, their northward view shut off by a mountain wall. These crinkle-edged crags, interspersed with high dark corries rising to rocky peaks, form the Cluanie Ridge, a famous traverse for mountain walkers with a lot more skill and daring than I possessed. As along the drove road, the high country to the north of Glen Quoich is all psammite and pelite, shot through with dykes of microdiorite, an igneous rock like a dark-coloured granite. There was one such dyke due north of Alltbeithe, running steeply up beside the cleft of Allt Beithe Mòr. Microdiorite is the end product of hot magma cooling instantly on contact with cold seawater. Here was a jolt to the imagination. I had to remind myself, as so often on this journey, that these ancient rocks were once buried deep beneath the sea.

Back on the heels of the Famous Highland Drove, I followed the road along the loch shore. The way led among miniature hillocks of gravelly rubble mounded up by the Ice Age glacier that pushed its way along the valley. Five hundred million years down to twenty thousand and then to the present day in the space between mountain slope, roadside mound and loch; a stretch of mental gymnastics between one footfall and the next.

Just beyond the dam, built in the 1950s and furnished with a remarkably functional and ugly set of powerhouses, there was a sharp reminder of the problematic relationship between modern man and ancient geology. The hillside here was disfigured by a long scar, the exposed rock shiny with rain, a memento of the day in November 2018 when a great slide of rock, mud and heather came crashing down the treeless slope, over the road and the dam spillway beyond. Pylons and cables were torn down. Twenty thousand homes far away in Skye and the Outer Hebrides had their power cut, and the one and only road to Kinloch Hourn was severed till the following spring.

I ran my finger up the map as far as the legend 'cave'. It marked a bluff 1,000 feet overhead, the source of the landslip. A fault runs through the bluff, a crooked line I could just about make out through binoculars blurred with raindrops. The slide was caused by a 'toppling failure', said Scottish & Southern Electricity's geologist, a recognized hazard where fractures such as the fault in the bluff dip into the rock face at an angle. A block had started to rotate and split away from the main structure, tumbling down towards the dam, bringing 9,000 tons of mountainside with it and disrupting the lives of people 100 miles away.

Now the road began its long and gradual descent east towards the Great Glen. After the drab miles through the hills, colour crept in. Mountain torrents painted white streaks high on the mountainsides. The rocks by the road were bright with orange, white and grey lichens. Bushes of rhododendron and gorse dotted the verges. Trees made a reappearance, silver birch and willow flanking the road and obscuring views over Glen Garry. A deer appeared from the trees and trotted springily away down the road before bouncing off into a meadow. Through the trees I glimpsed a handsome wooden fishing lodge down beside Loch Poulary, the slopes beyond clothed in forestry. The 5 miles from the dam of Loch Quoich had completed a transition from barren uplands to well-favoured river valley. Even the rain relented, allowing a watery sun to break through and turn the gorse to gold.

At Tomdoun, a handsome house of grey granite with red sandstone facings, civilization reasserted itself in the shape of a little church and a red phone box. Just beyond here the Famous Highland Drove had turned off the old road, seeking green pastures and a soothing footbath of formaldehyde for Matilda and her long-suffering consorts. A mountain path awaited cattle and drovers, a south-going track that would bring them out to cross the Great Glen at the tail of Loch Lochy. Thence their route lay 'in wildest Lochaber', as Irvine Butterfield styled it, south through the Grey Corries, past Ben Nevis to the old military road in Glencoe where I would cross their trail once more. For now, though, I watched in my mind's eye the tails of the shaggy Highland cattle disappear across the long girder bridge over Loch Garry. Then I turned to face the last miles through the trees to the Great Glen and the next step of the journey, down along the most dramatic rift in the body of the British Isles.

'Loch Quoich ... is one which seldom sparkles in the sun, the grey depths of its waters imparting a deadness to an already wild and inhospitable landscape.' – Irvine Butterfield

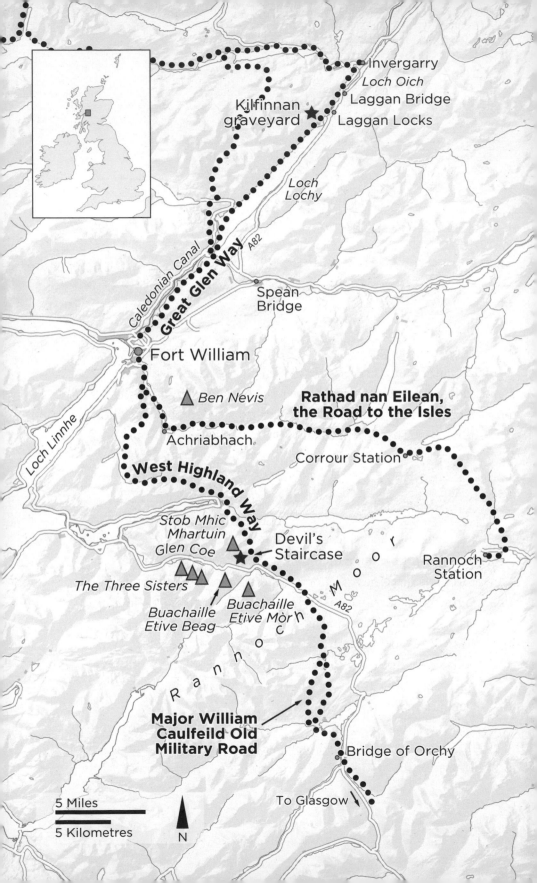

Invergarry

Loch Oich

Laggan Bridge

Kilfinnan graveyard

Laggan Locks

Loch Lochy

Caledonian Canal

Great Glen Way

A82

Spean Bridge

Fort William

Ben Nevis

Rathad nan Eilean, the Road to the Isles

Achriabhach

Corrour Station

Loch Linnhe

West Highland Way

Stob Mhic Mhartuin

Glen Coe

Devil's Staircase

The Three Sisters

Rannoch Station

Buachaille Etive Beag

Buachaille Etive Mòr

M o o r

A82

R a n n o c h

Major William Caulfeild Old Military Road

Bridge of Orchy

To Glasgow

5 Miles

5 Kilometres

N

Plutons, Volcanoes and Military Roads: Great Glen to Rannoch Moor

CARS WERE JAMMED UP NOSE to tail behind a motorhome as it trundled blithely at 20 miles an hour down the A82 towards Fort William. Half a mile away across the waters of Loch Lochy I watched the procession from a summit on the roller-coaster lane to Glas-dhoire. The potholed old road from Laggan is one only a local's car would brave. To a walker, though, it is one link in the chain of paths that constitutes the Great Glen Way, a very welcome alternative to the death-dealing verges of the main road on the far side of the long and narrow loch.

The Great Glen, ruler-straight in its 65-mile course between Inverness and Fort William, is an obvious choice for a journey across the Scottish Highlands through mountain scenery. It's not only drivers who benefit from its directness. Canoeists and boaters take advantage of its unbroken string of rivers, lochs and canals from sea to sea, as do fishing boats wanting to avoid the long and stormy passage round Cape Wrath from the Irish Sea to the North Sea. And since 2002 horse riders, cyclists and walkers have been able to follow the Great Glen Way, a succession of minor roads, forest tracks and hill paths that runs parallel with the A82 throughout the Great Glen.

Cyclists can whizz in a couple of minutes from Laggan Bridge to Laggan Locks, where the Caledonian Canal opens its jaws and expels narrowboats, yachts and kayaks into Loch Lochy. Walkers take a little longer. From the top of successive rises in the rough-surfaced lane I got great views southwest down the glen, its flanks rising sharply, its bottom spreading and flattening into a green strath where sheep and cattle grazed among the muck spreaders and silage clamps. From Laggan Locks the country lane idles on southwest beside the loch, undriven and unhurried. It's not a road conducive to great effort. At the old graveyard of Kilfinnan beside its green promontory I found a seat against the wall of the stark grey mausoleum of the MacDonnells of Glen Garry, and between sleepy yawns and blinks tried to focus my wandering mind on my notebook and what I'd scrawled there regarding the violence and upheaval, geological and human, that have been endemic to the story of the Great Glen since its formation some 400 million years ago.

The fault that contains the Great Glen is more than 300 miles long. It starts out in the North Sea beyond Shetland and heads southwest, slicing the main body of Scotland neatly in two and continuing through the Isle of Islay and on through the northwest of Ireland. The mainland section that forms the Great Glen runs for 65 miles from Inverness to Fort William in a remarkably straight line, and in fact the whole fault hardly deviates from this direct course from end to end. It is classified as a strike-slip fault, or more graphically as a wrench or tear fault, and those names give a good idea of what happened here in three phases over the last 400 million years or so.

The Great Glen Fault was formed during the Caledonian orogeny or mountain-building period between around 490 million and 390 million years ago, an age of violent geological upheavals as the tectonic plates carrying the two rock giants of Laurentia and Baltica pushed against each other. Their mutual shoving at this particular location was not head on, like two bulls in a pen, but at an angle. The pressure caused the Earth's crust to rupture vertically, creating two rock faces, one moving in opposition to the other in a sinistral or leftward movement to the southwest. The massive pressure caused the psammite and pelite rocks of the Northern Highlands Terrane to the northwest of the fault to be melted, folded and pushed sideways and upwards.

The second phase of activity along the Great Glen Fault took place at some time during the Carboniferous period (359 mya–299 mya). This time the fault was wrenched in the opposite direction, a dextral or right-handed movement to the northeast. Around the same time the fault region was cut across by dyke swarms of lamprophyres, dark igneous intrusions full of magnesium and iron. The crude and tiny geological map from the *Philip's Modern School Atlas* wasn't much help when it came to placing these little upward stabs of igneous rock. But by now I'd discovered the BGS's online Geology Viewer, a fabulous interactive map including both bedrock and superficial geology, with a telescopic zoom that let me get up close. It showed the lamprophyre swarm as a shower of olive-green worms, pushing in amongst weaknesses in the psammite. Among the green lamprophyre swarm wriggled other worms, purple ones representing intrusions of microdiorite, an older igneous rock I'd met along the Famous Highland Drove route from Kinloch Hourn. The geological map seemed to burst into life with these flecks of colour and their likeness to a mass of tiny creatures swarming across the land.

A final phase of movement came during the late Cretaceous and Palaeogene periods between 50 million and 80 million years ago, around the time of the great extinction event that wiped out the last of the dinosaurs. The North Atlantic Ocean was opening, and western Scotland was the scene of huge volcanic eruptions and magmatic intrusions as the sea-floor crust stretched, thinned and ruptured.

When the dust had settled on all the major movements of the Great Glen Fault, the rocks of the region had been displaced by more than 60 miles. The geological map shows a complete separation of one side of the Great Glen from the other, a sharp line with no cross-border correspondence between the splashes of colour denoting the individual rock types. No continuous colours cross the line of the fault on the map. Across the narrow strip of the Glen, psammite stares at sandstone, cataclasite (crushed) metamorphic rock confronts granite gneiss, flag-stone opposes conglomerate. Like a pair of sliding doors opened by an impatient giant, the action of the Great Glen Fault has wrenched the rocks in parallel, sliding them miles away from those which were once an indivisible part of them.

During the Ice Age the region lay deep beneath the ice. When the glacier that covered the fault receded at the end of the last glacial period some ten thousand years ago, it took the line of least resistance and followed the cleft, carving it deeper. Much of the bottom of the Great Glen lies below sea level. Its string of rivers and lochs is continuous. So the Northern Highlands and Hebridean terranes to the north and west of the Great Glen form what is effectively a separate island from the rest of Scotland.

A soldier in the summer of 1727, fighting off the midges as he laboured with pick and shovel on General George Wade's military road through the Great Glen, might have laughed sourly at the thought of anyone actually enjoying a journey on foot or by boat through what was then a hostile and ominous place. General Wade, Commander-in-Chief of North Britain, had been sent north in 1724 by the British government to survey the unmapped, roadless Highlands. These remote Catholic regions were still in a state of surly disaffection, if not outright revolt, some ten years after the defeat of the 'Old Pretender' James Francis Edward Stuart, son of the late exiled King James II. That had brought to an inglorious end the 1715 Jacobite Rebellion. The English were not wel-come here, their soldiery in particular. The clan system still held sway over the people of the Highlands, who were loyal to their clan chiefs and to the Old Pretender, but certainly not to a Protestant government issuing edicts from an unimaginable place called London far away beyond the mountains.

At the extremities of the Great Glen there already existed a pair of run-down forts dating from Cromwellian times, Fort William at the southern end where the glen reached the sea inlet of Loch Linnhe, Fort

George in the north where the River Ness debouched into the Moray Firth. General Wade recommended upgrading these strongholds and building a new one halfway between them at what is now Fort Augustus. More importantly, he proposed a road-building scheme such as had not been in seen in Britain since Roman times: 250 miles of well-engineered, well-drained, properly surfaced roads, 16 feet wide with easy gradients, ramifying across the Highlands, along which troops, supplies and messages could move rapidly whenever there was a hint of rebellion.

The force that laboured on the Great Glen military road, the first to be built, was between one hundred and three hundred men strong. The soldiers could work only during the few clement summer months each year. They had to be protected at all times. The locals feared and despised them, the midges loved them, army discipline brutalized them. They got the job done; they garrisoned the forts, and marched out as living proof that the Hanoverian regime had everything under control.

George Wade was a gregarious man, fond of a drink, unmarried but vigorously amorous, siring at least four children out of wedlock. Promoted to Field Marshal in 1743, during the national emergency of the second Jacobite Rebellion in 1745–6 he received the ultimate accolade due to a popular hero: a special verse dedicated to him and added to the National Anthem:

> Lord, grant that Marshal Wade
> May, by thy mighty aid,
> Victory bring.
> May he sedition hush
> And, like a torrent, rush
> Rebellious Scots to crush –
> God save the King.

Wade himself didn't crush the rebellious Scots, but the military roads he planned and built made it possible for governmental forces not only to confront and defeat Bonnie Prince Charlie's Jacobite army, but also to march throughout the Highlands after the victory at Culloden in 1746 and carry out the brutal suppression and punishment raids that became known as the Harrying of the Glens. More generally, throughout the Highland region the speaking of Gaelic was banned, as were the wearing of tartan and the playing of bagpipes. Clan leaders, stripped of their local influence, sent their children south for education in the wider ways of the world. And shortly after the smashing of the Jacobite army began a phenomenon that in its bloodless way was more destructive to traditional Highland life than anything else – the infamous Highland Clearances, lasting from the mid eighteenth century until the mid nineteenth century. Clan chiefs out of touch with their tenants, along with incomers who bought into the region, began the long-drawn-out process of clearing people from their lands to make way for more profitable sheep. The emigrant ships were filled as the glens were emptied. For the Highland clans, and for the whole of Highland society, there would be no coming back from that.

I completed my Great Glen walk to Fort William along the towpath of the Caledonian Canal, with the River Lochy snaking alongside on its short journey from parent loch to sea. The canal, linking the lochs of Dochfour, Ness, Oich and Lochy, was completed in 1822 under the direction of Thomas Telford. Although fishing boats did use it as a shortcut between the west and east coasts it was never the great commercial success the promoters hoped for. But as a tourist attraction through scenic country for slow boaters and strollers, and nowadays for competitive canoeists, it has more than held its own. The landscape through which the canal passes is hilly and beautiful, but from it one never quite catches a sense of the wildness of the Northern Highlands hinterlands, nor the drama of the Cairngorm mountains hidden in the broken country that steps eastward from the Great Glen.

The bulk of the Grampian Terrane lies here to the east. Its basement cover, the rocks that geologists classify as the Dalradian Supergroup, is as widespread and fundamental throughout the Grampian Terrane as the Torridonian sandstones are in the Northern Highland Terrane. Their story goes back 1,000 million years to the time when the supercontinent Rodinia was beginning to break up during its collisions with Laurentia. Eroding rocks settled as layers of sandstone and coarser-grained conglomerates. These sediments subsided into a succession of enormous basins, filling them up and eventually weighing them down to the point where they sank. As each of these basins subsided, the sea level rose and the shore lifted to higher ground. Along the east side of the Great Glen between Invergarry and Fort William the rocks carry evidence of this rapid change from a landscape of rivers to one of wide deltas and then a shallow marine environment as the reef basins overloaded and sank.

My present journey had brought me southwest along the straight line of the Great Glen to the unlovely outskirts of Fort William. There wouldn't be time or scope to include the Cairngorms, jewels in the Grampian crown 50 miles to the east. But I could picture those mountains easily enough, having walked them in hot summer weather and slept out on their Arctic plateau in a snowhole during a bitter winter. They owe their commanding height and stubby shapes to the second phase of development of the Grampian Terrane. Around 470 million years ago a volcanic island arc, a bow-shaped belt of volcanoes, formed in the Iapetus Ocean. As the ocean closed the volcanic belt approached the ancient craton of Laurentia and eventually collided with it, exactly in the location of the Grampian Terrane. In the ensuing upheavals of the Caledonian orogeny the region's rocks were pushed out of place, folded and reshaped. Superheated magma from shallow depths rose up, full of gas. It failed to escape at the surface, but instead expanded within the crust and pushed the whole Grampian Terrane upwards several miles into the air, a massive mountain-building process that left the region resting on the granites that formed when the magma cooled into rock. These plutons or intrusive tongues of igneous rock, mostly granite in the Grampian Terrane, are blobbed across the geological map in vivid scarlets and purples. Erosion has whittled down the Dalradian rocks they infiltrated, exposing the granite plutons at the surface in famous mountain locations such as the Cairngorms.

The most famous pluton in Scotland, however, rises just to the east of Fort William. It was next in my sights. Ben Nevis, ahoy!

'Intermittent rain over western Scotland, turning heavier,' gloomed the Met Office entrail-reader, 'with widespread thunderstorms, hill fog, low cloud . . .'

A typical Highland summer's day, then. I gritted my teeth, girded my loins in Gore-Tex and set out from the visitor centre in Glen Nevis in full expectation of a storm-battered slog up the notoriously ill-tempered Ben Nevis. By my side walked Andrew Bateman, a man whose origins in the flat country of the English Midlands have naturally lit a fire of ambition to climb any peak that's put in front of him, the tougher the better. This is a man dedicated to opening the delights of the rugged outdoors to all comers. He can dig you into a snowhole for a night out in winter while plying you with whisky and plastic bags of snow-moistened porridge, or accompany you from sea level to the top of Britain's highest mountain.

Up the flank of Ben Nevis we climbed the well-worn pony track. It rose through layers of rock that carried the stamp of massive heat and pressure. First came a band of clay mudstone metamorphosed by volcanic heat into a green-black rock called hornfels, followed by a grey section of speckled grey quartz diorite, an igneous rock dotted with flecks of quartz. That melded into a band of dark granite, hard and rough to the touch. The wide and accommodating pony track, pitched with stones, is rather patronizingly known as the 'tourist path', and we passed family groups in jeans and trainers walking the track along with sterner hikers in mountain boots. Ben Nevis is a popular mountain – everyone knows it's the Daddy at 1,345 metres (4,413 feet), and a good number of visitors to the Highlands set out to climb it. There's nothing technically difficult or demanding about the path. But a wise walker doesn't underestimate Ben Nevis. It's always cold at the top, and the mountain has a well-earned reputation for sudden storms.

We plodded through a rain shower, then into thick mist. Mountain torrents tumbled over the path and crashed on down out of sight. 'There's the halfway loch,' said Andrew, gesturing at a wall of mist. Beside the path the rock had turned a patchy pink that felt smoother under the fingers. This was the upper section of the plutonic granite that formed and cooled beneath the Dalradian sandstones and conglomerates during the Caledonian mountain-building era.

Now we tackled the second and stiffer part of the climb, crunching interminably back and forth up the graphically named Zigzags. The path rose into a zone of dark volcanic rock, magma that made it to the surface and spewed out across the pluton in a molten gush. Beneath it the trapped magma continued to exude heat and pressure until a mile-wide cylinder of rock collapsed into the chamber below. A giant crater or caldera was formed, with tuff or solidified volcanic ash piling high amid the lava rock. Once more I marvelled at how such ancient scenes of violent upheaval, devoid of their sound and fury and intense spectacle, could be stripped back so prosaically to these cold wet stones on the mountain where our boots were clinking.

Suddenly we popped out of the murk. Glorious blue ridges and mountain peaks floated away on a soft wool carpet of cloud. Soon we were looking down on them all from the heart of the caldera, the boulder-strewn summit of Ben Nevis. Especially striking was the view southwards across Glen Nevis to the disembodied tops of the Mamores range, their white quartzite crests gleaming in the sunlight as though streaked with snow.

Ben Nevis is no summit for solitary meditation. Hundreds were there with us, children seeming hardly big enough to walk, one tiny girl in pink plastic sandals. 'It's lovely here,' she breathed. Sandwiches were munched, pop bottles popped around the summit cairn. An ark-shaped shelter built atop a 6-metre stone pedestal vividly demonstrated how deep the snow and frost can lie up here in winter. 'Wonderful,' enthused

Andrew, like a man who had never climbed Ben Nevis before. 'The summit's been in cloud the last seven times I've come. Just look at that . . .' And he pointed to range upon range, fold upon fold of mountains, a mosaic of dinosaur ridges, ink-shadowed valleys and intensely blue lochs. We stood and gazed, tiny beings at the apex of Britain.

Along the southerly skirts of the mountain runs the narrow cleft of Glen Nevis. A minor but busy road from Fort William, a former droving track, heads south along the glen past the visitor centre before swinging east to reach the car parks where a steep gorge forbids further progress by wheeled traffic. I set out to walk the lower glen on one of those Highland mornings when the high tops have shawled themselves up in misty cloud after days of stair-rod rain, and only a fool, or a walker with X-ray specs, is headed for the summits.

The low-level stroll followed the glen where the River Nevis pours seaward from the mountainous heart of Lochaber. This brisk morning the river ran dimpling over pebbly shallows and round bushy islets in a tunnel of alder, sycamore and ash. A sandpiper darted upriver with silvery calls and a flash of white rump. On my left hand the flank of Ben Nevis rose into smoking grey cloud, great purple buttresses of granite cut with gullies where white strings of rain-swollen torrents came tumbling – Red Burn, Five Finger Gully, Surgeon's Gully. The forward view showed the river winding from its gorge under twin peaks streaked with pale quartzite – Sgùrr a' Mhàim and Stob Bàn, one of Scotland's greatest low-level prospects.

I passed an ancient graveyard, a square of immaculate sward inside mossy walls guarded by beech trees, silent and peaceful. The squelchy path dipped to Achriabhach Bridge past a bunch of Highland cattle with ferocious horns and mild manners. They put me in mind of Matilda the Highland cow and her fellow cattle on the Famous Highland Drove. The old road followed by those adventurers of 1981 took a more northerly route than the drove road through Glen Nevis, though the two tracks crossed each other a few miles to the east. It was the drove road that ran through Glen Nevis and on across Rannoch Moor, however, that was famed as Rathad nan Eilean, the Road to the Isles. Harry Lauder immortalized it in 1926 with his huge hit version of 'The Road to the Isles', as he sang, 'Sure by Tummel and Loch Rannoch and Lochaber I will go . . . as step I wi' my cromach tae the Isles.' Standing by the bridge at Achriabhach I recalled a walk I'd done some twenty years before from the remotest railway station in Scotland, 15 lonely miles east of here. My walking companion had been Irvine Butterfield, chronicler of the Famous Highland Drove of 1981, as we traced the route of the Road to the Isles across the wild wastes of Rannoch Moor.

Corrour Station lies right out in the wild at the northern edge of Rannoch Moor, a great expanse of unbroken blanket bog, loch and rough heather plateau that covers 50 square miles to the southeast of Ben Nevis. This is one of the wildest places in Scotland. Dalradian sedimentary rock underlies it, with a granite ceiling above. The mountains are ice-smoothed whalebacks on the skyline, carved during the last Ice Age by an ice cap more than half a mile high which scraped the granite and dumped it in loose heaps and drumlin hillocks. As the glacier receded, meltwater floods filled the scars and scrapes, creating a landscape very like that of the interior of Lewis, a wide boggy expanse of 'knock and lochan'. The fugitives David Balfour and Alan Breck Stewart, fleeing from the redcoats, were trapped on Rannoch Moor by their creator Robert Louis Stevenson in *Kidnapped*, and found it a desolate place:

The mist rose and died away, and showed us that country lying as waste as the sea; only the moorfowl and the peewees crying upon it, and far over to the east, a herd of deer, moving like dots. Much of it was red with heather; much of the rest broken up with bogs and hags and peaty pools; some had been burned black in a heath fire; and in another place there was quite a forest of dead firs, standing like skeletons. A wearier-looking desert man never saw . . .

Across this forbidding waste the drovers of yesteryear followed the Road to the Isles in reverse as they walked their cattle to Crieff market.

I was pleased to get an invitation to walk with Irvine Butterfield. My idea for a geological journey across the British Isles had been simmering for a little while, and so had the notion of following the route suggestions in *The Famous Highland Drove Walk* through the mountains of the northwest. 'We won't bother with that slog from the Grey Corries down past the Mamores,' Butterfield said decisively. 'Corrour's the place to start. We'll get the train from Rannoch to Corrour and walk back from there along the Road to the Isles. A good track, and you'll see quite as much of Rannoch Moor as anyone could want.'

The two-car sprinter train from Glasgow rattled north across the eastern edge of Rannoch Moor, sending a herd of red deer bouncing away from a loch beside the line. Cloud shadows dimmed the moor colours to sombre tans and olives; then the sun slid through again and spread thick gold across heather and grass. The light train gave a gentle spring every now and then, a reminder that when the West Highland Line was built across Rannoch Moor in the 1890s the engineers had to float the track on bundles of brushwood. The treacherous peat, up to 7 metres thick in parts, was far too yielding for any conventional hard foundation. Railway construction on shelterless Rannoch Moor could be a dangerous business. On 30 January 1889 a party of seven engineering experts, including the West Highland Line's contractor Robert McAlpine, set out on foot in their city suits to prospect the 40 miles from Spean Bridge southward across the moor to Inveroran. This was a brave if not foolhardy venture in midwinter. A pea-soup mist descended and the party became scattered, then separated. Benighted and lost, with one of the party helpless with exhaustion, they had to be rescued.

At Rannoch Station Irvine Butterfield hopped on to the train, stocky and badger-bearded, a Skipton-born man with a rich Dales accent and a deep well of knowledge of Scottish landscape and culture. As we pulled away for Corrour he leaned forward and pointed out of the window. 'That's Loch a' Chlaidheimh, the Loch of the Sword. Now there's a great story attached to that, about the time the Earl of Atholl agreed to meet alone at the loch with the Lochiel of the Camerons of Lochaber, to settle which of them the land belonged to. But both had men hidden in the heather. When these guys jumped up and faced each other, the chiefs declared a kind of draw. Atholl threw his sword into the loch, saying the land would belong to Lochiel and Lochaber until the sword should be found. Anyway, in 1826 a boy actually fished Atholl's sword out of Loch a' Chlaidheimh. But no one wanted to stir things up, least of all the Lochaber people to whom Atholl had ceded the land! So they took the sword off the boy and chucked it back in the loch again – where it still is.'

Corrour Station lies 15 miles east of Glen Nevis as the kyloe plods, and at least 10 miles from the nearest tarmac road. It has turned itself into a walkers' guesthouse, a beacon of light, warmth and good cheer in the vast wastes of Rannoch. Here we left the train and struck out along Rathad nan Eilean, the Road to the Isles, making for Loch Ossian in its cradle of mountains. As we crunched down to the lonely youth hostel by the lake, Harry Lauder in my inner ear was singing 'As step I wi' my cromach tae the Isles', and I was tickled to see that Irvine Butterfield was stepping with his own fine cromach, a long stick with a carved ram's horn crook.

The Road to the Isles left the stony lochside and ran parallel to it before plunging abruptly away southward, a firm green track founded on a bed of stone, with rough stone culverts carrying burns beneath it. This was no casual cattle path wandering through the heather, but a proper moorland road maintained down the centuries by the drovers who depended on it for their livelihoods. 'They didn't make these tracks because they liked playing about with cattle, you know,' observed Butterfield. 'They chose the best routes, the routes that would give them shelter for the night and keep them out of the bogs.'

I looked round the bleak scene: the rolling empty moor, stripped hundreds of years ago of the Great Wood of Caledon that once covered its 50 square miles; the inhospitable granite humps of the mountains;

ragged edges of lonely lochs. Where would the drovers find shelter for the night? 'Oh, there'd be stances along the way, rough-and-ready inns with a patch of grass. The men could have a dram and put their heads down while the cattle grazed. But when I see this landscape I think of winter and other travellers on the track, in a blinding snowstorm maybe, or lost in mist. Rannoch doesn't always look like it does today, you know!'

Rannoch Moor can be cruel. On a rock beside the track a plaque commemorated Peter Trowell, 29-year-old warden of Loch Ossian's youth hostel, who disappeared while wintering there alone in 1979. Somehow, he drowned in the loch, and was not found until spring when the lake ice melted. The carefree verse below the inscription only sharpened the poignancy of the story:

> I have a friend, a song and a glass,
> Gaily along life's road I pass,
> Joyous and free, out of doors for me,
> Over the hills in the morning.

As we followed the green ribbon of the Road to the Isles a wonderful panorama of mountains unrolled to all quarters – unknown peaks to me, but Irvine Butterfield had each of their identities nailed. Behind in the northwest stood the pale quartzite ridge of the Grey Corries, with Ben Nevis high and mighty in a blanket of snow beyond. Down in the southwest Buachaille Etive Mòr, the Big Shepherd of Etive, formed a right-angled cone over invisible Glencoe, and to its left rose snow-streaked Beinn Achaladair. 'Ah, that's a great name, meaning the Field of Hard Water. The mountain overlooks a place at the head of Loch Tulla that floods and then freezes every winter.'

The granite rocks along the track were speckled pink and grey. We passed the ruins of Corrour Old Lodge, a broken wilderness of grey stone walls, and drank from a spring beside the path, cold and sweet with a hint of iron-flavoured snowmelt. A tiny frog watched us from the mossy ground with unblinking golden eyes. 'Oh, that spring water's good,' said Butterfield, wiping the drips from his beard. 'Some walkers take it too far, though, drinking nothing but water, making a temple of their body. Well, bollocks to that. Make mine a pub. Mind you, I have to admit that I did overdo it once. Next day the sweat was pouring off me. I got to the top, the icy blast hit me, and pow! I went like a hare, up three Munros, one after the other.'

What about rock climbers, Irvine, those paragons famous for their iron control, for subsisting on hard tack and cold water? 'Crag rats? I'm amazed some of them in the old days could stand up in the morning, let alone climb. "They knew no fear", that was the image, but I suspect they were fearless because they were concentrating on not being sick. People weren't so bloody *earnest* about their walking and their climbing back then. There seemed to be more fun in it, I suppose.'

We came down the long track to Rannoch Station in cold clear sunshine. Ahead across the waters of Loch Rannoch, the lonely peak of Schiehallion ('Three thousand, five hundred and forty-seven feet,' intoned my companion automatically) stood islanded at the end of a breaking sea of mountain tops, its northern slopes white with snow. Butterfield paused and stared at it for a long time, leaning on his ram's horn cromach. 'You know,' he said, 'I'm thinking of that young man wintering alone in the Loch Ossian hostel, living hard just because he wanted to, because he loved the life out here. In the end, no one owns the mountains; they own you.'

If any mountains feel 'owned' by the public, it's those that hang so ominously and massively over the Pass of Glencoe on the road south from Fort William. Everyone with a taste for dramatic landscape heads for Glencoe. These close-packed mountains with their lofty, rugged profiles offer an instant hit of serotonin. They are famous climbing ground, fabulous walking terrain, peerless subjects for photography in snow or sunlight or threatening weather. The crocodile-backed ridge of Aonach Eagach bulks large to the north as you drive the A82 through the narrow jaws of the pass, while staring across at Aonach Eagach from the southern flank are the Three Sisters of Glencoe, Aonach Dubh and its neighbours Meall Dearg and Geàrr Aonach. They are rather troll-like in appearance, these triplets, with blunt face and simian brows, massive blocks of mountains. They slant back from the pass, but such is their presence that they give the impression of overhanging it.

Glencoe has not just an ominous air, but a bloody history. It was among these enormous crags that terrified men, women and children sought refuge in a snowstorm in the early hours of 13 February 1692 during the Massacre of Glencoe, a brutal piece of ethnic cleansing in which dozens of members of the local MacDonald clan were murdered by soldiers billeted in their houses. Gaelic-speaking Highlanders such as the MacDonalds, loyal to their clan chief Alasdair MacDonald, 12th chief of Glencoe, were seen as barbaric and incorrigible cattle thieves, and as traitors to the new political order under the Protestant monarchs King William and Queen Mary of Orange. 'They must all be slaughtered,' were the unequivocal orders to the officer in charge, Captain Robert Campbell of Glenlyon. 'Put all to the sword under seventy . . . have a special care that the old fox and his cubs do upon no account escape your hands.'

MacDonald's heirs did in fact survive the slaughter, though the chief himself was murdered. But the Massacre of Glencoe was the first of a succession of mortal wounds to the Highland way of life. Its death was finally confirmed some fifty years later in the punitive aftermath of the Battle of Culloden. The independent, traditional, cattle-raiding society of the clans was finished, and would never return. Bonnie Prince Charlie's romantic failure had sealed its fate.

The birth era of the Three Sisters and the other mountains of Glencoe was the Silurian (444–416 mya), when a hot, dry and mountainous Scotland lay 20 degrees south of the equator and felt the full tectonic force of the Caledonian orogeny. About 420 million years ago magma at depth rose up through faults in the overlying Dalradian rocks, forming sills which spread sideways through old river and lake sediments. The dark rock of the sills shows up very strikingly as columnar ledges, in particular on the north face of Aonach Dubh ('the Black Ridge') as it looks towards the Pass of Glencoe.

Capping the dark sills of Aonach Dubh is a layer of pink-grey solidified lava that poured out of volcanic vents above a huge magma chamber, measuring nearly 10 miles from side to side, during later violent eruptions. These enormous volcanic explosions were fuelled by masses of gas in the magma as it welled up through faults in the chamber roof and caused the rocks above to burst out at the surface, forming pyroclastic flows such as the one that obliterated Pompeii. Rock fragments, dust and pumice were hurled miles into the air and fell back to Earth, then instantly raced away from the vent like a shock wave and flowed down the steep mountainsides, at the speed of an avalanche but hot enough to melt rock and anything else in the way.

As at Ben Nevis, the roof of the magma chamber collapsed and the rocks above, together with the sills infiltrated by the previous plutonic events, sagged downwards, a process rather graphically called 'cauldron subsidence'. The oval depression or caldera they formed was filled with tuff (petrified ash) and volcanic glass, a granitic mixture known as ignimbrite rock that solidified out of the chaos of the pyroclastic flows. It is this hard, muddled rock that forms the brows and craniums of the Three Sisters of Glencoe. And the same story of fire and fury is repeated just along the glen in the craggy shapes of the Big and Little Shepherds of Etive, Buachaille Etive Mòr and Buachaille Etive Beag.

After a week of mist and murk, finally a day of clearing weather in Glencoe, the southwest wind less stormy and cold, the air less rainy than for the past few days. After wavering about which hill to climb, which path to take to get the best view of these magnificent mountains and their geological setting, I opted for the Devil's Staircase at the eastern extremity of the Aonach Eagach ridge on the north side of the road.

This fine zigzag hill path follows yet another very old drove road, probably dating back to the fourteenth century or earlier, by which the black cattle could be brought from the road to Skye over into Glencoe. After the Massacre of Glencoe and the Jacobite rebellions of 1715 and 1745, General George Wade's successor, Major William Caulfeild, was charged with constructing a military road north across Rannoch Moor. The course of the old mountain path was judged the best route across the Aonach Eagach ridge. Converting the old track into a military road involved hauling and carrying heavy equipment up and down the steep bends in the mountainside. The soldiers on the job in 1752 detested it, and nicknamed the zigzag climb 'the Devil's Staircase'. Some 150 years later, the construction of the Blackwater Reservoir dam to the north saw labourers camped out in winter conditions. These thirsty men would use the Devil's Staircase as a shortcut to and from the Kingshouse Hotel by the road in Glencoe. When it came time to walk back, full of beer, late at night in often atrocious weather, the zigzags proved too much for many; their bodies lay under snow until they could be retrieved with the following spring thaws.

Setting off from Altnafeadh beside the River Coupall, I followed the very well-trodden route of the West Highland Way as it climbed north up a steep slope towards the two opposing peaks of Stob Mhic Mhartuin and Beinn Bheag. Many thousands walk the West Highland Way between Fort William and Milngavie near Glasgow each year, and their boots have forged a stumbly, pebbly channel in the peat and heather, impossible to miss. The rocks were salmon pink and grey granite, most of it very rugged, breaking down into large quartzy crystals as coarse as gravel, shot through here and there with heavily striated dykes. Where the Allt a' Mhain stream sluiced across, the water rush had scrubbed and smoothed the rocks to a beautiful smooth orange, a very intense colour.

Flowers that love acid heathland spattered the hillside. Pure white lousewort grew in wet patches of the boggy heather. Flea-sedge with tiny primrose-yellow flowers, milkwort and little rowan trees a foot high flourished where no sheep's teeth could snip them off. Bilberry

was flowering, the tiny pink bells dangling. Bog myrtle released its res-
inous smell when pinched and sniffed. On the bright crimson sphagnum
grew branched lichen, as pale and delicate as the finest coral.

It was one of those 'watch your step, take the next zig and zag as they
come, how long to the top?' slogs up the Staircase, but at last the path
smoothed out at the bealach. Here I turned off the West Highland Way
on to a narrow stony track that rose across slippery slabs and squelchy
black bog to the modest cairn at the summit of Stob Mhic Mhartuin,
400 metres above the floor of Glencoe.

Looking across from this position, the Big and Little Shepherds of
Etive were revealed in all their glory, Buachaille Etive Beag on the right
looking east across the tight-squeezed glen of Lairig Gartain at her big
sister Buachaille Etive Mòr. Their main bodies lay back to the southwest
and couldn't be properly seen from my vantage point, but each mountain presented a northerly face rising to a peak – Stob nan Cabar (800
metres) on Beag, Stob Dearg (1,022 metres) on Mòr. The two Buachailles,
and the other peaks in view to the southwest, were dark with a semi-
molten look, as though liquid rock had been caught in the act of
flopping downhill.

The face of Buachaille Etive Mòr was a lumpish one, its lower half
hidden in scree, the upper section pale and wrinkled. Snow streaked its
gullies and the deep channels where volcanic dykes had once pushed
their fiery magmatic tongues in among the rocks. The mountain was
footed on pre-volcanic Dalradian sandstone and mudstone metamorphosed into psammites, semi-pelites and glassy quartzites. On top of
those lay a flow of shales and coarse conglomerates in an alluvial fan,
before three thick bands of rhyolitic or volcanic rock, separated by
bands where rivers had deposited sediments in the long periods
between the short, sharp volcanic episodes. Above that a blobby mass
of pale pink rock rose to the peak of Stob Dearg, a cap of ignimbrite
rock produced by the pyroclastic flows from the later, violent explosions of the Glencoe volcano. Tiny figures were moving cautiously
along up there, making for the summit cairn.

Buachaille Etive Beag, by contrast, showed a boldly slanting sequence of rocks tilted to the northeast. Here the lowest part of the mountain, a thick band of coarse glassy tuff or volcanic ash, was topped by a fan of alluvial stones and rocks, then angular blocks of Dalradian rock that sank into the caldera when the roof of the magma chamber collapsed. Capping those was nearly 250 metres' thickness of tuff, the upper section around Stob nan Cabar from a single massive pyroclastic flow.

I looked back at the scree slopes of Buachaille Etive Mòr. That great block of rock sticking out low down must be the Waterslide Slab. What did my notes say? 'Remains of fossil plants found here.' That was it. But somehow I knew this was a definitive moment on my journey. Up till now I hadn't given more than a passing thought to the dawning of life among the bones of Britain. Nothing had lived or died among the fabulous stripes of the Lewisian gneiss. There had been those billion-year-old stromatolites back on the shores of Coigach, but the microscopic organisms they represented were too hard to relate to. Of all the abundance of life in the Carboniferous period, the giant fish and steamy jungles, there had been little or no hint among the psammites and pelites of the Northern Highlands Terrane. But these bald words scribbled in my notebook, 'Remains of fossil plants', brought home to me a sense of separation. Here on Stob Mhic Mhartuin I was saying goodbye to the lands of fire and fury, the most dramatic landscapes in Britain, forged over thousands of millions of thunderous years when rocks clashed and melted, ripped at each other and rose into mountain chains the height of the Himalayas. Barren ages, back to the molten world of no oxygen and no life anywhere. Ahead of me there would still be volcanic episodes written in the rocks of this long journey through the geological body of Britain, earth movements, faults and upheavals that changed and moulded the landscape. But there would be abundant life, too, developing and ramifying; not simply the struggle for existence of single-celled blobs, but the magnificent expansion of life on Earth.

Geological tales printed in the rocks, so neat and orderly to read about in pamphlets and research papers, are naturally a damn sight harder to interpret on the ground. I sat and puzzled over the heavily lined and wrinkled faces of the two Shepherds of Etive. Glancing from Mòr to Beag, down to my laboriously copied notes, up to Beag and back to Mòr, I wondered if I would ever be like a proper geologist in the field. 'That's a rhyolitic extrusion, and *that* . . . is a lithic tuff with breccia lenses.' Such zen-like assurance seemed a long way off yet.

The cataclysmic era at Glencoe was drawn out over about 5 million years, with tens of thousands of years between some of the eruptions, long enough for lakes to form in the caldera and water to accumulate underground. Each event saw red-hot magma rising to meet cold water, supercharging the explosive energy of the ensuing eruption. Most of the upper rock formed by these later upheavals has been eroded away. But as the rocks within the caldera descended, and the magma continued to rise and intrude along the faults or cracks in the surrounding rocks, it solidified into a ring – more properly, an oval shape 9 miles long and 5 miles wide – of granitic rock around the rim of the caldera.

The ring fault that bounded the great Glencoe cauldron outcrops spectacularly just under the peak of Stob Mhic Mhartuin. I found the band of pink and grey rock, tilted on high and pushing in among the Dalradian quartzite of the mountain, a curiously smooth surface to run my fingers across as I stared across the glen. Tiny slivers of humanity were still inching up the ridge of Buachaille Etive Mòr. On this side of the glen walkers on the West Highland Way were crossing the bealach towards Kinlochleven. Up here on Stob Mhic Mhartuin in the wind and sun there was only me with the rocks of the ring fault under my hand, a ghostly imprint of the long-obliterated volcano in the landscape of Glencoe.

Next day threatened rain showers across Rannoch Moor. Writers have always enjoyed catastrophizing over the famously dour moor. 'An inconceivable solitude,' wrote geologist Dr John MacCulloch in *The Highlands and Western Isles of Scotland, in Letters to Sir Walter Scott* (1824), 'a dreary and joyless land of bogs, a land of desolation and grey darkness.' The authors of the West Highland Way official guide, Bob Aitken and Roger Smith, remark dryly of Rannoch Moor: 'In rain or snow with low cloud driving before a gale, it tends to promote the conviction that Hell need not be hot.'

The tiny, cheerful Walkers' Bar at the Inveroran Hotel, 10 miles south of Glencoe, was full of wet refugees from the West Highland Way, which very conveniently passes the door of this delightful and remote hotel at the foot of Rannoch Moor. The lobby was piled high with massive travel bags awaiting onward transportation in a Sherpa van. Down beyond the hotel at Victoria Bridge a cuckoo was calling from the trees by the river. A still afternoon, perfect for the hatching of the first hungry midges of the season. I sprayed on Smidge, and set out to trace the faint and mysterious thoroughfare known as the old military road.

General George Wade's name is forever associated with the making of military roads in Scotland, but Wade had retired by the time the road across Rannoch Moor was built by his colleague and successor, Major William Caulfeild. With the Scottish Highlands in a ferment after the defeat of Bonnie Prince Charlie and the Jacobite clansmen at the Battle of Culloden in 1746, the government decided that Wade's network of military roads had to be extended in double-quick time to get men and materials to wherever rebellion raised its head. Major Caulfeild's road across Rannoch Moor to Glencoe and Fort William was one of these. It was completed in two summers, 1751 and 1752, with three soldiers keeping guard against ambush for every squaddie labouring with pick and shovel on the road itself. It was built some way up the hillside, partly to keep high above the sucking bogs of the glen floor, partly because the mountain torrents could be crossed by means of cobbled fords rather than lower down where a proper stone bridge would have been needed, expensive in money, time and materials.

Nowadays Major Caulfeild's road has been more or less absorbed back into the moor grass, bog and heather. I had to get my eye in, letting my gaze roam across apparently trackless moor until it snagged on a darker groove in the grass, or caught the faint curve of an embankment.

The old military road is a well-engineered 'wild road', something you can appreciate once you learn how to spot it in the landscape, running in long straight stretches with occasional deviations. The cuttings are supported with stone walling and there are frequent side scoops showing where the soldiers quarried the rocks and pebbles needed for the road. I paced out the width between a cutting wall and an embankment slope at a point where the shape of the road looked fixed. It was about 20 feet wide, enough to allow one file of soldiers to march past another.

Rannoch Moor is floored with granite, with Dalradian sedimentary rock underpinning the southern section. Over this floor the glaciers laid a carpet of sand and gravel, and later rivers carved out terraces of pebbles and gravel, sand, silt and clay. These necessary materials for road-making, so providentially at hand, were brought up from the river plain by horse and cart for the works.

Looming over the old military road as I followed it northward was the craggy whaleback of Beinn Toaig. Up its flank started two red hinds, their white rear patches betraying them instantly as they bounced away. The ice carved and sculpted not only this fine hump and the taller sunset-red peak of Stob a' Choire Odhair behind it, but also the tremendous southern Grampian mountains that rose east of the Water of Tulla – Beinn Achaladair (1,038 m/3,406 ft) with its high corrie facing north, flanked by Beinn an Dòthaidh with a tilted top to the south (1,004 m/3,294 ft) and the serene weather-eroded cone of Beinn a Chreachain (1,081 m/3,547 ft) to the north.

Across the wilderness of the moor the old military road advanced. Direction-finding was helped by a series of sighting stones on successive skylines beside the road. Sprays of blotched leaves as slim as penknife blades showed where heath spotted orchids would be blooming come full summer. Rather surprisingly, there were also several clumps of wood anemones in full flower, a whisper from the long-vanished Caledonian forest that covered Rannoch Moor before man arrived to chop, fell and clear. At a block of forestry the character of the old military road changed from an open moorland track to a black peat sludge, both sticky and slippery. I forded a couple of streams and followed the forestry fence down to reach the West Highland Way in the valley below.

By the time Scots-born engineer Thomas Telford received a commission to build a new road across Rannoch Moor early in the nineteenth century, William Caulfeild's route was already half reabsorbed into the moor. Telford's brief was to build a proper level road, broad and stone-surfaced, crossing streams and rivers by solid stone bridges, with the roadway well blanketed in gravel and sand, kinder than cobbles on the hooves of the flocks and herds that the drovers brought along the new road. The old military road half a mile away up the hillside was never really used again, and Telford's road, completed in 1803, itself lost its importance when what is now the A82 was constructed parallel to the two older roads as a bypass across Rannoch Moor in the 1920s.

Telford's road languished, like its military predecessor, until in 1980 the West Highland Way was opened along its disused stretches. Now it's a very popular five- or six-day challenge. I squelched down to it, bog-hopping and stream-scrambling, and turned for home along the cobbled roadway. The raised stones burned and bruised my feet, now accustomed to the soft black peat mud of the old military road, as they had bruised the hooves of the cattle some forty years before on the Famous Highland Drove. I followed the ghosts of Matilda and her posse down the knobbly road. The mountains to the east stood in soft evening sunshine, their high corries shining. Wood anemones along the roadside were shutting their petals for the night, and a lone walker sat eating from a pannikin at the door of her tent beside the West Highland Way, comfortable Crocs on her feet, boots steaming beside her, the wide moor all around her, and a grin of pure satisfaction across her face.

Major William Caulfeild's military road arrows across the blanket bog of Rannoch Moor.

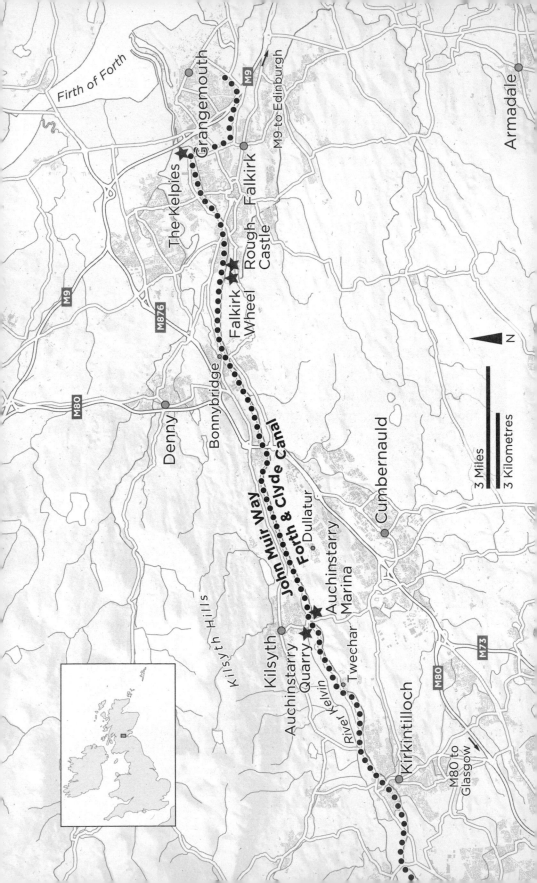

6

Whinstone and Wheel:
Kirkintilloch to the Kelpies

STONE QUARRIES ARE NOT REALLY places of great beauty. The old disused roadstone quarry in the Borough of Partick on the northwest outskirts of Glasgow had become a bit of an eyesore by the mid-1880s, when thoughts were beginning to turn to how best to mark the Golden Jubilee of Queen Victoria's accession to the throne. Why not celebrate the nation's good fortune by creating a new public park around the worked-out excavations? So the idea for the city's Victoria Park was born.

Public interest in rock gardens had been stimulated by the explorers of the British Empire and the exotic trees and plants they brought back to home ground. There was a taste for rational recreation, a little light learning to spice up a stroll, along with the necessary delicious smack of the wild, even if it was only an 'awful crag' hacked out of a quarry face. So a secluded retreat within the larger park was planned where you could walk between miniature canyon walls, under ledges dripping with ferns, beside water trickles and on beneath shady conifers to admire the colourful bedding displays. By 1887 the work was well under way. Artful little cliffs and grottoes were being shaped out of the old quarry ledges. One autumn day in that Jubilee year, workmen were forging a pathway through the rock when they came across a dozen cylindrical and oval objects a yard or so high, each about the size of an occasional table, sunk in the sandstone floor of the cutting.

What were the chances of the labourers simply destroying these obstacles with a few pickaxe blows? One of the gang must have responded to the incongruity of the round shapes and called the boss. Representatives of the Geological Society of Glasgow were summoned to the site and the objects were carefully excavated, revealing long extensions stretching away from their bases. They were the fossilized stumps of lycopods, primitive trees complete with their roots. They dated back to Carboniferous times, which made them around 325 million years old; and the only reason they had survived the decades of stone extraction at the quarry was that the quarrymen had been after the hard volcanic whinstone or dolerite, so useful for sealing road surfaces, and hadn't dug down into the much older layer of sandstone just below it where the fossil trees were rooted.

The lycopod forest of Victoria Park became a public sensation. A brick building with a slate and glass roof was erected over the stumps, and it became quite the fashion to end your perambulation round the fern grottoes and rock gardens of the ornamental park with a visit to the fossil trees.

On a changeable Glasgow morning I found Dr Neil Clark, Curator of Palaeontology at the Hunterian Museum, waiting for me with his wife Clare at the gate of the Fossil Grove. Rain and sun glanced in succession off the glass overhead as Fossil Grove Trustee David Webster unlocked the door to let us into the chilly, dank interior of the Fossil House. 'Damp and mouldy after ill-advised renovations in the 1900s,' apologized David briskly. 'The fossil trees don't take kindly to the condensation, which is damaging them, so if you can lay your hands on the eight hundred thousand pounds we need to do the work on the building, we'll be glad to have it.'

Large though that sum is, it seems a small price to pay for preserving the unique geological treasure that lies in the leaky old Fossil House. We stood at the brink of a dark excavation pit, rectangular and ashy grey. The sides were of dimpled dolerite above layered sandstone, sloping to the floor of the pit. Down there were the fossil stumps, a couple of feet tall with smoothed-off tops, each supported on branching tendrils. These lycopods were arborescent plants, organisms that resembled the trees of today. They grew on this spot when Scotland was drifting near the equator in a steamy tropical climate some 325 million years ago. Back then the sea level was fluctuating. This part of Scotland was sometimes under shallow seas, sometimes just above them on the shoreline of great estuaries. Giant insects flew and crawled in swampy forests where the lycopods supported themselves on bifurcated roots. These primitive plants were scarcely a metre in diameter, but up to 40 metres tall. Their diamond-patterned stems sprouted leaves all the way up to the forked branches from which dangled great conical fruiting bodies. Their trunks were hard-shelled, but their hearts were tender and pithy. Being so tall, slender and soft-centred, how they held themselves upright is a bit of a mystery. Perhaps the forest canopy high above formed an interlocking mesh of branches and foliage, so that the plants were part of a mutual support system.

Lycopods were the ancestors of trees, and they rotted and fell in enormous numbers over millions of years, squashing down into ever-deeper layers of peat that eventually metamorphosed into seams of coal. After the lapse of hundreds of millions of years it's incredibly rare to find a fossil example of an erect lycopod, let alone a cluster of upright trunks a dozen strong. So how did these specimens survive intact all this time?

'Probably,' said David Webster, 'the forest was hit by a massive flood, what miners call a washout, which toppled all the trees.' He pointed down from the viewing gallery to where a thick black spar lay crosswise between two of the lycopod trunks. 'That's the trunk of one that got completely knocked flat. But these lycopods we're looking at were snapped off short by the flood, a few feet above their roots. Of course, the soft centres of the stumps soon decayed, and the hollow trunk became filled with sediment, sand and silt, swirling around in the flood. That hardened into casts of the trees. The outer skin became a coaly rind around the cast, and that's how they stayed for the next few million years while layers of sandstone, siltstone and mud built up on top of them.'

Neil Clark indicated the rocks around the edge of the pit, several metres above the level of the stumps. 'Maybe thirty million years after the great flood, around two hundred and ninety million years ago, there was another great volcanic episode, the intrusion of a sheet of dolerite basalt – the Whin Sill, in fact, which you'll meet in several places quite dramatically in northern England. The dolerite spread and hardened above the silty and sandy layers in which the tree fossils were set.' He gestured across the grey stumps at the volcanic rock above. 'When it came to quarrying here, the quarrymen simply didn't dig down far enough. That's what saved the fossil lycopods.'

After dodging the weather among the mountains and tramping down the West Highland Way on sore feet, it had been good to spend a couple of days cooling my heels in Glasgow. But now I could feel the long road nudging at me again.

Somewhere towards the lower end of Loch Lomond I had crossed the Highland Boundary Fault that marks the southern edge of the Grampian Terrane. Here begins the fourth of Scotland's five terranes or geological areas. This Midland Valley Terrane, bounded on the south by the Southern Upland Fault, is an area of much younger rocks, but has no less striking a geological history than the three terranes – Hebridean, Northern Highlands and Grampian – that lay behind me. I was stepping down that chronological ladder now, getting in among the younger rocks as the journey continued to the south and east. It was still nothing like the orderly progression down the ages from A to Z that I had pictured when with great excitement I first sketched out the Bones of Britain route across the geological map torn out of my *Philip's Modern School Atlas*. But here I was, a lot older, maybe a touch wiser, and at least no longer afraid of the word 'orogeny'. That was something gained, wasn't it?

I couldn't hang around the Fossil House for long. It was time to get out of town and push on. I had a date with a lady up in Kirkintilloch, and I'd been warned not to be late. She was going to take me on a journey right across the neck of Scotland from Clyde to Forth. But not by road, track or path. The next few days we'd be cruising on Easy Street, the lady and I.

Maryhill lay moored in Southbank Marina at Kirkintilloch. What a trim and lovely little craft. She's a one-third-size replica of a Clyde puffer, the chunky pilot vessels that fussed about bigger ships on the sea approaches to Glasgow in days gone by. *Maryhill* is kept in good fettle by the dedicated members of the Forth & Clyde Canal Society. She grunts and rattles to and fro with her cargo of pleasure-seekers along the 35-mile-long Forth & Clyde Canal. This man-made slash of water connects the country's two great lowland cities, Glasgow and Edinburgh, and also provides a direct shortcut between the Irish Sea and the North Sea.

'I'm Dave Connell, and these are James, Bill and Christine.' The four members of the F & C Canal Society who stood waiting around the narrow deck of *Maryhill* were all retired; self-deprecating, humorous people, able to slip a round turn and two half hitches over a stanchion at a second's notice, but also capable of steering full tilt into a reed bed or the retaining wall of a lock during a momentary lapse of reason. We got along very well over the next couple of days, sitting on the cabin roof in relays and watching the world go by at a nice, civilized 4 miles an hour.

You soon get the basic geographical picture. The Forth & Clyde Canal, along with a tangle of roads and railways, traverses a long valley containing two rivers. The Kelvin flows west down to Glasgow, and from the watershed at Kelvinhead the Bonny runs east to the Firth of Forth. The valley is enclosed by hard knobbly hills, the Kilsyth Hills backed by the Campsie Fells to the north – these always in view as a tall ridge – and a series of gentler hills to the south. The canal snakes its way between the foothills of these modest ranges through a landscape of grazing pastures and boggy marshes interspersed with a scatter of small towns, many economically depressed these days. Only the presence of the occasional bing or mine waste heap hints at former activities here, when coal and iron, alum and stone were dug from pits and quarries all along the valley. It was the underlying geology that shaped the fate of these settlements, a series of events over the kind of long-drawn-out timescale that I was becoming increasingly used to by now.

Three hundred and fifty million years ago, when Scotland lay just south of the equator, massive volcanic eruptions and lava flows under the sea hardened into a basalt sheet up to 500 metres thick. For the next 50 million years, during the Carboniferous period, the basalt lay in warm shallow seas, along estuaries or under the kind of steamy tropical forests that sheltered the lycopods of the Fossil Grove. The sea level rose and fell. Limestone, sandstone, coal and ironstone were laid down in their turn, and along with that up to 2,000 metres' thickness of softer sediments – the remains of corals, crinoids and shells, settling in warm seas now rich in oxygen and teeming with life.

As the Carboniferous period closed, the great volcanic event that produced the Whin Sill intruded a sheet of dark dolerite in among the softer sediments, the same dolerite that lay above the fossil trees of Victoria Park. Earthquakes fractured rocks in a series of faults, including the Campsie Fault along the northern edge of the Kelvin Valley. That produced a 'downthrow' or downward movement, similar to that mighty displacement back in Coigach so coolly mentioned by Peter Drake as we walked the shores of Achnahaird Bay. This one in the Forth–Clyde Valley, nearly a billion years later, pushed the rocks down by well over half a mile, so that softer sediments now butted up face to face with hard lavas.

On the south side, the Kilsyth Fault dropped the ground another 300 metres. A dramatic illustration of these displacements was the efforts of miners in the 1880s to get at the valuable resource of coking coal. A seam of this material reaches the surface high on the slopes of Garrel Hill to the north of the valley. The Campsie Fault caused the seam to slip down 1,000 metres or more from its original position. The downthrow from the Kilsyth Fault dropped the seam another 300 metres, so that it now lay far beneath the valley. When the demands of steel-making made it profitable to mine the seam, a shaft of 413 metres had to be sunk from Dumbreck in the valley bottom, down through the doler-ite layer to reach it.

Over the following 300 million years the softer sedimentary rocks eroded. The hard lavas remained to the north of the Campsie Fault in the shape of the Kilsyth Hills, and on the south side of the valley the Whin Sill dolerite formed the hills of Bar and Croy. Soft deposits and allu-vium still floored the Kelvin Valley, however. During the great glaciation that commenced a million years ago, fast-flowing water under thick ice scoured these out to great depths, over-deepening the valley up to 50 metres below today's ground level in the Kilsyth area. What the Ice Age took out, it then gave back as rocks, sand and gravel were washed into the trench and filled it. Finally, as the ice melted and retreated, a thick blanket of boulder clay was smeared over hills and valleys.

The ice was succeeded by torrential rains, flooding across a bare land-scape of sand and gravel deposits. Gradually forests and bogs developed along the valley. Man arrived and prospered on the fertile soils. Hunter-gatherers gave way to farmers who cleared the forests and drained the marshes. Then came industry and its handmaidens: roads, railways, mines, quarries – and the Forth & Clyde Canal.

We set off from Kirkintilloch, *Maryhill*'s exhaust bubbling fatly. At every lock and swing bridge there was a cohort of gate openers and paddle raisers already in position, not necessarily members of the Forth & Clyde Canal Society, but all volunteers from Scottish Canals. This powerful and well-funded body oversees and carries out all the admin-istration of Scotland's canals, as well as the physical processes necessary at each lock: winding windlasses, shoving heavy gates, preparing the next lock to be either full or empty depending on the requirements of the approaching boat. It's a prescriptive and controlled set-up, utterly unlike the English and Welsh canals where you can hire and drive a narrowboat as long as a lorry without ever having set foot in one before.

'That throttle for fast and slow,' you are told, 'this tiller for left and right, *don't put your arm down that hatch.* All right? See you next week,' and you are off and running, free to ram the next bridge, lock or oncoming boat as soon and often as you like.

The F&CCS volunteers are universally elderly, retired, friendly and very involved. They are also worried about whether their society will survive very much longer. No one young is stepping up to take their place. Without their constant promotion of the canal, raising money and chartering the three boats they own, there would be little traffic west of Falkirk, where the ingenious boat lift at the Falkirk Wheel and the giant sculptures of horses' heads known as the Kelpies attract the visitors and the day cruisers. But retired volunteers can only take so much. 'We had our containers broken into,' said Stewart Procter, chairman of the F&CCS, when he joined us a little way along the canal. 'They took all our tools, everything we'd collected to do our work. So we got a new container, a better one. And they came back and stole that, too. Lifted the whole container with a crane, and went away with it. After that you might begin to think, well, is it worth it, trying to maintain this canal for the public good? Why don't I just give up and enjoy my retirement some other way?'

Yellow flags lined the canal. Water-lily pads and grasses bowed as we went by, the undertow swirling them around and ducking them under. Away to the north the Kilsyth Hills rose in a long dark wall of hard basalt rock clothed in high moor and low pasture. At intervals the handsome stone buildings of former canal-horse stables stood among the trees, blank windows and patchy roofs open to the sky.

A hoarse blast on *Maryhill*'s whistle, and we were admiring the twin pistons under Twechar Bridge as they slowly raised the decking to let us slide through. Just south of the canal rose a hillock of flaky dark rock, a hummock of mine spoil known locally as a bing. Deep in the layers of sandstone and limestone beneath the Kelvin Valley run seams of coal and ironstone, and Twechar was just one of many villages along the valley that sprang into being thanks to these mineral treasures far below the ground.

*

Coal and steel baron William Baird established his headquarters at Twechar in the 1860s, and for a century the village didn't look back. Baird himself was unpopular locally, seen as an autocratic figure who didn't care about his workers or seek to help them if they got into difficulties. But his mining enterprises for coal, for alum shale to be processed at Campsie for the textile trade, and for coking coal destined for Baird's Gartsherrie iron-smelting works on the eastern edge of Glasgow brought employment and a fierce sense of community to Twechar and the surrounding countryside.

There were workshops for painters at Twechar, for blacksmiths and engineers, joiners and electricians, roofers, wagon drivers and chimney sweeps, all working for William Baird. The works had their own private coal railway, but they also kept the Forth & Clyde Canal busy with their barges.

Some of the coal pits shut down in the 1960s after a century of production and employment. The railway closed, then the canal. The Miners' Strike of 1984–5 did for the rest of Twechar's pits. An excellent local study has collected oral accounts of what kept the miners at such a very dirty and dangerous job.* As ex-miner Andrew Bell expressed it:

> It was a horrible job, but it was a horrible job done by good men who cared about each other. You couldnae go doon the pit and no' care about the lad who was next to you. You had to care about him, because he had tae care about you, because you're three hundred and fifty feet underground . . . What it was, was good men doing a rotten job tae the best o' their ability.

From the bows of *Maryhill* I watched the landscape go by, a green valley where coal-mine and sand-quarry bings made small irregular hillocks with dark shaly flanks slowly eroding away. Patches of marshy

---

* twecharpitvillage.com.

wetland, now wildlife reserves, showed where subterranean mine workings had collapsed and flooded. The valley used to be notorious for its bogs and marshes. The thick sands, silts and clays that filled the great ice-gouged trench 100 metres deep below the valley floor were waterlogged, hence the wet ground. When the Irish and Scottish Highland navvies arrived with picks, shovels and wooden wheelbarrows to dig the course of the Forth & Clyde Canal between 1768 and 1773, many of the marshes were drained. Local landowners caught on and began to drain their own wet patches. The bogs became pastures and ploughlands. Then came William Baird and the coal mines, far more profitable to a landowner than agriculture. Coal-mining with its roadways, shafts, levels and tunnels proved a widespread cause of subsidence, and the bogs and wetlands began to form again, notably at Dumbreck Marsh southwest of Kilsyth where the shaft descended over 400 metres to reach the seam of coking coal displaced by the downthrow of the Kilsyth Fault.

Ironically, the former mine workings of the Kelvin Valley might yet be turned to good account once more. All those old collapsed tunnels bring rock and water into close conjunction under pressure, with possible potential for heat exchange and the production of deep thermal energy. This is only a wait-and-see idea at present, but it's a space well worth watching.

At Auchinstarry Marina on the south side of Kilsyth we pulled in and moored for a lunchtime break. I went off to look at the remarkable rock face of Auchinstarry Quarry just across the road. The Whin Sill, a wedge of hard dark dolerite that pushed in among the basalt of the Kilsyth Hills some 290 million years ago, stands here up to 100 feet high, a sheer curtain of ragged purple-grey rock, fractured across and cracked at a steep slant. Up close the quarried surface was iron hard, close-sheared, with a very fine granular texture like stubble on smooth skin. Climbing ropes dangled; pitons stuck out of cracks. Auchinstarry Quarry is famous among rock climbers. Beginners can get a feel for the sport here, and more experienced participants can still find a challenge. In the flooded quarry pit swim rudd, roach, tench and sizeable pike.

On the way back to the marina a couple of women stopped me to exclaim with pleasure at the sight of *Maryhill*. 'I remember being about fourteen or so in Glasgow on the west side and watching those little cluthas – we called them "cluthas" – puffing away as they guided big boats up the Clyde to the docks. All gone now, cluthas and big boats both. Nothing comes up the river to the city any more.'

Under way again we passed a long red and black bing, a spoil heap from the former sand quarry at Dullatur. Sand barges would be side-lined here, waiting to be filled via an aerial ropeway that spanned the canal. Next came a big bay, a sort of layby cut out of the canal bank. 'Before the railways came,' said James, 'you'd have high-speed boats rushing down the canal between Glasgow and Edinburgh, cargo boats that waited for no one. Every other boat had to clear out of the way, had to pull in and park up in the bay, to give the fast boats the passage. If they didn't give way they'd have their ropes cut.'

From Lock 19 to Lock 18 I took the tiny, spoked steering wheel and found that *Maryhill*'s bows took several seconds to respond to the wheel. One had to juggle queasily with it, trying not to over-compensate, to fight the impulse to spin the wheel frantically to starboard, harder and harder, as the bows dithered and *Maryhill* continued to head for the canal bank. It wasn't inadequacies in the art of steering that nearly brought disaster, however, but a knotting problem in Lock 17.

The rope that steadies the bows against the lock wall has to have a running or loose knot so the line can be paid out as the boat sinks in the lock. If it is a fixed knot the bows are held firm as the stern drops, leaving you dangling like a fish on the hook. Christine had demonstrated a fixed knot, a round turn and two half hitches, so neat and appealing that I decided I'd like to have a go. I fumbled it. Christine tried to undo the damage while *Maryhill* was in the act of descending the lock and the boat began to jib and hang by the bows. A furious shout from the lock keeper – 'NEVER do that in the lock! Christ!' – and he was winding the paddle like a madman to refill the lock and restore *Maryhill* to a level stance. Christine's blushes, but my bad.

At the halfway mark between Glasgow and Edinburgh we slid into a berth at Falkirk, right beside the Falkirk Wheel and just in time to watch the last action of the day. This fantastically shaped rotary boat-lift was opened in 2002 to raise boats the 79 feet from the Forth & Clyde Canal to the recently reopened Union Canal which runs all the way to Edinburgh. The boat-lift could have been made to slide vertically up and down in a boring old bridge shape, but instead the design of the Falkirk Wheel was bold and futuristic, while at the same time echoing the glories of the industrial past with its great cog wheels and massive cast-iron body. It resembles two enormous birds side by side, each with a hooked beak and a hollow eye. With a greasy whine and a series of loud clanks the bird heads tilted slowly backwards together, a pair of counterbalancing heads rose from under the structure with a faint groaning of metal under stress, and a big pink pleasure-boat, cradled on a long scoop between the heads, rose with slow dignity to the upper level where it chuntered off and out of sight.

Time for a leg stretch. I set off west along the line of a remarkable Roman monument, the Antonine Wall. This earthen rampart was built between the Clyde and the Forth across the narrow neck of Scotland in AD 142–3. The man who ordered it done, Emperor Antoninus Pius, was the adoptive son of Emperor Hadrian, who had built his own much more substantial stone wall across Britain from sea to sea twenty years earlier and 100 miles to the south to mark the northernmost boundary of the Roman Empire. Once the Romans had decided to reinvade the Pictish lands, the Antonine Wall was built, garrisoned and maintained for about twenty years, before the troops were withdrawn once more to Hadrian's Wall and the more northerly barrier was abandoned.

A mile or so along the valley, the Antonine fortification of Rough Castle lay ringed by trees. At the gate a notice exhorted dog walkers: 'Cura ut canis excrementum in receptacula in area vehiculorum posita deponas.' This dog Latin wasn't too hard to decipher.

An outer bank led into a deep ditch surrounding the central raised area of the fort. Grassed-over lumps and bumps showed the location of principia (headquarters), granary, barracks and commander's house. Thanks to the meticulous record-keeping of the Roman army we know that the commander at Rough Castle was Flavius Betto, the centurion commander of the VI Cohort of Nervians, a tough guy who had come up the hard way through the ranks of the XX Legion. The Nervians who made up the garrison that Flavius Betto commanded were Belgians, and a bit of a tasty bunch themselves – Julius Caesar in his *Commentarii de Bello Gallico* described them as 'a savage people of great bravery'. Analysis of the latrine waste at other Antonine forts suggests that the soldiers ate a good mixed diet of bread, porridge, bacon, shellfish, cheese and vegetables, along with imported treats such as coriander, figs, wine and olive oil.

Notwithstanding the good food, it was a hard life in harsh conditions, the weather often foul, the discipline rigid. The enemy could be expected at the gates at any time. Just outside the fort, on a neck of land to the north which attackers would be likely to pick as an easy approach, the ground is pitted with dozens of little oval depressions. These are *lilia*, or pit traps. A sharpened stake was fixed in the bottom, the pit filled with rushes or decayed leaves and the opening disguised with branches, and the charging attacker was left to stumble in and skewer himself. One doubts whether much sympathy would have been extended by the Nervians to Pictish victims of these Roman land mines.

Next morning *Maryhill* departed from the Falkirk Wheel and spent a long morning descending the sixteen locks that lowered her to sea level in the broad flat littoral that lies along the western fringes of the Firth of Forth. Under the grass and scrub the over-deepened valley runs ever deeper underground, then on out beneath the firth. Massive sand and gravel deposits and a good layer of glacial till pack the void, above which *Maryhill* cruised serenely towards journey's end at the Kelpies in the windy wastes of Helix Park.

The park was a project for the Millennium, and the idea was to give a much-needed lift to a large area of decayed industrial ground and a dozen deprived neighbourhoods around the oil refinery town of Grangemouth. The centrepiece of the new park had to be something striking, big and bold. A giant sculpture of two horses' heads by Andy Scott was chosen, a marvel of structural steel and silvery cladding that towers 30 metres high. One horse bends its neck gracefully down; its companion strains skywards, mouth open in a silent scream.

According to their creator, the Kelpies are intended to honour the heavy horses that helped make this region prosperous by pulling ploughs, wagons, canal barges and coal ships. Most critics loved the Kelpies, though the *Guardian*'s Jonathan Jones dismissed them acerbically as 'just a kitsch exercise in art "for the people", carefully stripped of difficulty, controversy and meaning'.

That was a bit harsh, I thought as I walked in their enormous shadows. Kelpies themselves are not cuddly beasts. They are mythical water horses with a malevolent habit of enticing people to ride them, then dragging them into the water to drown. Andy Scott's work catches that contradictory aspect of water horses and of horses in general, the submissive attitude of the downward-looking horse only a mask for the angry energy of its alter ego alongside.

Back beside *Maryhill* I said goodbye to her jovial crew. They would be making their way back west by gradual degrees, as would she. For me it was eastward ho! out along the Forth towards Edinburgh, with the refinery towers of Grangemouth falling behind and the Kelpies outlined against the cloudy sky for a long while until they sank out of sight to the west.

The Falkirk Wheel cocked for action.

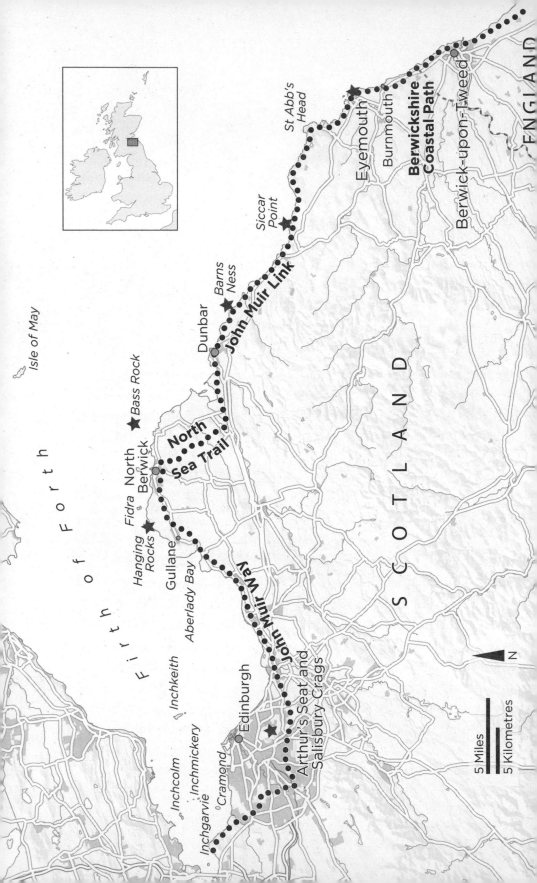

7

Volcanic Lumps and Lightbulb Moments: Edinburgh to the Border

'IF YOU CAN TEAR YOURSELF away from that mandolin,' said Dave Richardson as he laced up his boots, 'we'll go and take a walk on Arthur's Seat.' We were sitting in Dave's kitchen in Edinburgh, and I had just retrieved the lively hornpipe called 'The Steamboat' from the memory sludge of fifty years before, stimulated no doubt by thoughts of *Maryhill* and her predecessors, the Clyde cluthas. I'd always loved 'The Steamboat'. Dave used to play it when we were students at Durham University, long before he chucked in his nearly completed PhD in Botany and, as he puts it, 'ran away with the gypsies'. Cue long years of travelling the world playing Celtic and Northumbrian music with his band the Boys of the Lough and satisfying his restless curiosity about birds, plants, history and single malts in between whiles. Dave's enormous patience with my musical fumblings never ceases to amaze me. He'd already sat at the kitchen table and plonked out 'The Steamboat' many times over at funeral speed, and still I hadn't caught hold of those two bars at the start of the B part. Never mind – there was all the evening ahead to get them down and dusted. Better not waste a moment more of this breezy, spattering day in Scotland's capital.

The wind was rushing over the old city as we descended the Royal Mile. Just off Cowgate at the foot of the thoroughfare, the James Hutton Memorial Garden was looking a bit sad with fag-ends scattered on its gravel walk and empty cider cans among the bushes. A youth of peely-wally countenance and languid demeanour reclined on the bench, smoking a fat spliff while lazily skinning up another. He watched us examining the lumps of rock placed around as tributes to the eighteenth-century geologist's pioneering insight: some fragments of conglomerate composed of material from earlier rocks to demonstrate the never-ending cycle of erosion, deposition and formation; a couple of boulders shot with granite veins to illustrate how igneous matter could be intruded among layers of sedimentary rock.

'Aah . . . wha're ye doin'?' asked the lad as he deftly intruded igneous combustibles among layers of tobacco. Dave remarked on James Hutton's fame, and how the great geologist's house had once occupied this spot. 'Ah-hah? Yeah . . . uh, interesting,' slurred the youth. He puffed on his reefer and waved a limp hand around the tatty bushes and their basement layer of empties. 'Ah . . . nice view he had, annyways.'

It was a pretty nice view, at that, as we walked down Holyrood Gait towards the high dark curtain of Salisbury Crags. This great rampart of dolerite rock, 50 metres tall, stands at the edge of the Royal Park of Holyrood. It dominates the eastern end of the Old Town of Edinburgh, a backdrop for the stage setting of the Palace of Holyroodhouse in its neoclassical splendour and the ultra-modern Scottish Parliament building nearby. In the 1760s and 1770s James Hutton came frequently to stare at the crags and ponder their structure. He could see with his own eyes that the smooth dark basalt rock had not been laid down in regular horizontal layers like the sandstone around it. It stood up in vertical columns, sandwiched between sedimentary layers of sandstone above and below. In a little roadstone quarry just round the corner of the crags Hutton noticed that the layer of sandstone beneath the crags, itself composed of tiny fragments from some older source, had been bent upwards with great force at the point where it made contact with the volcanic rock. Not only that: it had been distorted and melted, evidently by the application of great heat. The fact that this layer of sandstone was beneath the dolerite showed that it had been laid down beforehand and was therefore older. The sandstone above the volcanic crags was evidently of the same age as the lower layer. So how could the younger igneous rock have been laid down as a horizontal layer between these two older sedimentary layers, as the geological consensus of the age dictated? Answer: it couldn't. It didn't fit with the notion of an orderly, unbroken succession of sedimentary layers. It must have come out of the depths and thrust itself into the middle of the already existing sandstone, forcing its way along a fracture or other line of weakness. And to have actually melted the sandstone it touched, it must itself have been excessively hot – molten, in fact, at the time of intrusion.

Further up the hillside, above the crags and sandstone, there were other layers of sedimentary rock and other volcanic crags. There was obviously a process going on involving unimaginable heat and pressure, endlessly repeated over aeons of time. But many reputable thinkers of Hutton's day subscribed to the theory that the world had been created in 4004 BC, as propounded in 1650 by Archbishop James Ussher of Armagh, Primate of All Ireland, in his snappily titled *Annals of the Old Testament, deduced from the first origins of the world, the chronicle of Asiatic and Egyptian matters together produced from the beginning of historical time up to the beginnings of Maccabees*. Others cleaved to the Biblical account, believing that it had all been accomplished within six days. Both theories had God swiftly depositing layers of rock, the oldest at the bottom and the youngest at the top, like a celestial brickie.

Of course it's all too easy to scoff as one looks back from the high ground of today's scientific knowledge. But to Hutton with his practical insights, even then these prevalent beliefs flew in the face of reason and the Enlightenment.

Dave and I wanted to get a good look at Hutton's Section, the little quarry where enlightenment began to tap James Hutton on the shoulder. But Radical Road, the driveway immediately beneath the crags, had been closed ever since a rockfall in 2018. Thanks to Health and Safety concerns, there was no realistic prospect of reopening it. How could anyone guarantee that no one would be hit by falling rocks, a process that has been occurring naturally for the past 350 million years? We found a rutted path that climbed the rim of the great dolerite cliff. It ran among wood sage, heath bedstraw and bloody cranesbill. The botanist in Dave Richardson took note, while I tried to refrain from endlessly humming 'The Steamboat' out loud. The earworm of that brisk little tune was driving me mad, and the B part still dangled tantalizingly just out of reach. 'De-dum-de-dum-de-*dum*', or '*dum*-de-dum-de-*dum*-de-dum'? I wished, not for the first time, that I'd learned to read music. Maybe I would now. Ah, there goes a flying pig.

Precious little can be seen of the face of Salisbury Crags from directly above – it's better to study it with a good pair of binoculars from below on Queen's Drive. Thanks to a curve in the cliffs near the apex of the path, however, I did get a glimpse of Hutton's Section while lying prone at the very edge and clinging on in the buffeting wind. Not really worth the palpitations, I had to admit. Just along the path we halted above a valley where many ways met. Ahead rose a double hump of high ground, the rugged Lion's Head declining southward towards the gentle slope of the Lion's Haunch to compose the striking miniature mountain of Arthur's Seat. Here above Hutton's Section is the place to stop and admire this iconic crag, a great lump of volcanic material that forms the centrepiece of the Royal Park of Holyrood, a carefully preserved swathe of wild land at the heart of Edinburgh. Lucky citizens, to have Arthur's Seat arising in their midst. Nice view they have, annyways.

The volcano that gave rise to this landscape of crags and lava flows burst into activity around 340 million years ago, after the start of the Carboniferous period. Sandstones and mudstones had been laid down as the sea rose and fell, and now from several vents over a long period of time magma poured forth on top of them in successive layers, sometimes in violent gas-enhanced eruptions, at other times in less explosive gushes of molten rock. From our standpoint the crown of the Lion's Head looked as handy a place as any for identifying what was left of the volcano and its basaltic vomit after all the subsequent years of erosion.

We climbed rock steps, polished green and red like serpentine by millions of boots, up the flank of Arthur's Seat where greenfinches gave their sneezy calls from the scrub bushes. From the saddle below the Lion's Head, stone steps with a chain link fence brought us easily enough to the summit. The wind was too strong and the rain too spattery up there to hold steady for a photo at the trig pillar, but no weather was wild enough to subdue the boisterous spirits of a crowd of German youngsters who'd made it to the top and were shouting over the city with primal glee. Here is one of those panoramas that has you gasping, a tremendous view north over the Firth of Forth to the hills of Fife beyond, south to the long line of the Pentlands, west across the city itself with the pillars of the war memorial and the preposterous tele-scope of the Nelson Monument on Calton Hill, and beyond that a skyline of spires broken by the volcanic wedge of Castle Rock with Edinburgh Castle draped along its summit.

Dave and I did our best to pick the bones out of this geological puzzle – like all such phenomena, quite easy to comprehend once you understand how to read it. Arthur's Seat, Calton Hill and Castle Rock are all products of the same volcanic upheavals. Castle Rock is a volcan-ic plug, a hard block of dolerite which became stuck inside the vent that opened over there. It's a classic 'crag and tail' formation, shaped by Ice Age flow from west to east, with a steep brow whose hard resistant rock forced the glacier to divide around it, and a long ice-sculpted tail running down to Holyrood Park. The ice also scooped out the hollow valley on the north side of the plug where Princes Street Gardens now lie. Calton Hill is a block of lava on top of volcanic ash or tuff. And nearer at hand the long slopes of Whinny Hill, a northerly extension of Arthur's Seat, are corrugated by ledges with rounded rims. These repre-sent a succession of lava flows from different sources, some from the Castle Hill vent, some from the crater on the Lion's Head. Below them in a humpy line run the Dasses, a tongue of intrusive basalt. Subsequent earth movements tilted everything 25 degrees towards the east.

As for the Lion's Head at the summit of Arthur's Seat, it is the top of a pluton, a vertical column of basalt that punched up from below into the crater. The Lion's Haunch is part of that crater, a great scoop filled with agglomerate rock from the eruption. At its maximum height the crater might have reached another 200 metres into the sky, before erosion began to whittle it away. Everyone who walks up Arthur's Seat therefore enters the very heart of the volcano and climbs across the plutonic rock that now lies exposed after all these millions of years – yet another testament to erosion and the new understanding of the geological cycle that James Hutton brought to the table.

Once we had reached lower ground out of the wind and hoisted in these basics with the help of Edinburgh Geological Society's excellent pamphlet, *Discovering Edinburgh's Volcano*, I took a moment to look around again. Over and above its geological significance is the breath of fresh air that Holyrood Park blows through the city. There's so much heather, bilberry and rugged rock here, so many rare flowers and birds, that it feels as though a sizeable slice of wild Scotland has been transported to this urban setting. The rock outcrop at the summit is slippery underfoot, shiny with the polishing of countless shoe soles, evidence of how much Edinburghers and their visitors appreciate this wild park.

Dave and I decided to descend by way of an unfrequented path across Whinny Hill. We soon saw why it was unfrequented. It fell abruptly away several hundred feet in a precipitous scramble among the spiky whin bushes, a path to be negotiated in reverse like a pair of sailors on a companionway ladder, then by sliding on our bottoms and snatching at grass clumps as they slipped through our fingers. We escaped with no worse damage than torn trousers, and I got the whin prickles out of my fingers in time to give 'The Steamboat' another damn good thrashing that evening.

From Arthur's Seat one gets a wonderful view over the Firth of Forth, with a tiny hint of a skeletal red humpback shape off to the left where the estuary disappears from view. This is the famous Forth Railway Bridge with its three diamond-shaped cantilever sections. The two road bridges beyond are hidden by a fold of ground. The railway bridge rests the tubular legs of its central cantilever on the low island of Inchgarvie, all that's visible of a submerged volcanic 'crag and tail' like nearby Castle Rock. I once had a LNER railway poster from 1937 that I loved to tatters. Captioned 'Over the Forth to the North', it encapsulated all the romance of a long-haul sleeper train, the legs of the railway bridge reflected in the calm waters of the firth, an indigo night sky behind, and contrasting with all these solemn blacks and blues a single splash of scarlet as the steam locomotive passes on to the bridge with the light of its firebox glowing off the cantilever arch. Such glamour! Many years after the poster had shredded away, I took a boat ride out to Inchgarvie and saw the massive bridge foundations and the wartime fortifications that litter the island.

That had been a fantastic day out on the choppy grey Firth of Forth, bouncing around in a rubber rib between Inchgarvie and its fellow islands. Inchmickery, another igneous rock, resembles a battlecruiser of First World War vintage – an illusion carefully fostered to deter potential raiders, with the building up of concrete towers and bunkers along the summit to suggest the superstructure of a warship. Inchkeith, slap in the middle of the firth and mostly of volcanic origin, has done duty at various times as a prison for political undesirables, a quarantine station (Black Death sufferers were dragged off willy-nilly to live or die here), a garrison fortified to the teeth and a hang-out for eccentrics. Slim-waisted Inchcolm, just off the northern coast of the firth, is greener and more fertile than the other islands, thanks to a geology of sandstone, shale and limestone as well as igneous greenstone. It boasts one of Scotland's finest preserved abbeys, built out here nine hundred years ago at the behest of King Alexander I as a thanksgiving for his deliverance from shipwreck in a storm.

A walker heading east away from Edinburgh follows the coast along the Firth of Forth by means of the North Sea Trail. What a fabulous concept this is, a single mighty path that runs with many loops and jumps along the shores of Norway, Sweden, Denmark, Germany, the Netherlands,

England and Scotland, 5,000 miles long from end to end. Or is it 2,700 miles? Or 4,000? No one seems to know. Here it runs as a waymarked path; there it's 'in development', or unmarked, or unheard of. Whether Britain's post-pandemic, post-Brexit section of the North Sea Trail will ever be fully up and running is open to question. But its wonky 'N' logo on posts and boards pointed me in shreds and patches, by way of the John Muir Trail, the John Muir Link and the Berwickshire Coastal Path, from the Scottish capital all the way round the coasts of East Lothian and Berwickshire to the English border.

Aberlady Bay lies a few miles east of Edinburgh. The last time I'd walked there with Dave Richardson had been in a winter's dawn. We'd left Edinburgh in the dark in order to be at Aberlady for the dawn flight of ten thousand pink-footed geese – or maybe twice that number; it was impossible to count them. Out on the sand flats they had become more and more restless and talkative as the light broadened, then as the sun appeared over the trees they rose in a long, long wave with a mighty roar of wings and babble of voices and flew overhead and on inland to search for their day's feeding. That early morning was memorable for two other reasons. Dave had been toting a melodeon in his backpack, and he sat down on a stone by the Forth against the sunrise to play me his self-composed tune 'A Walk Along The Shore', music and mood blending perfectly together. Later he sat down again, this time on the pie we'd brought for lunch, squashing it flat. Not so flat, though, that it couldn't be eaten with relish out at Gullane Point, henceforward forever to be known as Squashed Pie Point.

Today's expedition was a summer one, though. The geese were long gone to their breeding grounds in Greenland. Instead, we had chiff-chaffs and reed buntings in the scrub and reedbeds behind the shore path, and sedge warblers – beautiful little birds with bold pale eye-stripes under prominent black brows – unreeling their melodious songs as they grasped the reed stems and swayed with the wind.

If you're walking at Aberlady with a musician who's also an academic botanist and knowledgeable birdwatcher, sensations come at you thick and fast – the birds, the earworm tunes (yes, today's was 'The Steam-boat' again, curse that B part!) and the plethora of wild flowers in the grassy sand levels of the shore. Blue curls of viper's bugloss, yellow flags, early purple orchids. Bittersweet with psychedelically yellow stamens cradled in vivid purple petals. Ragged robin, pink and loose in the wet, calcareous grass. A feathery stem of hemlock; a golden carpet of scrambled eggs or bird's-foot trefoil nearer the sea. A riot of colours and shapes.

Out at Squashed Pie Point a big blunt cliff showed a thick band of grey mudstone below, seeded with little nodules of calcite, with a fat wedge of brown sandstone piled on top, all shot through with an orange-white net of calcite veins, like grog blossoms on a boozer's cheeks. Layers of ironstone as thin as pencils were squeezed into the mudstone, while the foot of the cliff was of rippled mudstone with lumps of quartz the size of baby teeth.

On past Hummell Rocks, an intrusion of flat hard dolerite from the volcanic upheavals west along the coast, where a glance back showed the unmistakeable sleeping-lion shape of Arthur's Seat above the battle-mented silhouette of Edinburgh's skyline, still dominated from this angle by church spires rather than skyscrapers. Patches of dark grey mudstone held interlocking teardrop shapes, the fossilized remnants of lycopod bark, last seen back at the Fossil Grove in Glasgow. Terns were circling with creaky cries like unoiled gate hinges before tipping over and spearing down into the water, to rise with a twinkle of silver at the beak as the fish writhed before being upended and swallowed.

The shorefront at Gullane was a strip of handsome Victorian seaside villas with candle-snuffer roofs, turrets and gables. Here Dave and I parted company, he to seek the Edinburgh bus, I to continue along the North Sea Trail through sandhills and woods at the edge of Muirfield championship golf course, where men in pastel sweaters, shorts and white socks swung and putted and squinted into the wind.

Opposite the long tidal islet of Eyebroughy the coast turned east again, with the volcanic plugs of North Berwick Law and the Bass Rock as prominent sighting points ahead. Just offshore lay Fidra, a basalt sill of vertical columns, with a natural sea-cut arch near the stumpy, yellow-topped lighthouse. Herring gulls in their thousands dotted its grassy slopes, and a little platoon of a dozen guillemots sat tight together like soldiers in black jackets and white waistcoats forming a square against the enemy hordes.

Crumpled cliffs at the back of the beach heralded the geological layer cake of Hanging Rocks. A jumble of boulders and rubble left by the glaciers formed the clifftop. Below that a grey lime-rich cementstone rested on a yellowish breccia or sedimentary rock full of fragments, which looked as though it might come away from the cliff face at any moment – hence the name of Hanging Rocks. The breccia in its turn was footed on a cream-and-purple tuff, a hardened volcanic ash with big chunks of what are engagingly known as 'bombs', splashes of thick lava spat out by the volcano that cooled into lumps before they hit the ground.

Two big caves sat side by side in the cliff about 20 feet above the beach, screened behind trees. If not for the British Geological Survey's notes and a small conical cairn at the foot of the sandy scramble to the caves, I'd never have known they were there. Both ran back between the breccia and the tuff, hollowing out into sloping hiding places hard to spot from below. The cave on the left still possessed a barrier of dry-stone walling that could have been built as much as 1,500 years ago. There was a flue to catch and funnel fire-smoke, and a curious squint giving a view on to the beach. Excavation here has unearthed a miscellany of objects including a quern-stone for grinding grain, a whorl for spinning wool, a spearhead and an iron knife, bone pins and buttons, glass beads from an armlet, and fragments of Samian ware dating from Roman times. Tantalizing hints of human ingenuity as people noted the weaknesses between the strata, then widened and dug them out. Millennia of occupation of these damp little burrows in the cliff, through times as tough as your enemies and hunger could make them, and other times when you felt secure enough to dress up in your jewellery and eat your dinner out of a fragile dish.

Beyond Hanging Rocks the shore was plastered with pink and purple jellyfish awaiting the next high tide. Plates of black basalt ran out to sea, and in the middle of this petrified lake of lava the remnants of a little plug of basalt lay in a raised oval of rock, formed when the plug cooled at the edges on contact with the existing rock.

Above the beach stood Marine Villa, a solid stone seaside retreat. Robert Louis Stevenson stayed here, gaining inspiration for *Treasure Island* from the dramatic view of Fidra just offshore. Beyond Marine Villa the little pristine bay had acquired a collar of enormous *Country Life/Coast* magazine-style buildings, great seaside palaces as big and impersonal as hotels, crammed up against each other with hardly room to swing a cat between them, stealing each other's light and air and privacy. Balconies and wood cladding abounded, each style individual, yet all similar in their discreet grey and cream colour schemes. Their huge picture windows looked seaward, but not on to the beach – each view was partly obscured by a line of pine trees between houses and beach. The expenditure, the uniformity and proximity of these mighty, empty houses was something staggering to contemplate. Of course it was stupid and unfair to compare them with the solid dignity of Marine Villa, let alone with the extraordinary cave dwellings back along the beach. But that's what I did, in spite of myself.

Three miles more along the North Sea Trail, and I was panting up the steep path to the summit of North Berwick Law. The big arch of whale's jawbone that crowned the volcanic plug when I was last here, twenty years ago, had been replaced with a fibreglass replica. A fellow climber explained it with a well-known phrase or saying: 'Health and Safety.' Whadda ya gonna do? as Tony Soprano would say.

Like the Bass Rock, its counterpart out in the Forth, North Berwick Law is a volcanic plug, hardened magma from the vent of a volcano long eroded away. And like Inchgarvie and Castle Rock back in Edinburgh, the hill was shaped by Ice Age flows into a classic 'crag and tail'. From up here this evening there was a superb view over to the Bass Rock, its slanted top and sheer east cliffs as formidable as ever. But something was missing from the scene. I got my binoculars focused on the Bass, and even then it took me a minute or so to catch on. The top of the great rock was pale brown, sparsely speckled with white. It should have been white all over – white with gannets, 150,000 of them nesting cheek by jowl, the world's largest colony of these big white seabirds with their dagger bills, black wing-tips and cold blue eyes. I'll never forget the day I landed on the Bass to spend an afternoon in their overwhelming company, watching the breeding pairs ecstatically intertwining necks, stabbing their deadly beaks at any other bird invading their space and patching their nests with seaweed and plastic bags, the whole scenario set to a roiling, boiling murmur of gannet voices and a most disgusting stink of fishy excrement. What on earth had happened to the Bass Rock gannets?

Enlightenment came courtesy of my neighbour under the fibreglass arch. Avian flu had struck the colony within the past month. Hundreds were being washed up, dead or dying, along the coast. Had I not seen them? Now that she mentioned it, I had caught sight from the coast path of one or two bundles of white feathers far off among the rocks. I'd assumed they were herring gulls, dead of normal causes. Over the next few days I kept an eye out, but the only gannets I saw seemed to be going about their usual business, flying parallel to the coast in line astern in groups of five or six as they looked for fish. It would take years for the full effects of this bird flu pandemic to be calculated, but it looked as though it was going to be the avian equivalent of Covid-19.

The Lothian and Berwickshire coast is a jumble of geological treasures and curiosities, often sitting on top of or next door to each other. Canted plates and crusts, volcanic bombs and plugs, showers of stone and ash, seams of coal, ancient forests, corals as beautiful as polished marble. Here James Hutton walked, sailed and pondered, wondering how horizontal sedimentary layers had set themselves on end, and the nature of the processes by which rocks of different types and ages got squeezed and jammed together. Hutton found what he considered incontrovertible proof of his theories at Siccar Point, 20 miles down the coast south of North Berwick. And halfway between the two, the Edinburgh Geological Society's trail at Barns Ness provides a succession of geological wonders for the most inexpert of eyes.

Barns Ness lies just southeast of the harbour town of Dunbar where John Muir (1838–1914), pioneer ecologist and Father of the National Parks movement, was born and raised. This little stretch of shore is where the largest limestone outcrops in central Scotland stretch seaward from the low cliffs. They were laid down at roughly the same time as the Fossil Grove rocks in Glasgow, during the early Carboniferous period when Scotland floated at low latitudes near the equator. Shallow tropical seas and swampy hot forests succeeded each other. All sorts of life forms throve in these conditions. The limestone sheets of Barns Ness, lying on top of sandstone beds, are themselves composed of the bodies of sea creatures, and are studded and packed with a tremendous array of fossils.

A big cement factory just inland demonstrates modern man's ingenuities in exploiting the limestone. The tip of its chimney, lazily coiling out fumes, was just visible above the cliffs as I set out east from White Sands beach along the low-lying shore under a sunny sky full of lark song. In the cliffs a limestone quarry and an old stone-built limekiln showed where men of former times burned lime-rich rock with coal, making fertilizer that would 'sweeten' acid ground for agriculture.

The pale limestone shelf ran seaward, with darker and coarser sandstone underlying it. In the sandstone I found a number of 'cock's tails', curved lines fanning out gracefully. These traces were made by sea creatures moving from side to side in the sand of the seabed. To place my fingertips in these petrified tracks and trace the wriggles made by some being of unknown shape and habits, to have its 330-million-year-old ghost under my hand, so to speak, was a powerful sensation.

The limestone held fossils of colony-forming corals caught in movement like dishrags swirling in unseen currents. Further out towards the sea were a series of dimples in the limestone, about the size of the *lilia* pit traps outside Rough Castle on the Antonine Wall. Instead of sharpened stakes, these depressions held stinking and half-rotted oarweed. They were floored with a greyish rock, the fossilized soil of an ancient wetland, often found just below seams of coal. Miners called it seat-earth. Stony sections of roots lay higgledy-piggledy in the seat-earth. It looked as though each depression marked the position of a tree in one of those steamy, swampy carboniferous forests.

I walked the shore, trying to distinguish the different geological layers. Pebbles and stones dragged here by the glaciers during the Ice Ages at the top, a thick ledge of mudstone underneath and under that a thin black seam of coal, only a few inches thick, outcropping in a line of tiny sharp zigzags like Norman dog-tooth decorations along a church wall. Under the coal a wrinkled, massively compressed layer of seat-earth, well squashed by the weight of all the rocks on top of it, and under that a floor of heavy iron-brown sandstone, dense and dark.

Near Barns Ness lighthouse lay something easier to understand and appreciate for its beauty. Half hidden among the scattered boulders were lumps of so-called 'Dunbar marble', mass graveyards of solitary corals and shellfish that died in their uncounted millions in those shallow Carboniferous seas all that time ago. White tubes and cylindrical sections, fan-shaped shells and strands, they lay hard packed, glimmering out of the rocks with a striking luminosity. If they all died and were buried in one single episode, it must have been a hell of a storm or mass poisoning to have exterminated them in such numbers. Back in the day there used to be a thriving local industry in collecting and polishing up slabs of this striking limestone. Nowadays the remnants of the coral necropolis lie unmolested for geologists and lucky beachcombers to enjoy.

Glancing back along the coast I suddenly realized that up around Dunbar I had crossed the invisible boundary of the Southern Upland Fault, and had left the Midland Valley for the Southern Uplands, the last and nethermost of Scotland's Five Terranes. Forty miles more along the North Sea Trail and I'd be crossing the equally invisible border into England. There was the same sense of closing one chapter and opening another that I had felt at the summit of Stob Mhic Mhartuin in Glencoe. Away to the west the Bass Rock lay diminished by distance in the Forth, a lump of volcanic rock that never bore a fossil, the vigorous life of its seabird colony suddenly under threat of disappearing. Here on the seashore at Barns Ness, under my feet and hands, was evidence of life abundant.

A blustery afternoon on the Berwickshire coast, a scant hour of light left in the day and a bank of rain blocking out the sea horizon as it moved in from the north. I followed a grassy path past the crumbling ruin of St Helen's Church, round the edges of the wet clifftop fields and on along the narrowing green finger of Siccar Point. Nothing to see here; nothing until I walked right out to the tip of the promontory and took a few steps down the slope, its grass skiddy with rain pearls, that steepened towards the cliff edge. There in the freckling rain and fading light I finally set eyes on what James Hutton saw from a boat off the coast on a clear day in early June 1788 – an electric-shock moment, even for me, with all the facts and all the story under my thumb at the touch of a smartphone keypad.

I was looking down on horizontal plates of red sandstone, their edges eroded and drooping like melted wax. Rising to meet them were vertical plates of dark grey rock, the meeting point of the two rock types a sharply defined but undulating line.

It was easy to see where the lower, upright rocks had come from. The outgoing tide had uncovered a maze of curved scars forming the bed of the bay. These were sandstone rocks laid down in Silurian times some 435 million years ago, but sandstones of a particular type known as greywacke, a name that for some reason I found peculiarly pleasing. The greywacke itself was formed of grains eroded from the ancient basement rocks of Laurentia, carried away by rivers and deposited in shallow marine waters, then pushed seaward by submarine upheaval or deposition overload to slalom down the steep continental shelf and spread out fanwise across the ocean floor. Interleaved with softer layers of mudstone and siltstone, this seabed rock was subjected to an irresistible squeezing force during the final closing of the Iapetus Ocean. The sandwich of rock strata was levered up till it stood on end, to be ground down alternately by the sea and by land erosion. It is the upper ends of these layered rocks, once horizontal and now vertical, that show as the black curvilinear scars in the bay at Siccar Point, and as the upright strata of the lower cliffs that have not yet been eroded and eaten by the sea.

As for the cap of horizontal red rock (actually very slightly tilted) that fits so snugly over the jagged upper ends of the vertical greywacke slabs, that is Old Red Sandstone laid down on land some 370 million years ago. The contact area between the two rock types does not present a flat neat surface. It's an undulating joint, mirroring the varying rates of erosion of the harder greywacke plates and the softer mudstone layers between them. The sandstone settled gradually on top, first filling in all the crenellations, then continuing upwards in its regular deposition of horizontal layers. But what of the joint between the two types of rock? There's no sign of any horizontal deposition along that surface during the 65 million years between the formation of the greywacke and the formation of the sandstone. This kind of gap or missing link in geological layers is called an unconformity, and this one at Siccar Point has become world-famous as Hutton's Unconformity.

From his boat in the bay, and during a scramble across the foreshore, Hutton freed himself finally from any doubts about the timescale and process of geological creation. His notions were confirmed by what was in front of his face. The greywackes must have been created, he realized, by infinitely slow horizontal deposition. Then a massive force had deformed and tilted them, just as had happened in the roadstone quarry at Salisbury Crags. That was followed by countless years of erosion clear of the sea in their upright position, with nothing being laid down above them, before the sandstones were deposited on top over more unspooling aeons of time. Deposition, uplifting and erosion: this process, Hutton saw, must have taken place over uncountable years, what we now call 'deep time'. Furthermore, it was, and still is, a cyclical process, whose infinitely gradual motions he could observe even as he stood there: greywacke and red sandstone in their turns eroding, grain by grain, falling or being washed away to deposition as fragments, the building blocks aeons hence of new layers of sedimentary rocks.

Standing on the cliff in the twilight I pictured James Hutton spluttering out these revelations to his companions on the trip, geologist Sir James Hall and mathematician John Playfair, Hutton's good friend and apologist. In a eulogy of Hutton in 1803 Playfair recalled that momentous voyage, and the effect of Hutton's insight upon the three of them, in words which catch all their excitement at making such a giant leap of understanding:

> We felt ourselves necessarily carried back to the time when the schistus on which we stood was yet at the bottom of the sea, and when the sandstone before us was only beginning to be deposited, in the shape of sand or mud, from the waters of a superincumbent ocean. An epocha still more remote presented itself, when even the most ancient of these rocks, instead of standing upright in vertical beds, lay in horizontal planes at the bottom of the sea, and was not yet disturbed by that immeasurable force which has burst asunder the solid pavement of the globe . . . The mind seemed to grow giddy by looking so far into the abyss of time.

From Siccar Point the North Sea Trail (or was it now the Forth to Farne Way, or the Berwickshire Coastal Path? The waymarks seemed unsure) turned east before its final southward dip towards the English border. On the crest of St Abb's Head I was visited with the memory of my son George, on a visit home from Australia one month before the first Covid-19 restrictions hit the UK, leaning against a blasting winter wind beside the lighthouse and yelling aloud for sheer exhilaration at being here and being alive. Around him stretched a landscape of explosive vulcanism, lava flows from a volcano triggered by the final closure of the Iapetus Ocean around 390 million years ago, still traceable in the pink and grey rocks of St Abb's Head. On either side was the greywacke coast, its strata stacked this way and that, above a sea flattened by the gale into a heaving grey mass devoid of waves or whitecaps. On this summer day, walking down to Eyemouth, I recalled the power of the sea that day and the way it had sucked so greedily at the tilted cliffs.

The fishing villages of this coast are set in coves burrowed out of the cliffs by the ceaselessly pounding North Sea. It is the sea that is master here. A cruel master, as they know only too well in Eyemouth, a dour stone fishing town huddled round a narrow harbour. During a long, wary relationship with the sea, Eyemouth fishermen and their families endured many losses in obscurity. But the calamity of 14 October 1881 hit the national headlines, when a hurricane-force storm fell on the fishing fleet out of a clear horizon.

'The sky suddenly thickened with dark, heavy clouds,' recorded the local minister, Revd Daniel McIver, in his book *An Old-time Fishing Town: Eyemouth*.

A fierce wind arose which was as wild in its fury as the calm was quiet; the sea began to heave its threatening bosom, like a man in whose heart passion was rising, and what with sudden darkness – it was then between eleven and twelve of the day – the shrieking of the hurricane as it drove at the creaking masts and ripping sails, and the thunderous roar of a boiling ocean, the poor fishermen thought that the Judgment Day had come.

So it had, for 129 of them. They drowned within sight of their families, gathered in horror along the shore, as their boats smashed on the Hurkers reef at the mouth of the harbour or piled into the cliffs. Five hundred women and children lost their loved ones and breadwinners. 'Eyemouth is a scene of unutterable woe,' lamented the *Berwickshire News* four days after the disaster. 'The town has received a blow from which it will be long ere it will recover; the fleet is wrecked, and the flower of the fishermen have perished.'

Eyemouth Museum preserves a tapestry created by twenty-four women of the town in 1981 to commemorate the centenary of the disaster. Embroidered with images of fishermen overwhelmed by the storm, the fish they gave their lives to catch, and a radiant sunrise as a symbol of new hope, it makes a very poignant and moving memorial.

Back in the eighteenth century this Berwickshire coast was a wild and lonely stretch of country. Eyemouth was not only a fishing town, but also a notorious smugglers' haven, thanks to its geographical position as the nearest Scottish port to the Continent. In 1753 local merchant John Nisbet, up to his elbows in the smuggling trade, commissioned architect John Adam to design Gunsgreen House on the harbour for him. Nisbet's cellars led directly to a landing place by the sea, and the capacious roof space of his new mansion was specially designed to store large quantities of contraband tea.

Eyemouth today feels like a place on the cusp of big changes. It's still a traditional sort of town where shoppers stub out their fag before going into the Co-op and light another as soon as they're out; still a fishing town with a long history of the craft and a harbour packed with trawlers from Lerwick, from Banff, from Port Seton and further afield in France and England. Yet smart eateries and stopovers have opened here recently, and there's a multimillion-pound redevelopment of the old fish market going on. Gentrification might be just up the road, Cornwall-style.

I walked out of town past ranks of trawlers moored along the jetties – *Stronsay Lad, Fear Not, Ella Mac, Still Game, Bright Ray* – sturdy boats with storm decks and many rust dribbles. A long green sausage of fishing net with floats attached had been unrolled along the dock. Gunsgreen House looked down on me from its grassy bank. Up on the cliffs beyond I stopped to look back over the rugged cliffs encircling the harbour, where the falling tide was beginning to expose the tips of the reefs just offshore against which the Eyemouth fishing fleet was wrecked during the storm of 1881.

The walking trail from Eyemouth along the cliffs to Burnmouth should be beaten wide and deep, because this is yet another spectacular stretch of Scotland's southeast coast. In fact, it's a scratchy and over-grown hike in many places, and my attention veered like a broken compass between the grass and bramble tangles snagging my boots and the coastal drama unrolling ahead.

From Agate Point onward the cliffs were contorted like layers of half-unfolded Swiss roll, most of the strata heaved up almost vertically. The barnacled plates of greywacke twisted and curved, each a few feet thick and separated by layers of more readily eroding mudstone. I followed one slender greywacke layer with my eyes as it fell away steeply towards the base of the cliff, executed a hairpin bend within touching distance of the sea, and reared up again a couple of hundred feet to the cliff edge. On either side of this downward fold or syncline, hundreds more mir-rored its bend and compression. Before my geological baptism I had thought of rock as essentially hard, stable, reliably immobile, but here in the cliffs between Eyemouth and Burnmouth the rocks had been bent and folded like a pile of handkerchiefs.

Down at Scout Point a white-haired sea angler had got himself to the outermost tip of the rocks. How he had negotiated the vertical strata, their barnacled surfaces as deadly to bare skin as any cheese grater, I couldn't imagine. His friend, a venerable Eyemouth fisherman to judge by his cap and salt-stained jersey, was waiting for him at the corner of the golf course, leaning on a post and watching the world go by.

A line of gannets flew past beak to stern. The sight caused me to remark on the toll recently taken among gannets and other birds by the incursion of avian flu. Oh, no, no, no, intoned the Ancient Mariner. That wasn't it at all. Something dreadful, some unnameable but truly evil thing had been dredged up from the ocean floor by an unfortunate trawler and was now killing all the birds – not just gannets but seagulls, puffins and albatross. The fish as well. And the dolphins. Whales, too, washed up dead around these islands, all the way down to Wales and Cornwall. Having delivered himself vehemently and sincerely of this End-Of-Days scenario, the old man smiled quite unexpectedly and hoped I would enjoy my walk.

At Hurker's Haven the cliffs took an inland dip, a bite out of the coastline. Perched high above I found the hollow concrete shell of a wartime observation post. Inside, two tiny tables and chairs, a couple of framed pictures and a toy sailing boat faced the view out to sea. This was some enterprising small person's eyrie. A razorbill flew purposefully out from the folded cliff, turned and scurried back as if it had left something vital in the nest. Out at sea fifty guillemots formed a circle on the water like a fleet of tiny fishing boats with white hulls and black upperworks. Orchids and bird's-foot trefoil grew in profusion in the grassy areas between wheat and rape fields. All nature seemed to go about its business along the tortured landforms of the coast.

The tripartite fishing settlement of Burnmouth came into view, gradually and piece by piece – first the few dwellings of Ross, isolated to the south by a bulge of cliffs, then the long crooked arms of the main harbour, a stronghold for fishing boats with a large building behind; finally a line of houses rising up the cliff. Just north of Burnmouth Harbour, around Partanhall, the grey crags are Silurian greywackes laid down around 435 million years ago, roughly the same time as those at Siccar Point. But those to the south, exposed on the shore and at Ross Point, are composed of alternating layers of sandstone, siltstone/mudstone and cementstone, deposited in Carboniferous times a hundred million years after the adjacent greywackes, and tilted almost vertical during the Variscan orogeny or mountain-building episode 200 million years ago. They form a sharp-toothed shore that's fiendish to walk across, as many a local can tell you.

Burnmouth was one of the local sites, early this century, where palaeontologist Stan Wood and his colleague Tim Smithson unearthed the fossils of tetrapods, four-legged amphibians which were pioneers on land. The discoveries of Wood and Smithson bridged a 25-million-year gap (known as Romer's Gap, after palaeontologist Alfred Romer who first recognized it) in the record of the tetrapod species. The fossil hunters concentrated on rocks that were laid down during the first 15 million years of the Carboniferous period, and eventually came across tetrapod fossils that filled in the missing link. These creatures lived partly immersed in water, partly in the shallows and partly on dry land, and the sturdy ribs of the fossils showed that their lungs were well developed enough to be serviceable out of the water – a vital step in the long transition of living, moving beings from the sea to the land.

Burnmouth's main harbour was strengthened in 1879, and should have been storm-proof, but twenty-four men from the little village were lost along with their Eyemouth neighbours during the epic storm of 1881. Down by the harbour a Burnmouth man grimaced as he recalled his experiences during the more recent Storm Arwen in November 2021. 'The sea was absolutely furious. Really frightening, with waves rolling up the harbour and crashing high. I've never known such a storm. There's no defence in Burnmouth against weather like that. The worst I've ever seen. Hope to God I'll never see anything like it again.'

His words stuck with me as I looked back from the cliffs beyond. Today in the sunshine, with a pale blue sea lapping quietly at the distant harbour piers, it was hard to imagine a malevolent force of nature attacking this quiet settlement. But nature, of course, is not malevolent, or greedy, or violent, or any of the other adjectives of moral degeneracy that one can't help reaching for when contemplating the catastrophic consequences of its processes on the lives and viewpoints of human beings. It simply rolls on and around, as James Hutton recognized, and very effectively too.

I came to the border just south of Ross. All Scotland lay behind me now; all that drama, the aeons of time and the giant upheavals stamped into the mountains and coasts of the Five Terranes. Here at the border there was no fanfare, no line across the path. Just the hiss of sea wind in the gorse bushes, and the rattle of a train heading south along the cliffs into England.

Sandstone and mudstone piled up in the cliff at Gullane Point.

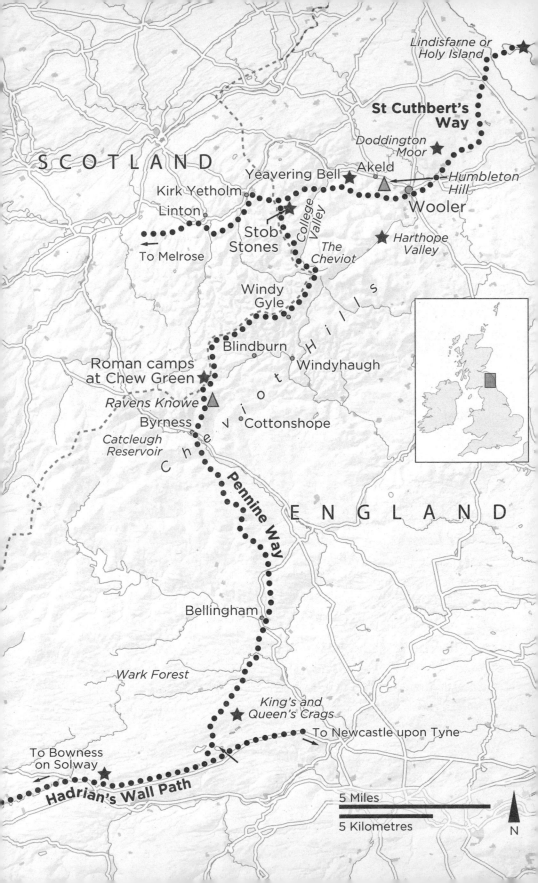

Lindisfarne or
Holy Island

**St Cuthbert's
Way**

S C O T L A N D

*Doddington
Moor*

Akeld

*Humbleton
Hill*

Yeavering Bell

Kirk Yetholm

Linton

Wooler

Stob
Stones

College
Valley

*The
Cheviot*

*Harthope
Valley*

To Melrose

H
i
l
l
s

Windy
Gyle

Blindburn

Windyhaugh

Roman camps
at Chew Green

t

Ravens Knowe

Byrness

o

Cottonshope

*Catcleugh
Reservoir*

v

Pennine Way

e

C
h

E N G L A N D

Bellingham

Wark Forest

*King's and
Queen's Crags*

To Newcastle upon Tyne

To Bowness
on Solway

Hadrian's Wall Path

5 Miles

5 Kilometres

N

8

Saint, Dragon and Devil:
Holy Island to Hadrian's Wall

'DAMN YOU! THE CASTLE IS ours.'

On 10 October 1715, at the height of the Old Pretender's rebellion, sea captain and ardent Jacobite Launcelot Errington relieved Samuel Phillipson, Master Gunner of Lindisfarne Castle, of his temporary command with these economical words. Phillipson, a part-time barber, had had Errington exactly where he wanted him only an hour or so before, sitting in a chair at a perfect disadvantage, head tilted back and throat proffered for a shave. If the master gunner had only known it, he could have pre-empted his humiliation with one quick slash of the razor.

The Jacobites had hoped to land a Franco–Spanish force on the Northumbrian coast. Lindisfarne, or Holy Island, is only a stone's throw from the Scottish border, and for a brief moment the remote island seemed a fair choice for an adventure of this sort. A body of soldiers might have been able to land there and remain undetected until they had marched ashore and met up with Scottish rebels coming south to overthrow the hated Hanoverian regime. But Lindisfarne sits isolated in the North Sea at the end of a tidal causeway, and it would have been a complicated business for incomers to garrison and retain the island against organized opposition.

As things turned out, Launcelot Errington and his nephew Mark waited in vain for reinforcements. They only held the castle for twenty-four hours before being captured on the run and clapped into Berwick Gaol, from which the resourceful pair soon managed to tunnel to freedom. Lindisfarne Castle itself, however, might have withstood quite a siege in its lofty position on top of Beblowe Crag on the south side of the island. The 23-metre knoll is a dolerite dyke that fed into the Whin Sill, and as Neil Clark had promised in the Fossil House back in Glasgow, this mighty magmatic feature with its dark cliffs and crags was to create a dramatic backdrop to my walks through northern England in at least four places – here on the Northumbrian coast, all along Hadrian's Wall, at High Cup in the rim of a classic glacial valley, and down in Upper Teesdale in tall cliffs and riverbed steps.

Looking south along the coast from the ramparts of Lindisfarne Castle on a bright summer morning, I made out the flat shapes of the Farne Islands low down on the southeastern horizon. These shelves of dark rock in the sea are more manifestations of the Whin Sill, a great upwelling of molten material some 290 million years ago. The hot viscous magma rose vertically from the depths and then pushed sideways between layers of limestone and sandstone laid down during the Carboniferous period some 40 million years before. It forced its way both vertically and horizontally between the weakened Carboniferous layers, deforming some rocks and folding others with its heat and energy. Gradually it cooled and set below the surface, and there it remains, a massive sheet of dolerite with an average thickness of 30 metres, more than twice that beneath Weardale and Allendale, outcropping in many places, a leviathan looming invisibly below more than 1,700 square miles of what's now northern Britain.

Gazing out at the Farnes in balmy weather before setting off to trace St Cuthbert's Way along the English border from east to west, I pictured the saint, Prior of Lindisfarne monastery and a famous solitary, decamping from Holy Island in AD 676 to the bleak rock of Inner Farne Island to live there as a hermit. For companions Cuthbert – or Cuddy, to give him his affectionately bestowed nickname – had only the seals and the eider ducks, known locally as Cuddy's ducks. The Farne Islands in their isolation resist the sea, but the sea has not yet done with them. Twice a day it reduces the twenty-eight dolerite shelves of the archipelago to fifteen, drowning the others before uncovering them again at low tide; and slowly, inexorably, as sea levels rise it will reclaim them all.

St Cuthbert's Way leaves Lindisfarne across Holy Island Sands. You can cross to the mainland on foot by way of a mile-long causeway road, but let's face it, that's cheating. It was off with the boots and socks for me, and out into a flat tidal world with ribbed mud and sand under my bare feet and a line of tall poles to mark the original old pilgrimage path that runs straight for 3 miles between mainland and island monastery. I passed the tidal refuges, a couple of barnacled boxes on stilts where unwary travellers caught by the tide can wait high and dry, if not in any comfort, till the sands are uncovered once more. A crowd of pattering sandpipers were my heralds as I came ashore at Beal Point. No saint was there to wash the mud from this pilgrim's feet; I had to do it myself, and that night I found plenty I'd missed between my toes.

The following morning I looked for the green line of the Cheviot Hills and found them beckoning 10 miles off on the southwestern sky-line. St Cuthbert's Way makes more or less straight for the Cheviots, and I followed it through the forests and over the wide grasslands of the Kyloe Hills. Big brown cattle roaming there raised their heads from their grazing to stare and snort and stamp their forefeet in quite a marked manner. In spite of the name of their home territory they were far from kinship with the little black kyloes that once walked the Scottish drove roads, and they looked pretty far in temperament from docile Matilda and her fellow Highland cattle of the Famous Drove of 1981. I was glad to get away from them into the trees of Cockenheugh and take a break at the mouth of St Cuthbert's Cave. Whether the hermit saint ever took refuge under the great sandstone overhang, or whether the monks who fled the Danes and carried his body away from Holy Island in AD 875 rested his carved coffin here at the start of their long wanderings, no one knows. Such is the power of belief and a good story that the legend has never been forgotten hereabouts.

It's tempting to think of rock laid down in the Carboniferous era as all limestone, the remains of tiny creatures floating and drifting gently down through the warm tropical seas to settle on the seabed. But sandstone is in the mix too, and this early Carboniferous rock forms a high crest to Doddington Moor, just north of St Cuthbert's Way as it approaches the market town of Wooler and the outer foothills of the Cheviots.

On a fantastically blowy morning I set out to climb to the crest of the escarpment. Across the sunlit farmlands the Cheviot Hills stood up proud on the southern skyline, rounded and bosomy, fold behind fold, an eye-catching patchwork of green and orange. Climbing the country lane from Doddington up to the moors, I kept turning round for another stare.

A noble view – was that why our distant ancestors chose Doddington Moor as the site for so many of their stoneworks? Up there they placed practical ones such as field enclosures and settlements, ritual creations in the form of stone circles, and most notably a scatter of mysterious cup-and-ring markings, rounded depressions the size of a teacup surrounded by a doughnut ring, gouged in the surface of flat rocks. Northumberland is rich in cup-and-ring sites, and Doddington Moor is one of the best places to find them.

I followed a hill track past a congregation of droopy-horned bullocks who jostled up to stand and stare like rude young men in a pub. The path led me around a vast field of oats that swung and hissed in the wind, and then by map, compass and the pricking of my thumbs to stumble suddenly on a fine cup-and-ring-marked rock, a sandstone slab looking east towards the coastal hills and dimpled with man-made hollows surrounded by concentric rings. Was it sited here to face the rising sun? There's no telling now; but the slab still held power and presence today for this twenty-first-century man.

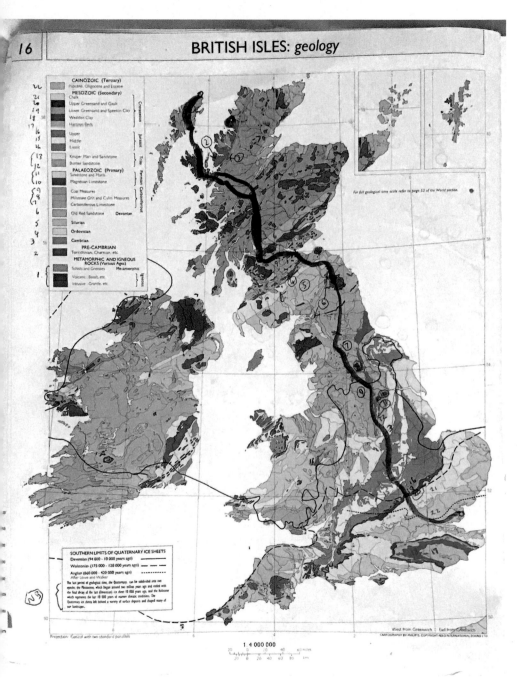

The *Philip's Modern School Atlas* geological map that started it all.

Above Swirly 2-billion-year-old gneiss at the Butt of Lewis.

Left The cliffs of Dail Mòr, striped with gneiss 3 billion years old.

Below Witchy figure by the tomb at the heart of the Stones of Callanish.

Above Skye contrast: smoothly weathered granite. Red Cuillin (**left**) versus jagged gabbro Black Cuillin (**right**).

Left Coigach time freeze: a lump of asteroid ejecta peppered with lapilli.

Below Beauty of psammite, pelite and quartz exposed beside the road from Kinloch Hourn.

Above View from the Devil's Staircase across Glencoe to the volcanic pile of Buchaille Etive Mòr.

Below Marker stone on the forgotten military road through the blanket bog of Rannoch Moor.

Above Clyde puffer *Maryhill* on the Forth & Clyde Canal.

Right The angry energy of the Kelpies.

Below Pioneer geologist James Hutton's inspiration: the volcanic landscape of Salisbury Crags and Arthur's Seat, Edinburgh.

Above Coal sandwiched between mudstone and seat-earth, Barns Ness cliffs.

Right Dunbar marble, a graveyard of carboniferous corals and shellfish.

Below Greywacke folds in the cliffs near Eyemouth.

Above Housey Crags, the roof of a volcanic chamber, stands out above the Harthope Valley in Northumberland's Cheviot Hills.

Below Hadrian's Wall, surfing the Whin Sill above Crag Lough.

Above left South Tyne Valley: green in-bye land rising to rougher grazing, then moorland with lead-mine scars.

Above right Iconic view towards Lune Valley from High Cup, a perfect U-shaped glacial valley.

Below High Force in Upper Teesdale, a 20-metre step in the dolerite of the Whin Sill.

The wind trembled the harebells and fought me like a foe past the tattered, seething firs of Kitty's Plantation. It was the kind of wind that gets under the tails and into the heads of cattle. In the field I meant to cross, a posse of thirty heifers, big and bulky, were busy pushing down the flimsy fence by sheer weight and numbers. Whatever their motive, it was driving their neighbour, a giant bull in the adjacent field, absolutely wild. He was shaped like one of those Andalusian fighting bulls, all chest and attitude. He slavered as he roved up and down along the fence, roaring and pawing, great silver gouts of saliva flying as he tossed his head. I let discretion be the better part of valour and found another way.

'Stone Circle (rems of)' said the map, and here it was: a big rough king stone the size of a man, crusted with lichens and deeply grooved by rain and weather, lording it over a circle of recumbent stones. I followed a path through head-high bracken along the escarpment, shoved and elbowed by great blasts of wind. Before descending into Doddington village once more, I sheltered by lonely Shepherd's House on the brink of the moor and tasted that mighty Cheviot prospect to the full, the rounded domes of Yeavering Bell, White Law, Gains Law and Humbleton Hill standing high across the valley above their deeply cut stream clefts. Whoever lives here is monarch of what must be one of the finest views anywhere in Britain.

Before tackling the Cheviot heights I spent a day fossicking about in their depths. The Harthope Valley runs straight and true into the northern flank of the Cheviots, shadowed by a country road that leads southwest out of Wooler and threads its way between the hills. Harthope contains only a couple of farms and a single bumpy road, but birdwatchers and walkers relish its quiet air of remoteness from the everyday world rushing by beyond the borders of Cheviot.

The remarkably straight valley follows the line of the Harthope Fault, which cuts right into the heart of the Cheviot massif, the remnants of a volcano that was once as tall as Mount Etna. This huge volcano erupted around 395 million years ago towards the end of the Caledonian orogeny, with one of its vents along the Harthope Fault. It was an explosive event. Andesite and rhyolitic lava, fine, light and fast-cooling, continued to pour out, eventually covering up to 230 square miles. And after the final grumblings of the volcano had died away, a tongue of granitic magma licked up and out through the andesite layers, intruding into the bowels of the volcano and forming a pluton or subterranean cone of granite. Much of the volcano was eroded away and the granite unroofed before the beginning of the Carboniferous period around 360 million years ago. After nearly 400 million years of erosion, what is left above the Harthope Valley is a line of tall crags, the roof pendant or remnant of the rock that was violently invaded by the granitic magma.

Curlews were fluting far away on the moors as I climbed a steep grassy path towards the skyline tor of Housey Crags. On the path a clot of sheep dung the size of a golf ball appeared to twitch into motion, then froze as I leaned over it. I stayed as still as I could, and was rewarded when a fat black dung beetle poked its head and shelly back out and continued its tasty feast.

Housey Crags and the neighbouring tor of Long Crags resemble castles of rock, all that remains of the baked and metamorphosed roof of a volcanic chamber. The columnar rock, coarse to the touch, subtly variegated in pinks and greys, carried a crusty lichen coat of many colours. Hollows and cracks were lined with dripping mosses and staghorn lichens. Life was not so much clinging to as thriving on these rough outcrops.

All around lay bleak moorland bubbling with curlew cries and red grouse cackles, dominated by the great bald dome of Hedgehope Hill, with a distant smear of grey misty sea back in the east. I could see Harthope Valley's straight cleft laid out in its southwestern course, the banks of the burn eroded into scoops by landslips and floods, with the long back of The Cheviot rising in the west. Crags, moor, hills and valley looked so solid, so settled, when set against the imagined violence and upheaval of the volcano, the clashing of continents and the opening and closing of oceans when Northumberland was a tropical slab of weltering rocks.

The map showed a path that dropped by degrees into the Harthope Valley. That mapmaker was a cruel joker. The bracken wilderness he led me into would have baffled a bloodhound. I gave up, turned back and went straight down the slope to Langleeford Farm in the valley bottom. Here in 1791 a young Walter Scott came with his uncle to stay 'in the very centre of the Cheviot Hills in one of the wildest and most Romantic sites'. He shot a crow to provide himself with a writing quill, and enjoyed drinking the goat's whey served to him in bed, early in the morning, by 'a very pretty dairy-maid' (*The Letters of Sir Walter Scott (1787–1832)*, ed. Sir Herbert J. C. Grierson).

Beyond Langleeford the road roughened, turning to a burnside path as it passed the sheep pens at isolated Langleeford Hope. In a miniature gorge beyond, the waterfall of Harthope Linn came chuting down over a dark granite step of rock. The gentle hiss of the fall, and the persistent call of a cuckoo, were the sound of the summer afternoon in this enchanting valley.

St Cuthbert's Way hurdles the northern outliers of the Cheviot Hills. The River Glen and the Bowmoor Water together form a very definite northerly boundary to the range, the hills descending steeply to the water. This was all disputed land in medieval times between neighbour and neighbour, as well as between the Scots and English, and near Akeld I marked the first appearance of a fortified farmhouse known as a bastle, a solid two-storey stone building with its entrance high in the wall of the upper floor. You could pull up the ladder and defy raiders, if you were lucky and they didn't burn you out.

The path led round the back of Humbleton Hill. Here on 14 September 1402 a Scots army under the Earl of Douglas, returning home after rampaging through the English borderlands, was brought to bay by Henry Percy, Earl of Northumberland, and his impetuous son Harry Hotspur. Douglas had the advantage of the high ground, where he ordered his men to form schiltrons, ranks of soldiers with 15-foot pikes levelled at the enemy. Unfortunately for the pikemen, the English archers simply stood off and shot them down in droves. Even an act of absurd bravery such as by Sir John Swinton of that Ilk could not turn the tide. The exhortation put into Swinton's mouth by Sir Walter Scott in his 1822 play *Halidon Hill* still brings a lump to the throat: 'O my brave countrymen! What fascination has seized you to-day, that you stand like deer to be shot, instead of indulging your ancient courage, and meeting your enemies hand to hand? Let those who will, descend with me, that we may gain victory, and life, or fall like men.' One hundred Scotsmen, fired up by Swinton's plea, rushed down the hill with him in a wild charge and were slaughtered to the last man. As to the outcome of the battle, that too was set out in William Shakespeare's play *Henry IV, Part I*, where the king gloats over the victory:

> Here is a dear and true-industrious friend,
> Sir Walter Blunt, new lighted from his horse,
> Stain'd with the variation of each soil
> Betwixt that Holmedon and this seat of ours;
> And he hath brought us smooth and welcome news.
> The Earl of Douglas is discomfited:
> Ten thousand bold Scots, two-and-twenty knights,
> Balk'd in their own blood, did Sir Walter see
> On Holmedon's plains. Of prisoners, Hotspur took
> Mordake the Earl of Fife and eldest son
> To beaten Douglas; and the Earls of Athol,
> Of Murray, Angus, and Menteith.
> And is not this an honourable spoil,
> A gallant prize? ha, cousin, is it not?

A side path led up to the crest of Yeavering Bell, 'hill of the goats', where a rough stone wall, 10 feet thick in places, enclosed an enormous Iron Age hill fort. Within the fort an inner stronghold had been dug out of the rock. Circular hut foundations lay scattered across the hill slopes, but I could make out only a couple in the thick grass and bracken. What a centre of civilization this must have been when the Votadini clan peopled the lands between Tweed and Tyne, from the Iron Age until after the Romans withdrew from Britain. Walking on along St Cuthbert's Way I remembered the master musician and composer Alistair Anderson playing his haunting air 'Yeavering Bell' in the church on Holy Island a few years ago, at the end of my previous pilgrimage along St Cuthbert's Way. But today, try as I might, I couldn't bring 'Yeavering Bell' to mind. Just as well; however beautiful the tune, it would have been an earworm even more persistent than 'The Steamboat', and I'd only just got rid of that.

A side track took me up to the Iron Age hill fort on Great Hetha with its double wall and views south down the College Valley into the heart of the hills. The Cheviots are wonderful walking ground, their granite and andesite rock pleasingly rounded by erosion, not too steep, riddled with secret hollows and valleys that can sneak you in, wind you round and push you out in quite another direction to the one you thought you were taking. Through these beguiling hills I followed St Cuthbert's Way down to Kirk Yetholm, where the Way meets the terminus of the greatest and oldest of England's long-distance paths, the Pennine Way, at the end of its leg-wracking traverse of the Cheviots.

On a saddle of ground just before reaching Kirk Yetholm I recrossed the Scottish border at the Stob Stones, two boulders from a dyke of speckled porphyry, markers of the traditional coronation site of the Kings of the Yetholm Gypsies. These hills are soaked in blood and history, and also in traditions never quite forgotten. Witness an ancient yarn from Linton, just along the valley from Kirk Yetholm, that tangles with my own family's story. In an addendum to his classic 1806 collection *Minstrelsy of the Scottish Border* Sir Walter Scott quotes the following splendid account, which he came across in a 'genealogical MS' of the seventeenth century:

In the parochen of Lintoun, within the sheriffdom of Roxburgh, there happened to breed a monster, in form of a serpent, or worme; in length, three Scots yards, and somewhat bigger than an ordinarie man's leg, with a head more proportionable to its length than great-ness. It had its den in a hollow piece of ground, a mile south-east from Lintoun church; it destroyed both men and beast that came in its way. Several attempts were made to destroy it, by shooting of arrows, and throwing of darts, none daring to approach so near as to make use of a sword or lance. John Somerville undertakes to kill it, and being well mounted, and attended with a stoute servant, he cam, before the sun-rising, before the dragon's den, having prepared some long, small, and hard peats bedabbed with pitch, rosett, and brimstone, fixed with a small wyre upon a wheel, at the point of his lance: these, being touched with fire, would instantly break out into flames; and, there being a breath of air, that served to his purpose, about the sun-rising, the ser-pent, dragon, or worme, so called by tradition, appeared with her head, and some part of her body, without the den; whereupon his ser-vant set fire to the peats upon the wheel, at the top of the lance, and John Somerville, advancing with a full gallop, thrust the same with the wheel, and a great part of the lance, directly into the serpent's mouthe, which wente down its throat, into the belly, and was left there, the lance breaking by the rebounding of the horse, and giving a deadly wound to the dragoun; for which action he was knighted by King Wil-liam; and his effigies was cut in ston in the posture he performed this actione, and placed above the principal church door of Lintoun, where it is yet to be seen, with his name and sirname: and the place, where this monster was killed, is at this day called, by the common people, who have the foresaid story by tradition, the Wormes Glen. And fur-ther to perpetuate this actione, the barons of Lintoun, Cowthally, and Drum, did always carry for crest, a wheel, and thereon a dragoun.

Not only the barons of Lintoun, Cowthally, and Drum – the heroic old legend gave the whole Somerville tribe its family crest of a dragon on a wheel (actually a two-legged beast, and therefore a wyvern), a device which features in old bookplates and on cufflinks and spoons passed down as family heirlooms.

What gave rise to these accounts of dragons that one finds all over the country? Were they just a damn good device to scare the kids round the fire of a night? Could they have been a folk memory of the dragon-headed ships of the Viking raiders that ravaged the land with fire and sword? Or might the cave of the Linton Worm have been a way to explain the geology of the area southeast of Linton, the igneous rocks and ancient cairns on Grubbit Law and Wideopen Hill? Good questions to chew on as I sat in the Border Hotel at Kirk Yetholm, deciding how to tackle the first and worst section of the south-going Pennine Way.

It was Tom Stephenson, Secretary and later Chairman of the Ramblers' Association, who in 1935 dreamed up the idea of a great long-distance footpath along the watersheds and high moors of northern England. It took him and like-minded colleagues thirty years and endless negotiations with farmers and landowners, water board representatives and foresters before the opening of the Pennine Way in 1965. Running for 270 miles between Edale in Derbyshire and Kirk Yetholm just across the Scottish border, it was – and remains – a formidable challenge involving boggy peat moors, steep hill climbs, long miles of lonely country and the notorious Pennine weather. On my journey through the bones of Britain I was going to tackle the Pennine Way in reverse, following the excellently waymarked trail south through the rugged Northumbrian uplands and the dales of Durham and North Yorkshire, down to the gritstone country of Derbyshire's Dark Peak. Most folk, though, do it south to north, as Tom Stephenson intended. When you reach the end of the trail, your only reward is your own satisfaction at having risen to the challenge and made it to the finish. That was the case, at least, until in 1968 Alfred Wainwright published his hand-drawn, handwritten, splendidly acerbic *Pennine Way Companion*, in which he promised to stand a glass of beer to everyone who limped into the Border Hotel at Kirk Yetholm having completed the trek.

'Limped' is the *mot juste*, because the north-going Pennine Way adventurer, footsore, blistered and bone-tired after some 240 miles of rough walking, is required to face a final day that involves a slog of 29½ miles from Byrness in Redesdale across the mires and featureless uplands of the Cheviot Hills, following the border between England and Scotland so often and so bloodily disputed in times past. There is no escape, except for walkers canny enough to have arranged in advance for transport to meet them on some lonely road several miles off the Pennine Way and drive them to overnight accommodation. You have the border fence as a guide through mist and low cloud for much of the way, but it's a weary and dispiriting trudge from dawn to dusk in the best of conditions, and you rarely get the benefit of those in the clouded Cheviots. In the early days before stone slabs and boardwalks were laid down across the squelching peat, before mobile phones and GPS pinpointed your position, Pennine Way lore was rife with tales of lost hikers, bog immersions, maps that blew away, boot soles that detached themselves and other final day travails. I wasn't going to get into any of that. Instead, I opted for a series of excursions, demanding enough in themselves, that climbed from the valleys on either side to sample the border path as it snaked along this most punishing section of the Pennine Way.

The first few miles of the walk followed the border steeply south from the Stob Stones, over White Law and below the aptly named, peat-smeared Black Hag, up to the round hill of the Schil, from where at a height of 600 metres there was nothing whatsoever to see, thanks to a wet mist or cloud that sidled up and wouldn't let me go. Back down a farm road to Kirk Yetholm with the Halter Burn as a cheerfully chattering companion and guide, to shake the mist out of my hair and head in the bar of the Border Hotel. Alfred Wainwright died in 1991, having forked out about £15,000 over the years to fulfil his 'beer for completers' promise. Much to their credit, the hotel and their brewery sponsors still honour Wainwright's commitment. There was a moment's temptation – 'Phew, just done ten miles, what d'you reckon, just a splash?' – but the barman didn't look as though he'd be amused. I bought my own pint of Border bitter. Very tasty, too.

For the central stretch between Byrness and Kirk Yetholm I went a few miles south and based myself in Upper Coquetdale. A network of ancient byways – Clennell Street, Salter's Road, Gamel's Path – crosses the border from this beautiful and mostly unfrequented valley that snakes into the hills from the east. The first excursion was the longest, up from Windyhaugh past circular bields or sheep pens to the remote farm of Uswayford, then on up the wide track of Salter's Road, passing the waterfall at Davidson's Linn where 'innocent' (i.e. tax-free) whisky was once concocted at Rory's Still. Here I totally lost my way. I'm still not quite sure how I arrived at the reassuring sight of the border fence and the paved path of the Pennine Way, but the scramble there involved plenty of heather-hopping and peat-plodging. After that it was gold-star walking southwest along the border to where the big sprawling stone heap of Russell's Cairn tops the gentle knoll of Windy Gyle. Or should the place name be 'Split the Deil', as the map suggests? The Devil certainly had plenty to do hereabouts in the bad old days. Russell's Cairn actually dates back to the Bronze Age, but it more or less marks the spot where in July 1585 the Wardens of the Marches and their representatives convened to try to settle issues of law and order along the border. As so often in these debatable lands, blood was soon spilled. The victim in this case was Lord Francis Russell, the Earl of Bedford's son, and the perpetrator – so the English side had it – was the staunch Roman Catholic and supporter of Mary Queen of Scots, Thomas Kerr of Ferniehirst. Kerr was imprisoned in Aberdeen's Tolbooth, where he very conveniently died early next year, ridding Queen Elizabeth of a powerful enemy.

From Russell's Cairn the Pennine Way bent across the hillside to rejoin the border fence and descend to the Street, another of those broad grassy tracks that carried drovers, reivers, whisky makers and pedlars between England and Scotland. It carried me off now, an exhilarating march away from the Pennine Way, over Black Braes and south along the ridge between the Carlcroft and Rowhope burns in their deep valleys, heading down to where the Barrow Burn hurries to join the River Coquet and the valley road to Windyhaugh. Just the sight of the name on the map sent another earworm tune burrowing into my brain, 'The Barrowburn Reel', and another memory sparked up the synapses: Dave Richardson at the Falun Folkmusik Festival in Sweden back in 1995, patiently leading a workshop on Northumbrian music. I was there as his guest and pupil, and by God how we all sweated over the Barrowburn Reel. We ended up playing it in front of hundreds of proper, actual musicians, just to prove we damned well could, scraping it out on fiddles, nyckelharpas, flutes, squeezeboxes, and in my case a blues harmonica. What a week that was, a blur of images: a Native American in full feathered headdress yelling in a phone box; the Boys of the Lough's piper waltzing with his girl to a wind-up gramophone in the open air; a trio of Sami dancers cutting impossible shapes in mid-air in a deserted concert hall; the band Väsen playing their gorgeous 'Josefins Dopvals' at six in the morning in a hotel lobby while guests descended in their dressing gowns to lodge objections with the manager and stayed to dance round the reception desk. 'One golden rule for this festival,' said Dave, 'don't go to bed. You'll miss something special.' He was absolutely right.

Further up Upper Coquetdale stands the farm of Blindburn, from where I spent a day wandering the old tracks round the remote upland area called Buckham's Walls. The ancient country custom of cheese-making was reflected in the name of Yearning Hall farm, a ruined house a mile below the border. The insectivorous plant butterwort – in Scots vernacular 'earning-girse' or curdling-grass – was used to curdle milk for cheese-making, and masses of butterwort leaves formed lime-green rosettes along the peaty banks of the Blind Burn below the ruin. Just above Yearning Hall, where the Pennine Way dips south from Lamb Hill, stood one of the two refuge huts which have very sensibly been provided along the route of the massive last day.

I went south at a good pace along the border fence, among the dark peat hags on Blackhall Hill and past an outcrop of wacke or decomposed igneous rock, dark and soft, that popped its head up on Greystone Brae. Down near the valley road two great Roman marching camps overspread the slopes of Chew Green, with a fort incorporated and a couple more strongholds as part of the complex, all built in these remote hills two thousand years ago at the orders of Governor Gnaeus Julius Agricola, a man who would have brooked no lawbreaking whatsoever.

The Pennine Way follows the green road the soldiers built, part of the mighty highway of Dere Street that arrowed from Eboracum north to the shores of the Firth of Forth. Subject to ambush, cold and over-stretched, often soaked and always on the lookout, how the Roman conscripts must have grumbled and groused on their long marches at the outermost margin of civilization. Today it was the curses of a pair of Pennine Way walkers that floated on the Cheviot winds as they limped up Dere Street, bruised and blistered of foot, towards the Scottish bor-der on the last day of their 270-mile ordeal. 'Good luck!' I chirped as I caught up with them on the Border fence. 'Luck you!' they snarled back – or something like that.

A side path led off east over the rounded backs of the Dodd and Deel's Hill (the Devil again! How he gets about!), with the deep valley of the young River Coquet out of sight in its cleft below. It was a won-derfully exhilarating march, a cold wind out of Scotland bowling me along, with far views across bleak treeless hills whose pale grasses seethed and raced as if stirred by invisible spears.

Down at Buckham's Bridge I dropped on to the valley road and turned back past the lonely farmhouse of Fulhope where half a dozen farmers had gathered to dip their sheep. A quad bike on the steep slope above carried a shepherd and an ancient collie, while the junior dogs crouched and raced at shouted commands: 'Left! Left! That'll do!' The sheep wheeled and scampered in a panic, the shape and direction of the flock skilfully managed to funnel them down to the dipping bath.

Back at the car park I met a squad of soldiers, doubling up the lane at the end of some ferocious exercise on the Otterburn Ranges. The lad in the rear – he couldn't have been more than seventeen – was wincing with every step. 'Hop in,' I said, opening the car door. He grinned, sheepishly, and hobbled on after his mates. Julius Agricola would have recognized the spirit.

The last circuit I did was from the little forestry hamlet of Byrness in Redesdale. The narrow road up the Cottonshope Valley had a red flag fluttering and a severe notice forbidding entry on account of soldiers training with live ammunition. But a call to the nice people at Otter-burn Range Control* elicited a courteous 'That'll be OK today, they won't be over your way at all.' A rattle of machine-gun fire and the pop of a rifle sounded occasionally from some far-off valley, interspersed every now and then with artillery fire much further away, a curiously feeble and hollow sound, like a giant punching an empty biscuit tin. I walked up the road to the lonely farm of Cottonshope, where a faint path climbed through rough grass pastures, swerving in and out of the boundary of Otterburn Ranges, up to meet the Pennine Way on Raven's Knowe.

Here was the place to stop and take stock. To the northeast the rounded bulk of Cheviot, its granite bulk lifting gently to the cloudy sky, the sandstone flanks rolling and tumbling down to where I stand. South and west the Scottish border plain, cruising smoothly out of sight. West, hills and forests running to the border. To the east, the barely perceptible path dropping out of sight into the Cottonshope Val-ley. South from Raven's Knowe it was all forest, great swathes of the coniferous cladding that has adhered to the Redesdale hills since the area was planted between the world wars of the last century. Having felled countless acres of prime broadleaved woodland, parkland and private tree cover to meet an instant need for props and boards in trench construction during the First World War, Britain was determined never to go short of timber again. Redesdale, Wark, Kielder and other swathes of border hill country were considered just the ticket for massive con-iferous planting. The sloping peaty ground would drain well, and best of all these acres held only a small, scattered population of hill-farming families who would not baulk if offered a decent price for their land.

What was not taken into consideration was the ecological

---

* 01830 520569.

impoverishment brought about by such a heavy monolithic intrusion – conifers, conifers, and nothing but conifers. They grew well and quickly, undeniably, but once harvested, the hillsides looked more like the mud-baths of Passchendaele than anything green and fertile. Rainwater flooded down the parallel alleys where the trees had been, sluicing chemical fertilizers into the watercourses. The birds and flora of the uplands departed the moors and valleys. And people who had expected to farm the hills like their forebears found themselves dispossessed and in many cases relocated.

From Raven's Knowe my gaze jumped over the trees and landed on the southern skyline, the long grey-blue lift of the northern Pennines. Somewhere between here and those hills ran the great dolerite upthrust of the Whin Sill, carrying on its crest the most extensive monument to the Roman Empire still standing in Britain – Emperor Hadrian's great declarative Wall. That was for another day. Now I just wanted to get down off the Cheviots and into a nice hot shower.

The wind pushed hard and cold from the west, hissing in the conifers below. I headed down from Raven's Knowe on a boggy track partially paved with flagstones from old mill floors, by way of Houx Hill, Green Crag and Saughey Crag. These outcrops of sandstone were of weath-ered grey, but in places the surface had been knocked away by Pennine wayfarers' boots and sticks, exposing a sandy-coloured stone winking and sparkling with crystals.

Catcleugh Reservoir came into view, a wedge of steely water among the trees. The Pennine Way descended among tuffets of bilberry and sphagnum, before suddenly tumbling precipitously down a staircase of rocks. I slid down boulder faces, hopped the rocks and grated my ankles. You weren't expecting that? Brace up, you clown – this is the tough old Pennine Way, remember?

Down in Byrness village beside the roaring A68 road I found the lit-tle Church of St Francis, built in the 1790s for a very small and very scattered congregation. In the south wall a stained-glass window of pre-Raphaelite realism was dedicated in 1903 to the memory of sixty-four men, women and children who died during the building of Catcleugh Reservoir. The reservoir, fed by the River Rede, was built between 1884 and 1905 to supply drinking water to the far-off but rapidly expanding Newcastle-upon-Tyne and Gateshead. Workers flocked to the area and made it buzz as never before. When they left, Byrness reverted again to its quiet isolation. Apart from the reservoir itself, the chief memorial to those workers is the plaque with the names of their dead fellows on the church wall, and the stained-glass window alongside, featuring half a dozen labouring men. Trousers hitched up with string, capped and hatted, their bearded and moustachioed faces are bent intently on their work with pick, shovel and wooden wheelbarrow. At their feet sits a child, clasping a bundle. She has come out to the works, dangerous place though that is, with her father's lunch wrapped in a cloth. In the background, bearing down on the men and the little girl, is a puffing railway engine and a line of wagons, soot black and ominous.

In the Forest View guesthouse that night I dined among walkers. All were northward bound, with a hellish long day to face, or maybe two, before journey's end at the Border Hotel in Kirk Yetholm. They were tired, aching and modestly proud of their achievement in coming the 240 miles from Edale on foot and in one piece. I didn't envy them the Cheviot tribulations that would be their lot tomorrow, and in their somewhat daunting and non-nonsense company I pretty much kept my own counsel.

At a communal table I rested my unbooted feet, ate a pasta bake out of an enamel dish, and drank excellent draught beer. I hadn't expected that either, but on the Pennine Way you learn to take the smooth with the rough.

From Byrness to Bellingham is a straight 16 miles, at first through smothering conifers at the eastern edge of Kielder Forest, then slippy, sloppy moorland and rough pasture going south over Padon Hill. It's dour country, especially under continuous cloud; so much so that I found relief for my mind and eyes in the stony spoil heaps around the long-closed Hareshaw Colliery, where domestic and steam coal was dug from the 1850s for the next century. Old limestone quarries and ironstone workings around the ferny gully of Hareshaw Linn were evidence of pickings, if not exactly rich ones, to be won from the Tyne Limestone Formation of Carboniferous rock underlying the moors. Turning to the geological map, I found it had gone wild in curving tiger-stripes of milky green and caramel. They represented layers of sandstone and limestone folded during the Variscan orogeny of the late Carboniferous period, when mountain-building far to the south exerted pressure on these northern strata.

Among the grey-pink sandstone shops and pubs of Bellingham, walkers going south on the Pennine Way rest their bones for the next day's trek of much the same length across much the same sort of country. The track winds as farm roads and obscure depressions in the moor grass to cross Wark Common and Warks Burn where thin layers of Carboniferous rock lie exposed. Soon I plunged in among the massed trees of Wark Forest, a daunting experience. It's a curious fact that open moor, be it grim and bleak, lacks the chill factor at the heart of a coniferous forest. There's something about the pitch blackness under the trees, the absence of light and colour and movement, that raises atavistic fears along with the hairs on the back of my neck, making me think of witches and wolves, of woodcutters' children and how beastly their stepmothers are to them.

It was only a step through the edge of Wark Forest, a couple of corners to cross, but I was pathetically grateful when the black trees fell away and I found the blue-grey eye of Greenlee Lough winking cheerfully ahead. To the left the ground rose abruptly into long sharp outcrops, the Queen's and King's Crags. Poor King Arthur – slumped in dejection on King's Crags, cuckolded and despised, he couldn't get Queen Guinevere to talk to him. Seated on Queen's Crags a few hundred metres away, she was combing her hair and daydreaming of Lancelot. In his frustration, Arthur picked up a rock the size of a wagon and hurled it at his consort; but she saw it coming and deflected it with her comb. And there it lies to this day, jagged and uptilted, its red sand-stone scored with scratches from the teeth of Guinevere's comb, or possibly from the abrasions of an Ice Age glacier. Nearby I found the sandstone hollows that the Romans had quarried. On a smooth rock face at head height some literate members of a military work party had carved the names of their three overseers, centurions Saturninus and Rufinus, and the optio or second-in-command Henoenus. These names were crudely cut to begin with, and two thousand years of exposure to Northumbrian weather in a north-facing position have not improved their clarity. You have to take their lettering as an article of faith. But as ever, the graffiti of the past brought a scene vividly to life: the grumbling squaddies at their toil, the bullying officers, some ingratiating toady who thought he'd get into that bastard Henoenus's good books by inscribing his name. 'Oi, Legionary Servius, what the f---! Spoiling my nice smooth rock face! I'll deal with you when we get back to barracks, lad! Right! Quiiick . . . *march!'*

The Pennine Way crossed a rise of ground, and there, dead ahead, was what I'd come a long and weary way to see: the great dolerite ram-part of the Whin Sill stretching away to left and right, with the long grey line of Hadrian's Wall riding the crest of Cuddy's Crags. Standing by Broomlee Lough and looking up at the Whin Sill cliffs I thought of St Cuthbert – 'Cuddy' to Northumbrians – and his hermit retreat on an island of the same dark dolerite out in the North Sea 100 miles back along the way, another link through time and distance in the ever-lengthening chain of this journey.

St Cuthbert's Cave in the sandstone crags of Cockenheugh.

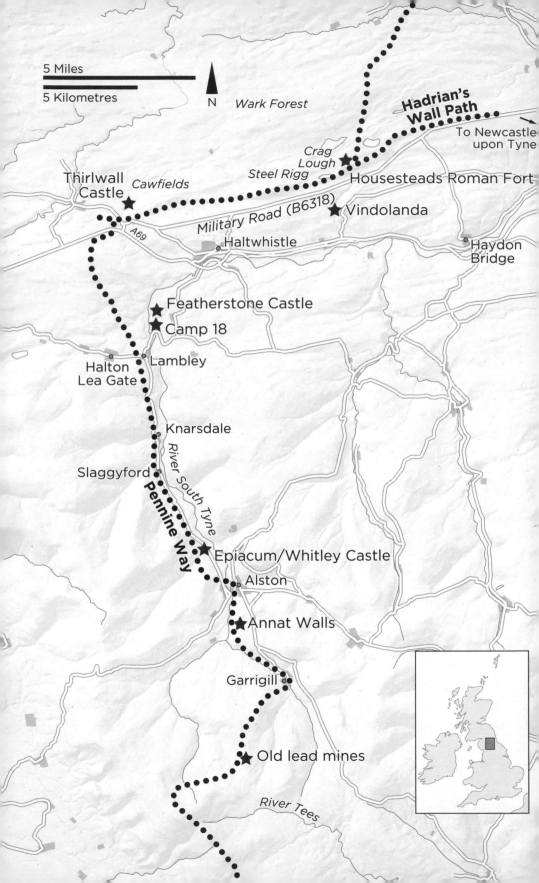

9

Mighty Wall, Black Nazis and Leaden Ore: Hadrian's Wall to Garrigill

'BRIGHT BUT COLD,' SAID THE weather forecast notice at fog-bound Housesteads. I allowed myself a smile as I climbed the slope towards the Roman fort. Today there was nothing but a white wall of mist up ahead, until suddenly the stark stone walls of the stronghold shaped themselves out of the vapour, with Hadrian's Wall a ghostly line along the ridge beyond. I wandered through the paved courtyard of the Commanding Officer's house, down to the soldiers' multiple-seater latrine with its floor drain and stone water tanks. The eight hundred men of the First Cohort of Tungrians, the original garrison troops at Housesteads two thousand years ago, must have known many a cold, foggy summer's morning such as this.

At least the men would have been used to it; they were not warm-blooded Mediterraneans, but tough recruits from the Low Countries. In point of fact, most of the soldiers who built and garrisoned Hadrian's Wall were northern mercenaries. Nevertheless, the Wall must have been one of the most unpleasant postings a soldier of the Roman Empire could hope to avoid, especially in the long northern winter. Bitter winds, blizzards, interminable nights and the threat of attack by Pictish warriors or hungry wolves were added to the endemic military miseries of boredom, barracks life and bullshit. Yet Emperor Hadrian had ordered the Wall not only to be built, but to be permanently garrisoned. And what the Emperor ordained was written in tablets, or rather in foot-square blocks, of stone – 330-million-year-old sandstone quarried, shaped and put in place by the soldiers themselves.

After their successful invasion in AD 43 under Emperor Claudius, the Romans took several decades to subdue the British natives. The northernmost boundary of the area over which they had control moved slowly up the island. It took around eighty years and several changes of emperor to settle the border along a line between the mouth of the River Tyne and the Solway Firth. Emperor Hadrian, coming to power in AD 117, decided that this northern boundary of his empire needed fortification. Canny and yet fortunate Hadrian! On a visit to Britain in AD 122 to inspect the coast-to-coast route, he saw that the central section comprised a massive cliff, 100 feet high in places, with its reverse slope running back into Roman territory and its sheer face turned due north towards the barbarian lands. The material of the cliff was too hard to use as building stone, but there were plentiful beds of easily quarried and shaped sandstone nearby. This section of the Whin Sill's tongue of dolerite, exposed as a great east–west cliff, couldn't have been better positioned if Hadrian had designed it from scratch himself.

The Emperor initially ordered a stone wall 10 feet wide and some 15 feet high to be built from the Tyne westward for 45 miles, with a turf wall 20 feet wide at the base to run the rest of the way to the Solway. However, the Wall as actually built between AD 122 and 128 was a slimmer construction, with a 'milecastle' or fortified gateway at each mile, and two observation turrets between each milecastle. A ditch 10 feet deep ran on both sides of the Wall, which was garrisoned by twelve major forts. Hadrian's Wall was not simply a means of keeping the Scots barbarians at bay; it was a grandstand for viewing what the northern savages were getting up to, and a launching pad for aggressive military action against them. And it was as much a symbol and political statement as a defensive structure. Hadrian's Wall said, loud and clear: here is the northernmost boundary of our mighty empire. From this point outwards, the writ of civilization and the rule of law cease to run.

Everything south of here is Roman, civilized and good; everything north is the outer darkness of barbarian country.

When the Romans withdrew from Britain at the beginning of the fifth century AD to shore up their crumbling empire further south, the civilian villages that had grown up under the protection of the forts continued to flourish. But the abandoned forts themselves decayed, along with the now ungarrisoned and useless Wall. Over the centuries the structure was continually plundered for its conveniently cut and shaped stones. All the way along the course of the Wall the handy stones with their distinctive square shapes can be spotted set into field walls, barns, churches and houses. Hadrian's Wall could have disappeared completely. Large sections of it actually did so when, after the Jacobite Rebellion, a military road – now the B6318 – was built on top of what remained. But even back then, antiquarians were interested in the remarkable old monument.

The miracle is that there was enough of the original building material left to allow the restorers to do such a good reconstruction job on the Wall. Gradually a preservation movement took shape. It culminated during the last century in the takeover of several sections of the Wall by the National Trust and English Heritage, and the declaration of the monument in 1987 as a UNESCO World Heritage Site.

Trying to walk Hadrian's Wall in the dim and distant past was a chastening experience. This 2,000-year-old national treasure was inaccessible for long stretches. Such sections of footpath as did run beside the Wall itself became a boggy quagmire after rainfall of any kind. Waymarking was poor, ancillary path marking non-existent. You struggled and sweated, puzzled over the map, detoured along many miles of remote byway, and generally had plenty of cause to pollute the clean Northumbrian air with foul language. Nowadays, praise be, the Hadrian's Wall Path National Trail runs from Wallsend in Newcastle-upon-Tyne to Bowness-on-Solway on the other side of the country, and it rides the Whin Sill roller coaster right next to the Wall on the most dramatic section of the dolerite rampart.

*

As I set off west from Housesteads a breath of wind swirled the mist into rags. The Whin Sill cliffs emerged ahead, an iron-coloured tsunami breaking northwards into a white sky. The old house and barns of Hotbank Farm lay huddled on the slope of Hotbank Crags, their walls much patched with Roman stones. The square-cut blocks of the Wall itself were blotched white with lichen and bearded with mosses. The stepped path of the National Trail swooped up the crests and down into hollows of the dolerite sill, passing the foundations of the milecastles and turrets where conscripts from the Low Countries paced and shivered and looked out into the debatable lands to the north from which the wild Picts might come screaming at any moment. I halted above Milecastle 39 and stared out from the Wall to the looming black line of Wark Forest and the steel-grey humps of the Cheviot Hills beyond, picturing those young men on Wall duty, sulkily clutching their cloaks around them and wishing they were down in Vindolanda where the latrines ran with clean water and the stew came hot to the table.

The Romans wrenched the history of Britain masterfully into a new course, and their influence still pervades our language, education and expectation of the civilized amenities of life. Yet it's not their enormous political significance, but the tiny details of their everyday lives that fascinate us most. Excavation and archaeological science have given us a remarkable account of the Romans of Hadrian's Wall – their hopes and fears, their clothes and appearance, their correspondence, wine bills, curses and prayers. On a previous walk along the Wall I had visited the excavated fort and settlement at nearby Vindolanda, just south of Milecastle 39. Here lie the foundations of the town the Roman soldiers and their families lived and loved in, its houses, temples, wells and paved streets. I had been spellbound by the finds on display in the museum: tough hobnailed shoes and exquisitely crafted leather sandals, decorated drinking bowls, earwax cleaners and nose-picks, delicate rings and brooches, combs with fine boxwood teeth, and a wealth of official and personal correspondence scratched on wax blocks. A floor tile on display carried the prints of a pig's incurving toes, as sharp today as the hour they were dinted two thousand years ago in the drying clay (how the potter must have sworn!), a tiny incident caught in the long movement of time like the lapilli trapped in a splash of ejecta on the shores of Coigach.

The Wall and its path dropped away into Sycamore Gap where a shapely sycamore tree stood at the very bottom of the hollow, a much-photographed icon of symmetry. At Steel Rigg a pale sun glanced through the cloud, dispersing the mist for long enough to allow me a look back at the Wall riding the bulwark of the Whin Sill above a gleam of grey water in Crag Lough. Then it was on by way of Winshields Crags, Bogle Hole and Thorny Doors, with wonderful views north over rough grazing rumpled into corrugations by the tilting of the layers of limestone, mudstone and sandstone that underlie these borderlands. From Cawfields onwards, the Vallum's vast open V of a ditch was my companion through this lumpy, wild and lonely country, with views out towards the dark line of the forested northern horizon.

Above Walltown Farm the Wall stood up to 10 feet high, a broad band of stone laid across the ribboning landscape. Yet, as elsewhere, many of its stones had been stolen and were plain to see reused in the field walls and farm buildings. And it was the same at Thirlwall, parting point of Hadrian's Wall and the Pennine Way. A ruined fourteenth-century pele tower stood alone and proud on its green mound beside the Roman wall. It was obvious at first glance that John Thirlwall, builder of Thirlwall Castle, used whatever materials lay nearby, and those were mostly the little brick-shaped sandstone blocks so handily quarried and dressed by his Roman predecessors. The castle walls, however ragged, stand 30 feet tall, the symmetrical Roman stones giving the ruin its strength as well as lending the walls an incongruous air of neatness.

Just beyond Thirlwall I followed Hadrian's Wall across the B6318. The former military road, which had been running parallel to the Wall on the south side in long straight stretches, now received a promotion to become the A69 highway towards Carlisle. As though relieved of its duties, the country road immediately bent this way and that, wriggled its shoulders and wandered away across the Northumbrian uplands in the general direction of Scotland. Here the Pennine Way and I unpeeled ourselves from the Wall and turned off south. A dip through the Vallum, a heart-stopping dash across the rush and grind of the A69, and off over the upland moors. When I paused on Blenkinsopp Common and trained my binoculars back the way I'd come, I could still distinguish the outline of Hadrian's Wall surfing the dark waves of the Whin Sill.

The igneous drama of the border country, the granite upthrust of the Cheviots and the dolerite tsunami of the Whin Sill, can bedazzle the south-going walker. They seem like the last sparks from the geological firework display that constitutes the bare bones of Scotland, exposed so definitively in the stark mountains, the ancient lava fields and squeeze-box cliffs one has walked so far. What England has to offer as it stretches ahead is no less remarkable, but it feels more low-key and looks less obvious in the landscape. As a rough and ready distinction, you could say that turning south from Hadrian's Wall marks the point where the fire and the fury of the early world are largely behind you. Ahead lie landscapes shaped by the Carboniferous era with its richness of life, its teeming seas and sticky heat, forests and swamps, and its creatures moving from the water to the land as the sea flooded and retreated and flooded again – the proof of all this to be read in fossils and footprints and thick horizontal bands of limestone, mudstone and sandstone, although these have still been bent and folded on occasion by faults and shakes and magma intrusions with their injections of minerals from far below.

As a rule of thumb, one can take a stance from the Tees–Exe line, an imaginary but handy mark stretching across England from northeast to southwest between the mouths of the rivers Tees and Exe. The rocks west of this line tend to be older, harder and more prominent in the landscape, with more igneous and metamorphic content forming higher and more sharply defined hills and uplands – for example, the northern and central Pennine Hills along which runs the Pennine Way. To the east of the line the rocks tend to be sedimentary, lying flat or gently tilted and folded; younger rocks forming softer and lower-lying landscapes which have retained more of the glacial sediment left as a legacy of the Ice Age, like the rolling clay and limestone cornfields and pastures of the Midlands.

As far as my journey through the bones of Britain was concerned, there was still a fortnight's travel over and through the Pennines in front of me before I would cross the Tees–Exe line on the borders of West Yorkshire and Greater Manchester nearly 200 miles away. Trudging south across the low moorland of Blenkinsopp Common I had the route ahead unrolling in my mind's eye: the upland moors and lead-mining hillsides, the Ice Age flora of Upper Teesdale where I'd rendezvous with the Whin Sill once more, the beautiful limestone dales of North Yorkshire, the harsh black gritstone of the bleak south Pennine moors.

What a difference the changing geology makes to one's perspective. I stopped on the low bumps of Black Hill and Wain Rigg to stare around at dark rushy moorland where sheep were selecting their mouthfuls with fastidious care among the sedge clumps of these sodden pastures. Under the black squelchy peat and a thin layer of nubby glacial rubble lies a band of coal-bearing rock. The Pennine Way winds its course by Greenriggs and Batey Shield, past old coal-mine shafts fenced off among the rushes. But the industry is far from consigned to history. Half a mile to the west lies Halton Lea Gate opencast coal mine, active from 2012 to 2020, during which time nearly 150,000 tons of coal came out of the ground. When I first came here fifty years ago it was bleak green fields. Next, a massive hole in the landscape, its edges almost up to the nearest houses, diggers and tipper trucks grunting in the depths. Finally, bankruptcy for the mining company, HM Project Developments, the abandonment of the mine and of the company's pledges to the community, and the long-drawn-out process of filling in and restoring the damaged land.

Eastward the land dips gently to the narrow north-going cleft carved out by the River South Tyne. Road, river and former railway run in close company, squeezed tightly together by the converging valley sides.

I detoured downhill to Lambley Bridge and hung over the parapet for the pleasure of watching the river. Down below hardy youngsters were towelling off after a dip in the clear, fast-flowing water. A dipper bobbed its white bib on a stone mid-river, a brief curtsey every couple of seconds as though encountering a never-ending line of royalty. Nearby a grey wagtail flashed its bright yellow bathing drawers. Sand martins dashed in and out of their nesting holes in the riverbank. All was lively and animated in a 10-foot airspace above the water.

A few years before, walking the east bank downriver from the bridge, I'd come across a line of buildings, stark brick boxes black-windowed and sinister, reminiscent of 1940s photos of concentration camps. This clutch of derelict huts left to rot in the quiet river pasture was all that remained of the wartime site known as Camp 18. American troops readying themselves for D-Day were housed here in 1944. After they'd gone, Italian prisoners-of-war took their place. At the conclusion of the war the Italians in their turn were replaced by German captives, among them many who had been labelled 'Black Nazis' – officers whose fanatical allegiance to Nazi ideals had never wavered. Their political views may have been despicable, but their captors decided, in a remarkably far-sighted initiative, that the intelligence and leadership qualities they obviously possessed could be vital assets in a Germany that needed all the help it could get to raise itself again from the ruins of the war. If they were capable of being 'de-Nazified', the authorities decided, they could be repatriated to help with defeated Germany's reorientation.

Camp 18 in its heyday was a sprawling affair of two hundred huts in four compounds. It held four thousand officers and six hundred orderlies, whom those running the camp were determined to treat not as a bunch of evil no-hopers, but as possessors of bright, articulate, creative minds that could be channelled into better and more constructive courses. To cater for their bodies and souls there was a chapel, a library, a theatre, a medical centre, a bakery. The inmates formed three separate orchestras and published a newspaper, *Die Zeit am Tyne*, the *Tyne Times*. By night they were treated, or perhaps subjected, to lectures by dons brought in from Durham and Newcastle universities. It was all rated a big success. By 1947 the barbed wire surrounding the camp had been removed and the inmates were out each day working on local farms. The following year Camp 18 was closed and the Germans were repatriated, the scales fallen from the eyes of many if not all of them. The brick huts by the river were abandoned to the slow working of time and the Northumbrian weather.

Poking around among the empty buildings I came across a memorial stone erected in 1982 by former inmates of Camp 18. It was inscribed in honour of Captain Herbert Sulzbach, OBE, the camp interpreter, who 'dedicated himself to making this camp a seed-bed of British–German reconciliation. Our two nations owe him heartfelt thanks.' This memorial was the gift of men who had been true-believing Nazis before the interpreter and his colleagues worked to change their minds. Herbert Sulzbach was one of three senior officers in charge of Camp 18, and what was truly remarkable was that all three of these brave and selfless men were Jewish.

Back on the Pennine Way I walked south into a gently rolling landscape underpinned partly by limestone, partly by gritstone, often labelled 'millstone grit'. This coarse, abrasive rock, perfect for use in the manufacture of grinding stones for corn mills, was laid down during the later Carboniferous period when wide tropical rivers spread layers of sand between layers of mud around 320 million years ago. Up on the surface of gritstone country the exuberant limestone flora disappear, disliking the acid conditions. Field and house walls are dark, but they sparkle in the sun. On this warm afternoon wading birds were about their nesting in these lonely upland pastures. The birds do not care what rock lies below. Their only concern is for an absence of disturbance and an abundance of insects. Oystercatchers called shrilly, *pik, pik!* Lapwings complained about the intrusive stranger as they tumbled clownishly overhead with their creaking *eee-wit!*, and curlews emitted that haunting, fluid cry so evocative of northern moors in springtime, a musical bubbling of increasing urgency that fades suddenly away as the bird descends to land with a shiver of wings and a dip of its long, down-curved bill.

Now the Roman presence revealed itself once more in the upland landscape as a fine straight track that carried the Pennine Way arrowing across the skyline. The way crunched firmly underfoot, a relief after the boggy miles I'd trudged since leaving the Wall. Marching troops tramped this Maiden Way two thousand years ago on their way south from Magnae Fort near Thirlwall to join the cross-country road between Eboracum (York) and Luguvalium (Carlisle). After the legions left and their wonderfully engineered road system fell into disuse, herdsmen and drovers took over the well-drained and levelled old route and pursued it across the uplands for the next thousand years. These days it's the blistered and peat-splashed footsloggers on the Pennine Way who share the Maiden Way with farmers on sheep patrol. 'All reet,' grunted the ruddy-faced quad rider who passed me in a blue haze of oil and petrol smoke, and 'All reet,' I responded.

Down below Slaggyford the Maiden Way passes the Roman fort of Epiacum or Whitley Castle, built around the same time as Hadrian's Wall. The ramparts and ridged lines of defensive outer ditches lie clearly marked out on the hillside. It's a curious shape, not the usual strict rectangle but a squeezed-out lozenge, as though the engineer had had difficulty focusing. The Romans were mining these hills for lead and silver, and Epiacum was well placed to guard the workings and associated transport, as well as providing a handy stopover point along the Maiden Way.

Hereabouts the south-going Pennine Way seems to be wincing and feeling its feet. It quits the uplands and drops down to the riverside meadows opposite the old lead-mining town of Alston. The town with its pubs and chip shops perches high around its sloping cobbled market square. It's exactly the place to stop and get the bloody boots off, and that's exactly what I did.

*

Next day was a doddle. I left the Pennine Way in the valley of the River South Tyne and followed a hillside lane that ran in parallel above, a narrow old way, dead straight and walled handsomely on both sides. At High Physic Hall a tiny lad was trailing down the lane behind his mum, shoulders bowed under a weight of troubles, for all the world like Shakespeare's boy 'creeping like snail unwillingly to school'. Ahead the postie's little red van went bouncing along on creaking springs towards Fairhill Farm.

The dale side fell away to the river, rushing over flat platelets of flood pebbles. It has cut deep down through an alternating succession of limestone, sandstone and mudstone laid down some 332 million to 320 million years ago. These layers form this particular area of what's known as the Alston Block, a massive horst or slab of raised rock nearly 2,000 feet thick in places. The floods of millennia have exposed and terraced the uppermost face of this rock. Looking across the river I saw the dale side rising in broad shelves sitting far back. The green pasture near the river – the 'in-bye land' – sloped up to the farmsteads whose white-walled houses looked out among shelter trees. Above rose the rock terraces sculpted by flood, frost and enormous ice intrusion.

When the glaciers of the last Ice Age came inching across this high ground from the southwest, the tough Alston Block limestone, footed on hard granite, resisted their advance. It had already withstood subter-ranean pressures that had buckled weaker rock, and now the ice behemoth diverted to find a way through and onward by pushing its head into the cleft of the South Tyne Valley. Just upriver from Alston, Park Fell on the opposite side of the dale showed a fine example of the crushing yet subtle power of this advancing ice. Coming down the val-ley from the southwest it squashed the first part of the rock bench it came to, the southern end, forcing it steeply up and squeezing it together; then the forward motion pared away the rest of the bench as the ice pushed past, forming a narrower, streamlined end that tapered away to the north. Just ice did that: frozen water, with an engine of infinitely slow motive power behind it.

These dale sides are scarred with pale fans of lead-mine spoil, and the dark holes of disused adits lead back under the rims of the rocks. Above the lead-mine levels, stone walls climb the slopes in a mesh of dark lines wriggling towards a clean-cut moorland skyline unbroken by any tree.

Beyond Fairhill, a proper old Norse longhouse built all of a piece with its barns, I came to Annat Walls. There was something both formidable and familiar about the construction of the two conjoined barns by the lane. Where had I seen this sort of thing before, not too far away – the thick stone walls, the tiny slits of windows, the door sills several feet above ground level? It took a little while before I recalled walking in Allendale, a few miles across the fells to the east, and stumbling across a roofless ruin at Old Town with the same characteristics. The penny dropped – here was a pair of bastles. The farmhouse, too, looked as though it had been built as a bastle, as much for security as for domestic comfort.

The datestone over the low lintel said '1707'. You don't have to go back very far, only three hundred years or so, to enter an era when this countryside, so close to the much-disputed Scottish Border, was a lawless and dangerous place. Hereabouts might was right, and a prudent householder made sure that raiders and reivers could be kept at bay. The rule of law and the Crown, distant and often feeble entities, came a poor second best to the swords of the most powerful Border families and their kin and allies. People who lacked such connections, living in isolated areas such as this, built bastles if they could afford to – fortified farmhouses with cellarage at ground level, and the living quarters and front door above on the first floor. Access from outside was by means of a ladder. If you lived at Annat Walls and news came that a gang of reivers was heading your way, you could draw up your ladder and bolt your door and be reasonably sure that your enemies couldn't get at you. They would find it hard to burn a solid stone bastle. They might break into your storeroom, pinch your tools and seed and drive your cattle off, but at least you and your family would live another day.

Now big grey clouds came jostling up from the west. They sat low on the moors towards Cross Fell, the next and toughest leg in this stage of the journey. I put my head down and skeltered on along the roller-coaster lane. In the miniature ravine of Nattrass Gill liverworts shone among the mosses. A dog was barking its poor heart out in the barn up at High Nest. The deep cleft of the Black Burn across the river rose away towards the clouded hills that hid Cross Fell. Hawthorn and rowan had been newly planted at immaculately done-up High Sillyhall. And the empty house and byres of Low Craig stood silent and dank, with jackdaws in the chimney and a big conical mine tip just up the bank.

Apart from the back-breaking and dangerous work of extracting and processing lead ore from the bowels of the Earth, the lead-mining business had a visibly pollutant effect on the landscape. Mine spoil was tipped as close to the workings as possible; there was no point in barrowing the heavy waste any further than absolutely necessary. Hundreds of hummocks of spoil, pale and fan-shaped, project out from the hillsides and line the riverbanks all across the North Pennines. Standing on the bridge over the River South Tyne at High Redwing, just north of Garrigill, I looked up and downstream. Here the river has cut down through the deep litter of gravel and cobblestone left behind at the glaciers' eventual retreat, and further on down through the Tyne Bottom limestone that floors the dale. Huge boulders litter the fringes of the river, testament to the power of its winter floods. A series of shallow terraces shape the dale sides, terraces of alluvial stones and rocks spread by the river in its rages. The stones and gravel of the lowest terrace have the pale tinge of the bankside lead spoil heaps from which they have been washed out by the vigorous, flood-prone River South Tyne.

Back from the water's edge, among the greys and browns of this lead-contaminated rubble, minuscule clumps of flowers were showing. I scrambled down, looking along the lowest shelf of flood material for metallophytes – a shiny new word I'd only recently come across, signifying plants that can grow in ground impregnated with heavy metals. Soon enough I had spotted an old favourite from these parts, the delicate white five-pointed stars of spring sandwort. The tiny loose flowerheads of alpine pennycress, the subtlest of pale pink, evaded my search, but the website of the North Pennines Area of Outstanding Natural Beauty told me they shared this precarious foothold with the spring sandwort.

Metallophytes are picky about where they take their stand. Spring sandwort can grow in various unpromising soils, and among these it's able to tolerate any heavy metals, such as lead, that might be present. This is a rare enough characteristic among plants. But alpine pennycress is even more unusual, in that it can't survive in the absence of such pollutants. These beautiful and highly specialized plants find the habitat they need here in the North Pennines where competition is sparse thanks to the mineral-infused ground, the protracted winters, and the cold, rain and wind endemic at these relatively high altitudes. Even little Garrigill, sunk in its dale bottom, is 1,100 feet above sea level, and the top of Cross Fell, wrapped in mist, rain and ferocious winds most days of the year, is almost 2,000 feet higher.

Down in Garrigill it was as silent as the grave. It's hard to credit that this immaculate little hamlet with its tree-lined green, its neat hillsides and clear river, was crammed, foul and noisy with lead mines and miners two hundred years ago. The Industrial Revolution was in full swing back then. During the first half of the nineteenth century the populations of Sheffield, Leeds and Manchester swelled threefold. Cities all over Britain were rapidly expanding. They needed water pipes, guttering and lead-based paints. A thousand men, employees of the London Lead Company, burrowed the dale sides around Garrigill for lead ore at Whitesyke and Bentyfield, at Tynebottom and Ashgill and Cross Gill. Thousands of others lived and laboured just over the hills at Alston, Nenthead and Middleton-in-Teesdale.

The ores of lead had been injected into cracks in the Carboniferous limestone around 290 million years ago, 50 million years after the rocks themselves had been laid down. Hot fluids infused with minerals rose and crept along the splits and fissures in the limestone, gradually slowing, losing heat and solidifying into veins of lead, or more rarely zinc and copper. Finding the precious lead veins was a puzzle in itself. They were never very long, a few miles at best, often fractured or shot across with intersecting veins, and frequently only a few inches wide, though they could be deep. The lead ore was mixed with what was called gangue, worthless material that had to be discarded. Between 5 and 10 per cent of lead content in a seam was considered a rich deposit. And the deposits themselves were generally patchy and intermittent 'ribbon deposits'. When you got to the end of the ribbon, that seemed to be that – the seam had run out. But experience taught the mine engineers that the seam might resume just a few yards away, if you could gauge or guess exactly where. So the whole enterprise was an imprecise science, which mine owners tended to apply as much in hope as in expectation.

Up to the start of the Industrial Revolution the life of a North Pennines lead miner was by any reckoning nasty, brutish and short. Lead poisoning, accidents at work, a diet of bread and tea, insolvency, and large families living cheek by jowl in damp and insanitary dwellings were all unconducive to living happily ever after. By the early nineteenth century the London Lead Company had bought out most rival companies and owned what amounted to a monopoly on lead mining in the area. But the LLC was no rapacious Moloch. It was interested in financial profit, but it was also run by Quakers with philanthropic ideals. Its policies were long-sighted, and it had broader aims for its land and people than the mere production of lead. The mining families had to be able to feed themselves. The company built houses for its employees up the dale sides, with enough land attached to be run as smallholdings. It invested in draining and fertilizing the rough pastures and the moorlands above, and also in planting larch and pine trees to provide timber for the company's mines. Not only that – as Quakers, the board members felt a sense of Christian obligation towards their fellow men. Libraries and reading rooms were provided for the miners' education, chapels for their spiritual improvement, baths for their corporal cleansing, schools for their children, a mill to grind their corn, and shops with large stocks that broke the miners' historic reliance on credit traders with crippling rates of interest by selling goods strictly for cash only. Desperate working conditions, a bleak northern setting, a nice moralistic backdrop – what a subject for Charles Dickens to have passed up!

Last time I set foot in Garrigill was over forty years ago. My father and I had come limping down from Cross Fell towards evening, sodden with rain and sore of sole, bewitched and bamboozled by a 19-mile *via dolorosa* from distant Dufton. In pea-soup mist and drizzle we had slogged up and over the highest point of the Pennine Way, a route not long opened and, to say the least, ineffectually waymarked. We had got hopelessly lost, straying far off the path down the steep and unforgiving banks of the Cash Burn, adding many miles of splattering bog-hopping to the miserable day. My God, how glad we had been to sit down outside the George and Dragon Inn, gingerly unpeel our bloodied feet from our stinking boots, and get in out of the rain to a plate of stew, a couple of pints and the lumpy beds in the room we had booked. I'll never forget the look on the faces of a pair of girls who had failed to reserve any accommodation. They sat heads down by the stair door, too weary even to debate what to do next. I wonder what happened to them. Maybe a brace of *verray parfit, gentil knyghts* gave up their beds to those two disconsolate maidens, or offered to share with them, perchance. It certainly wasn't going to be us. Cheery cries and guffaws came up through the floorboards of our room from the bar where Pennine Way hikers with their beds safely booked drank and gave thanks for another day's survival. But sleep drowned all that out in a nanosecond.

Annat Walls bastles or defensive strongholds, reminders of wild days
along the lawless Border country.

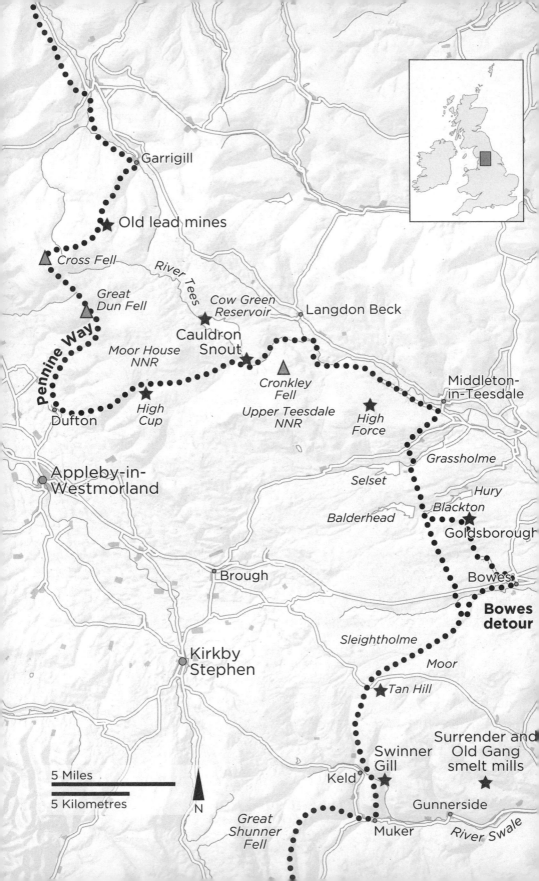

10

Hill Farmers and Teesdale Jewels:
Garrigill to Tan Hill

A STONY LANE WINDS UP from Garrigill through moor and bog, mine levels and mining spoil, for 8 long miles to the summit of Cross Fell. It's a relatively easy ascent for a Pennine Way walker. Not so much for the funeral parties who are said to have shouldered coffins along this ancient corpse road in times past. Before St John's Church was built at Garrigill in 1753 the village had no consecrated ground, and anyone wishing to bury a relation or friend according to Christian principles was obliged to carry them all the way up and over the flanks of Cross Fell, then down to the graveyard at St Laurence's Church in Kirkland, 10 miles to the west.

Or so the stories say. There's a simple question to be posed: wouldn't anyone with an ounce of sense convey the corpse down the dale to St Augustine's Church in Alston, 4 short and easy miles away? Probably a few natives of Kirkland were borne over the hill to burial in their home parish. But to abandon the romantic notion of the black-clad processions climbing the old corpse road with their grisly burdens would mean letting go of all those rich and meaty folk tales of hauntings and demonic apparitions, of burial parties driven back by storms and corpses discarded by the wayside. And where's the fun in that?

Whatever the truth about the corpse carriers, the old lane across the fells certainly saw a tremendous amount of traffic by boots and hooves, sleds and carts during lead-mining days. Everywhere you look as you gain height there's evidence of the lively, busy industry that scarred the moorland of Long Man Hill and Backstone Edge. Tunnels with well-crafted arches bore away into blackness – dry levels to admit the miners, wet levels to drain the constantly flooding workings. The track is spattered with tiny gemstones of fluorspar, pink, blue and purple, cast aside as a useless by-product of lead mining, but later used in steel-making to help prevent the cooling metal from cracking. Broken cottage walls, the ruins of the bothies where the miners sheltered, stand on the rare aprons of level ground. And the open mouths of air shafts, some but by no means all fenced off, pockmark the ground side by side with the natural subsidence hollows called shake holes that resemble so many velvety navels in the earth.

At last you turn south off the (putative) corpse road and climb a final grassy bank to reach the cross-shaped shelter at the summit of Cross Fell. How I would love to join the thousands who have eulogized about the prospect from this, the highest point along the Pennine Way at 2,930 feet (893 metres), with views that stretch north and south across endlessly unrolling moorland towards, respectively, the Scottish Border and the Yorkshire Dales, and west across the broad green farmlands of the Eden Valley to the bewitching outlines of the Lake District fells, 25 miles off against the sky. Two things militate against this: one, I have never been here in anything other than thick mist, or rain, or both; and two, I suspect the gentle and widespread convexity of the fell at the very top hides all views to every quarter. In fact, the Carboniferous limestone lump of Cross Fell, with its gritstone capping and screes, is a hill that walkers look to, rather than from. You can see its crouching beast shape from viewpoints for hundreds of miles around – from Lakeland, from the Scottish Borders, from the Eden Valley to the south, and in rare clear weather from the peaks of Snowdonia.

But what one sees in the mind's eye – the 'skull cinema', as supreme walking writer John Hillaby termed it – counts just as much when you are piecing together the jigsaw, or better still the ever-fracturing and reshaping kaleidoscope of impressions, that makes up a long journey on foot. Sitting in the mist with my back against the shelter walls, rain dripping off my spectacles and nose, I shut my eyes and pictured the gigantic peaty sponge that blankets the upland massif of Cross Fell, leaking and squeezing out the trickles of water that within a few miles of their sources reach their valleys and become the northern-flowing River South Tyne, the River Eden heading northwest for the Solway Firth on the Scottish Border, and the snaking, south-going River Tees. Cross Fell sits towards the western edge of the Alston Block, and imagination, if not my actual eye, delved down off the fell into the fault valleys that form three sides of the roughly square horst of raised upland – Tyne Valley to the north, Stainmore Gap to the south, and the beautiful Eden Valley out west.

Heading east off Cross Fell summit, somewhere around Crowdundle Head I stepped across an invisible border into the North Pennines' last remaining wilderness. The 22,000 acres (8,900 hectares) of upland that officially carry the somewhat burdensome title of Moor House–Upper Teesdale National Nature Reserve lie high and wild between the valleys of Eden, Tees and South Tyne. The reserve is split distinctly between Moor House, the great swathe of uninhabited upland that slopes south and east off Cross Fell, and Upper Teesdale, the lower portion where the River Tees flattens out and carves itself a course through rugged farmland and over a series of spectacular falls. Cow Green Reservoir separates the two halves of the nature reserve, and is included in neither, for reasons to be explored a little further down the way.

The habitats of the NNR as a whole range from the flowery hay meadows, rough grazing and juniper woods of Upper Teesdale to the higher limestone grassland, heath and blanket bog that takes its name from the lonely scientific research centre of Moor House in a cleft near the infant Tees. Rare black grouse and golden plover mate and nest in the high uplands. Ring ouzels with their white parsonical collars and chirruping, interrogative calls find safety among the rocky streams and craggy outcrops far from human habitation. Nowhere in England is there one location comprising such a variety of delicate and endangered habitats. And nowhere in the world is there an upland site so carefully warded, so meticulously studied and so thoroughly understood. The weather and hydrological research station at Moor House has been recording data since the 1930s. The minutiae of climate change are read and studied all over the NNR, from changing river levels in spring spates to fragile plants migrating up the slopes in search of cooler and less competitive locations. Drainage channels in the bogs, known as 'grips', are blocked to preserve the peat as a storage medium for organic carbon. Sheep are excluded from the limestone grassland where the rarest plants are found. Yet all these admirable advances in research, education and conservation stand or fall by the level of cooperation that exists between the field officers of the NNR, Natural England, the governmental officials of DEFRA (the Department for Environment, Food & Rural Affairs), the Raby and Strathmore Estates that own the land, and most importantly, the hill farmers who tend the land and have to make a living out of it. That is asking a great deal of human nature.

For the walker following the sometimes obscure course of the Pennine Way across the western flank of Moor House NNR between Cross Fell and Swindale Beck, the reserve is simply a wonderfully lonely and beautiful piece of unspoilt upland. The source of the Tees is very near Cross Fell's summit shelter. You can squelch aside to find its marker stone and rock-filled hollow. And you certainly can catch sensational views to the west and south, if you choose a day of fair weather instead of today's murk and mizzle, as you march the long ridge, a natural grandstand, southward from the summit of Cross Fell by way of Little Dun Fell and Great Dun Fell with its hilltop litter of air-traffic-control radar centre and transmitters. Away from the path down from Great Dun Fell runs Dunfell Hush, a man-made gash 100 feet deep, its steep sides scabbed with a grey pebbly scree. This was what the miners did when all the indications were that there was lead in the vicinity, but they were uncertain of exactly where the seam ran: they dammed up a hill stream, then released the head of water to crash down its gully, scouring away all the vegetation, soil and loose rubble to expose the rock below and reveal that elusive, gleaming hint of lead and, if doubly lucky, silver. Beyond this mighty piece of industrial vandalism, the Pennine Way falls gradually southward across the springy turf of Knock Fell, skirts the deep scar of Knock Hush and the miniature gorge of Swindale Beck, and drags itself at long last across the ancient clapper bridge over Great Rundale Beck and down the length of Hurning Lane to Dufton. Hardy yompers in good weather might positively enjoy these final 5 miles down from Great Dun Fell, but with a front-row seat in the skull cinema at a private viewing of 'Dufton Awaits' – the quiet little village at the fell foot, the Stag Inn, the wide green with its resting benches – this footsore Pennine Way adventurer descending from the mists and winds of Cross Fell would rather have them over and done with.

Having diverted far to the southwest to reach Dufton, the Pennine Way swings abruptly back eastward for another long moorland crossing, 15 miles across some of England's wildest and remotest country. As I left Dufton on a fine morning after a rainy night, it was hard to believe such arduous walking lay ahead. A crowd of day walkers was setting out to do the first 4 miles, climbing the walled lane up Dod Hill and then following a good grass path, aptly named Narrow Gate, at the rim of a spectacular cleft. They were heading for a sublime picnic spot with an iconic view, a hollow in the upland moor known as High Cup, where one can very pleasantly idle away an afternoon before descending to Dufton once more.

As we climbed the lane a sudden roar to the right made everyone jump, a low ripping scream as though sky and earth were being torn apart like Velcro. There was nothing to see. A little further up a musical *clink, clink!* sounded not far away. It came from behind a wall where a farmer was sorting through a great pile of stones. 'Did you see that bloody jet?' he said, gesturing towards the invisible cleft half a mile away. 'I were down on the edge just now and saw it. He made a turn and went right up t' dale below the rim, and popped out over High Cup Nick. Must have caused a few heart attacks up there, cheeky bugger!'

The pile of stones was destined for mending the boundary between the farmer's pasture and his neighbour's. Few if any of the sheep that had kicked the wall down to ground level were his, he said, but he was philosophical as he gauged the task with his eye. 'Oh, I might have a couple of hundred yards of this walling to do, but I'll have my lad up here to help me tomorrow. How long will it last? Well, I reckon it were last done properly two hundred year ago, and I hope it'll stand another two hundred.'

The east wind blew cold from the moors. The farmer, heated by his physical labour, was working bareheaded in shirtsleeves, and he couldn't help grinning as he saw me zip my fleece up high and plunge my hands in my pockets. 'Aye, hope we get a bit of warm – we could do wi' it. Long winter it was. Didn't affect the lambing, though. Most of the farms had twins – about seventy per cent, I should say. It's a tough job to keep going up here, mind. Lamb's not fetching the price it should. Wool? You've to pay to get it taken away. And they're not letting us graze the sheep on some of the high tops any more.' He shook his head sardonically as he reached for another stone. 'Summat about preserving the wild flowers, they say. And *I* say – what about preserving us hill farmers, eh?'

The rim of High Cup's dramatic valley is whinstone, or dolerite, the same stuff that forms the Whin Sill outcrop I'd last seen carrying Hadrian's Wall across the neck of Britain 40 miles away. The dolerite stands upright in clusters of hexagonal columns or dark towers, a thick band of hard rock along which I walked, fording a couple of rain-thickened becks, skirting slippery patches of lichens, to come to where the valley sides converged in a chute of scree and a perilous rock staircase rising to flat ground at the pointed head of the dale.

You can't walk on regardless. You just have to stop and gaze, as handfuls of other walkers were doing, propped on their day packs or leaning on their walking poles. High Cup is a geological and scenic phenomenon. It's not the only U-shaped glacial valley in the country, but it's the finest and most geometrically perfect one in England. Glacial action shaped it, scooping out softer layers of Carboniferous sandstones and limestones, mudstones and silt, leaving the tough dolerite all around the upper edge. The sides fall opposite one another in symmetrical concave curves of scree, plunging down from the belt of dolerite exposed at the rim. They meet some 600 feet below at the bottom of the cleft in a tumble of dolerite blocks loosened from the rim by the frosts of millennia. Below the valley floor, under the scree and the glacial rubble, lies older and more resistant limestone, piled in its turn on ancient slates and sandstones some 460 million years old. Dead centre in the picture, the glinting little river of High Cup Gill sinuates away towards the flat green pastures of the Eden Valley, with a hint of taller fells beyond towards Shap and the eastern outliers of Lakeland. It's a view you never forget, and a place you are drawn back to.

I sat with Wainwright's *Pennine Way Companion* open on my knee, looking along the northern rim for the tall free-standing dolerite pillar called Nichol Chair, or sometimes Nichol's Last, drawn in fine pen-and-ink detail by the Master. There it was, standing out from the plum-coloured shadows under the cliffs. If Mr Nichol, a Dufton cobbler, ever actually scaled that slender rock pillar and mended a pair of boots while seated on its sloping top, he must have been very nimble and possessed a hell of a head for heights. It was all to win a bet, apparently, and he had his name and feat immortalized. But what about High Cup Nick, the cleft where a little beck rushes over the northern crags near the apex of the rim? How did that insignificant notch or nick, meticulously labelled by Wainwright with its own little arrow pointer, come to have its name usurped? Everyone speaks of High Cup Nick in reference to the whole valley, but *everyone is wrong*! There! Clever boy! You're in the right. Here's a pat on the back. Feeling better?

I snapped Wainwright shut, gave myself two slaps (one for pedantry, the other for smugness) and set off east between the moorland slopes of Dufton Fell and Murton Herds, heading for the promised land of Upper Teesdale with the stony rustle of Maize Beck for a guide. Last time I came this way it had been a freezing day with banks of mist just about keeping their distance, direction-finding made less problematic by a hand-held Satmap GPS device. Forty years before that, a boggy and poorly marked slog with the splashy fording of many becks, my father puzzling over a 1-inch map and haphazardly swinging compass, myself brushing rain spots off the *Pennine Way Companion* and its doomy annotations ('... the worst section of the crossing: a featureless moor with only a few cairns and posts and no natural landmarks'). Today there was gloriously clear sky, a well-trodden path, an actual footbridge over the bronze and purple waters of Maize Beck, and not another soul in sight or earshot. The southern boundary of Moor House NNR ran well to the north of the Pennine Way, beyond the quaking slopes of Dufton Fell, and the nearest border of the Upper Teesdale reserve was another 4 miles or so of moorland walking to the east. I went at it slowly, with curlew whistles and the chatter of successive becks for company, savouring the rush of cloud shadows across the fellsides and hollows of this no man's land outside the strictures of the nature reserves.

Towards the start of the Carboniferous period some 359 million years ago, this region of Britain was a scattered bunch of islands near the equator. They finally went under a warm tropical sea around 335 million years ago. Turbid water deposited mud on that ancient seabed; then the water cleared and began to teem, at first with life, and then with death as the shells and body parts of unimaginable trillions of minute organisms piled up on the muddy seabed to form hard Carboniferous limestone over the mudstone. As the seas receded, giant river systems across the exposed land deposited thick belts of sand; as the waters advanced to drown these new shores, so more calcium carbonate from more dead sea organisms drifted down and was transformed – slowly, over aeons – into more thick layers of grey Carboniferous limestone over the coarser and ruddier sandstone. It was a cycle interrupted by the clashing and churning of tectonic plates as rocks were squeezed upwards into mountain ranges, only to erode away again; a surging this way and that of waters and silts over hundreds of millions of years.

It's the familiar cyclic succession of limestone, mudstone and sandstone of the Carboniferous period that underlies most of the Moor House–Upper Teesdale National Nature Reserve. When some 50 million years later the Earth's crust suddenly stretched and split, a rude tongue of boiling magma arose and thrust its way at 1,000°C into whatever cracks it could find in the rocks of what is now northeastern England. As it cooled, the magma formed the Whin Sill, the great subterranean bar of hard, dark dolerite that outcrops at the Farne Islands, along Hadrian's Wall and around the rim of High Cup. As for Upper Teesdale, into which I now began to descend along the Pennine Way, the erosion of the Whin Sill has produced one of the most dramatic and beautiful river valleys in Britain, the place I'll always associate with nature's annual springtime uprising in all its glory.

Where the land begins to fall eastwards towards the valley of the River Tees, the first big slab of the Upper Teesdale dolerite ventures west. Maize Beck makes a sudden right-angle bend to the south here, cutting across limestone and sandstone bands as though impatient to hook up with the volcanic intrusion. Once united with the dolerite, the beck follows it all the way down to its tumultuous junction with the Tees at the foot of the rocky fall named Cauldron Snout. The Pennine Way is cannier; it leaves Maize Beck to its detour and cuts straight across the moor to the old lead mine at Moss Shop. What a bloody awful peaty purgatory this section of the path used to be, devoid of waymarks and rich in the waterlogged muck that Wainwright terms 'slutch'. The only time during our Pennine Way adventures back in the 1970s that I can ever recall my father actually admitting to being baffled was here, as we gazed around hopelessly with rain freckling on Dad's spectacles and bog water lipping over the tops of our boots.

A few white flowers of spring sandwort starred the long spoil hummock behind the ruins of Moss Shop lead mine. The track, now stony and definite, swung down the slope to the lonely sheep farm of Birkdale, cradled in a soft limestone hollow between two fat arms of hard dolerite. From here I turned a corner and dropped down the hillside with Maize Beck for company far below, until the mountain river curved away to plunge and dissolve in the jabble of white water that marked the outpouring of the River Tees from the foot of Cauldron Snout.

Cauldron Snout is more a series of cascades than a true waterfall. The Tees, its peat-infused water a creamy yellow, comes bouncing and leaping down a rugged staircase of gleaming black volcanic rocks some 300 feet high and more than twice that in length. The volume of the water is rarely lower, rarely higher than this steady flow. The sound is constant, too: a battering hiss. It's not till the Pennine Way swings off with the farm road to cross the top of the cascade that you see why Cauldron Snout is so steady a spectacle. Belying the drama of the fall, up here the Tees itself flows quietly under the bridge after issuing at a regulated pace from the base of a great grey wall of stone blocks that fills the head of the valley a couple of hundred yards to the north. It's the dam of Cow Green Reservoir, from which the flow rates of both river and fall are carefully controlled. A minute more up the road, and you gaze down on the reservoir itself, 2 miles long and nearly 1 mile wide, filling a quiet hollow in the hills.

The basin of Cow Green is floored with hard dolerite, and flanked with permeable limestone. It sits high, 300 feet above the valley floor of Upper Teesdale, and some 1,600 feet above sea level. Winters are long, springs cold and rainy. Water is superabundant, leaching mineral nutrients from the soil. The intrusion of the tongue of boiling magma some 290 million years ago baked and metamorphosed the Carboniferous limestone rocks around it, producing a coarse crystalline marble called sugar limestone. This weathers into a coarse thin soil, rich in minerals. There are plenty of ledges and cliffs around the basin that can't be reached by grazing animals. A tough, harsh, unwelcoming environment for most plants, but ideal conditions for the community of arctic-alpine flora that colonized these uplands shortly after the last of the ice sheets melted away ten thousand years ago. Wonderfully beautiful and ecologically irreplaceable, the Teesdale violet, the tiny pink bird's-eye primrose and the royal blue spring gentian have clung on in Upper Teesdale ever since. Botanists and Moor House field officers were among those most vehemently opposed to the building of the Cow Green Reservoir when it was first mooted in the 1950s, because Cow Green was one of the prime strongholds of this rare arctic-alpine flora.

The trouble was that post-war British industry was still trying to get on its feet again, urged on by Prime Minister Harold Wilson with his vision of a Britain 'forged in the white heat of the scientific revolution'. Imperial Chemical Industries at the mouth of the River Tees, 80 miles to the east, consumed 35 million gallons of water every day. For the tens of thousands employed at ICI, that water meant the difference between employment and redundancy. Upper Teesdale had the rainfall and the wide-open spaces. Reservoirs at various locations in the region were planned and vigorously opposed, but it was the proposed flooding of Cow Green that drew the most vocal and widely publicized opposition. The Northumberland & Durham Naturalists' Trust raised a fuss. The Botanical Society of the British Isles drew attention to the significance of the impending loss. A letter to *The Times* in February 1957 was signed by fifteen botanists, including Durham University lecturer David Bellamy, already an *enfant terrible* of his discipline.

But representations to the Tees Valley Water Board fell on deaf ears. Awareness of ecology and the importance of conservation, while slowly gaining traction, had not yet spread far outside academic communities. And engineers and water board officials were pretty robust and no-nonsense about what they thought. In April 1961 the *Daily Express*, reporting on the commencement of construction at Balderhead Reservoir, recorded the exasperated words of Alderman Charles Allison, Chairman of the Tees Valley Water Board: 'All this fuss is a lot of tommy rot. It is sickening to think that a little flower is more important than the future of Teesside. Who cares if the gentian disappears – it is no good to anyone.'

That attitude prevailed, and the building of Cow Green Reservoir went ahead, a four-year project culminating in its opening in 1971. In the intervening years the water needs of industrial Teesside have subsided. Cow Green is now used mainly to regulate and control the levels of water in the Tees in times of flood or drought. Assiduous plant hunters can still find the Teesdale violet and the bird's-eye primrose around the edges of the reservoir, though greatly diminished in numbers. But the reservoir itself, despite its bird and aquatic life, remains outside the two halves of the NNR, which withdrew their boundaries rather pointedly, Moor House to the northwest, Upper Teesdale to the southeast, like two ladies of refinement whisking their skirts away from an ill-mannered gatecrasher at the conservation ball.

Beside Cauldron Snout I descended a steep and slippy rock staircase to where the River Tees swung away in the great northward curve it has carved for itself under the dark dolerite cliffs of Falcon Clints. It was stumbly going, picking a way across and between the boulders that had fallen from the cliffs to litter the riverbank. When I first came this way in the 1970s I took a nasty toss among these rocks, bashing and bloodying the point of my elbow and damaging the nerve, an extremely unfunny-bone that still tingles from time to time. At such moments at least I can claim kinship with Sherlock Holmes's faithful companion Dr Watson, who at the Battle of Maiwand 'was struck on the shoulder by a Jezail bullet, which shattered the bone and grazed the subclavian artery', leaving the doctor with a damaged shoulder (or possibly leg – Watson himself seems unsure) that 'ached wearily at every change of the weather'.

Others were picking their way gingerly under Falcon Clints. 'We've seen the dipper,' enthused a woman in boots spectacularly scuffed by the hard going, 'and a black grouse in the rocks just along there.'

'And a grey shrike,' put in her husband. 'And you've seen the peregrine, have you? And the ring ouzel ... Honestly, this place is such a paradise for birds.'

As if on cue, a slate-backed peregrine went darting out across the river from the crags, twisting like an acrobat before hanging in the sky on an invisible step. Staring up, I stumbled backwards. My elbow buzzed, a psychosomatic twitch. Come on – just concentrate on putting one foot safely in front of the other!

Soon enough the Tees went rushing on beyond Falcon Clints and began to wind its way eastward once more under the rounded back of Widdybank Fell. The Pennine Way, suddenly wider and more negotiable, passed below the white-painted buildings of Widdybank Farm, headquarters of Moor House–Upper Teesdale National Nature Reserve. And here I sat down for a drink and a think, casting back some fifteen years to a long walk started from Widdybank, a memorable springtime wander around Upper Teesdale with the NNR site manager, Chris McCarty. It was one of those days when it's borne in on you that sentiment without science is shadow without substance – that, much as you may feel a landscape in your heart, much as you may invest it with all manner of enchantments and glories, knowledge burnishes all things. Learning to appreciate, even if only in glimpses, the mutual operations of soil and plant, temperature and altitude, water and rock, man and the very specific environments with which he interacts, adds weight and substance to what you see as you walk about. Chris McCarty proved generous with his time and his knowledge, and seeing Upper Teesdale through the lens of his practical understanding was a breakthrough moment for me.

Lapwings were flying above the Teesdale meadows, flapping against the stiff spring wind as they cried their folk name of 'peewit' in creaking voices. Seeing the two humans approaching, they had become agitated on account of their newly hatched chicks crouching in the lee of sedge clumps. 'Holding their numbers well this year,' remarked McCarty as I steadied my binoculars against the gusts. 'And listen! That's a sound you don't hear too often.' Through the incessant *pee-wit! pee-wit!* of the lapwings I made out a background whirr. 'See the snipe? He's drumming – look, just over our heads!' A small dark body was diving groundwards, long bill extended like an aeroplane skid. 'It's all to impress the ladies,' said McCarty. 'He holds out his tail feathers, and the air drums them as it rushes by.'

McCarty was in no doubt which season shows Upper Teesdale at its best. 'It's got to be now, late in the spring. You get all the ground-nesting birds in full cry, the river's generally spectacular at this time of year after the snow's melted, and then of course . . . well, have you ever seen the valley's arctic-alpine flowers in full bloom?'

Actually, I had never yet seen the full string of these pearls of Teesdale, rare and delicate survivals from a post-Ice Age Britain. It was John Hillaby's recounting of his 1966 Land's End to John o'Groats walk in spring, *Journey Through Britain*, that first infused my imagination with the magic of this unique flora. 'No botanical name-dropping', he wrote, 'can give an adequate impression of the botanical jewels sprinkled on the ground above High Force . . . In this valley a tundra has been marvellously preserved; the glint of colour, the reds, deep purples, and blues have the quality of Chartres glass.'

Under the spell of John Hillaby's account and the observant prompting of Chris McCarty, I found myself wandering from one floral treat to the next like a kid in a sweetshop. The rain-sodden meadows beside the Tees, at first glance composed of thick grass, revealed themselves on closer inspection to be a rich salad of herbs. The leaves of great burnet and pignut, said McCarty, were a good indicator of unspoiled ancient grassland. Left unblasted by chemicals and allowed to lie uncut until their flowers have set seed, the riverside fields yielded pink and white milkmaids, shiny marsh marigolds, slender purple spikes of early marsh orchid, acid yellow balls of globeflower. Beautiful, one and all. But where were the arctic-alpines, the stars of the show?

Like most northern dales, Upper Teesdale lies separated into very distinct agricultural strata. The green valley-bottom pastures known as in-bye land give way to rougher 'intake' land further up the dale sides. This higher land is boggy and wet, lightly grazed to prevent its cover becoming too wild and dense – ideal conditions, in fact, for wading birds such as snipe and redshank to nest and rear their young. Beyond the intake wall, darker and rougher country rises to the open heather moorland that rolls away across the fell tops.

As McCarty and I crossed the racing Tees by way of the footbridge at Cronkley and began to climb towards the intake land, we started to spot the delicate arctic-alpine flowers on ledges and in crevices of the rocks among tattered juniper bushes. I got down on my knees and crawled along, transfixed by their fragile beauty. Mountain pansies, boldly coloured purple flowers with a broad lower lip of yellow striped with black; one or two tiny Teesdale violets; and then the bird's-eye primroses, delicate, deep pink, with an intense yellow eye glowing at the heart of each flower. Why had they clung on here, in this particular dale? Chris McCarty, kneeling alongside, ticked off the points on his fingers.

'They're here essentially because of lack of competition. The climate in Upper Teesdale stays cold and wet till late on in spring, which most British flowers can't deal with. And the formation of the Whin Sill is the other main factor. Think of it as a horizontal volcano about 300 million years ago, squeezing its lava like toothpaste between two bedding planes of Carboniferous limestone. It baked the surrounding limestone into crystals full of minerals – sugar limestone, we call it, see?' He scooped up a palm full of glittering crystals like dull little diamonds. 'It's got exactly the balance of nutrients that the arctic-alpines need. A huge slice of luck, and it gave us this fabulous flora.'

We climbed the well-worn track of the Pennine Way above Cronkley Farm, then turned aside along the ancient cattle-droving road called the Green Trod. Beside the hissing jet of White Force waterfall we ate our sandwiches under the ruined gable of an old mine 'shop', a bothy where lead miners once sheltered between their shifts in the levels of White Force's mine. Then it was on up the long nape of Cronkley Fell to the fenced and walled enclosures on the broad summit. These squares and lozenges of separated land are in fact 'exclosures'; their fences exclude the sheep and rabbits with their close-crop nibbling. White stars of the metallophyte spring sandwort spattered the sugar lime-stone. Thousands of bird's-eye primroses trembled in the wind. And seeded here and there in this rich pink carpet, trumpet-shaped spring gentians of a deep celestial blue vibrated as if blowing a silent paean to spring. This is what much larger swathes of the Upper Teesdale uplands might look like if the nibblers were banned, the vegetation oh-so-carefully controlled, the landscape preserved for the precarious little flowers. But then, as the farmer mending his stone wall on Dod Fell had remarked to me by way of valediction: 'What about preserving us hill farmers, eh?'

Down by the roaring Tees Chris McCarty and I turned back along the riverbank. The squeaky cries of the nesting lapwings recalled the creaking of fracturing ice, and on the walk back to Widdybank Farm I pictured the mountainous glaciers melting away northwards ten thousand years ago, in retreat from the deep valley they had gouged, leaving behind a naked landscape ready for seeding with the ancestors of Teesdale's gentians and violets, pink primroses and exquisite mountain pansies.

From my seat by the river below Widdybank Farm today I threw small pebbles idly into the shallows and soaked my feet, sore from skidding on the knobbly dolerite clitter under Falcon Clints. A good dozen miles from Dufton, and a couple more to go before I could ease off the boots and stretch out in the garden of the Langdon Beck Hotel. OK, might just close the eyes for a second or two . . .

The course of the Pennine Way downriver from Langdon Beck is characterized by Alfred Wainwright as 'a pleasant and popular walk'. If ever a phrase damned with faint praise, that was it. These 8 miles are on chunky limestone, shelves of sandstone and limestone stepping up the northern dale side, the elongated finger of the dolerite rock along the southern flank outcropping here and there as it points down the dale towards Middleton-in-Teesdale. This is some of the best river walking in England, a beautiful path that crosses the Tees at Cronkley Bridge and climbs the juniper-clothed hummock of Bracken Rigg for a high-level view of the snaky bends below the bridge. The view also takes in the giant hollow of Force Garth Quarry on the opposite bank, where up to 300,000 tons of dolerite or whinstone are removed every year to be crushed up for aggregate. That is a staggering amount of stone to be taking, and it illustrates again the Cow Green conundrum, the irreconcilable aims of landscape and wildlife preservation with economic growth and expansion.

From Bracken Rigg I descended to reach a crescendo of noise and drama at High Force. On occasion I have found High Force skimpy and diminished after a rainless period. Really, though, it begs to be seen as I saw it today, the river swollen with a winter's rain and spring's snowmelt, crashing over and down the Whin Sill, a 70-foot step of gleaming black dolerite, in a forked sluice of creamy yellow water, a furious rush out and then downward, trampling with crushing power into a basin misty with spray and brilliant with miniature rainbows. The sudden contrast with what has preceded High Force is remarkable. The peace of the long walk at river level through a wide valley studded with junipers is cracked by the shock and awe of the thunderous fall, and in the blink of an eye the Tees has dropped away as though sinking through the earth, to resume its journey in a steep-sided gorge far below.

High Force effectively marks the eastern boundary of all the breeding bird activity, the arctic-alpine flora and the hard volcanic scenery that one associates with the 'special ground' of Upper Teesdale. Low Force follows, an elongated cataract with rapids between whose short columnar rocks the Tees swirls before running the last few miles through quiet pasture into Middleton-in-Teesdale.

Middleton lies under green fellsides. The hills are quiet these days, but when the London Lead Company came to Teesdale in the mid-eighteenth century the lead mines raised a row in both the town and its surrounding countryside. The Teesdale rocks were full of lead, and the miners, like those in the South Tyne dales around Alston and Garrigill, were a rough lot who lived a rough life. The LLC, true to its Quaker philosophy, aimed to put its Teesdale workers in the way of salvation as well as employment. The company insisted on strict temperance and set up a Sunday School; it gave the men the weekend off, and opened allotments where the miners could grow vegetables for a healthier diet. And it developed Middleton as a model industrial village in stone and slate.

Today Middleton is a local shopping hub and a tourist town, the gateway to Upper Teesdale, its stone buildings of locally quarried sandstone sternly handsome in mottled grey and orange. You could easily think that's all there is to the town. But this is still sheep-farming country, and the auction mart on Station Road on the outskirts of town is where the farmers come to do their dealings. Weary enough from my Pennine Way footslogging, I couldn't resist going along once I found out the sheep sales were on. It was salutary to realize that I was the only inquisitive outsider present. Everyone else was there on business.

The sheds clanged and rang with clashing gates, rattling sheep hooves, anxious bleating and the kicking of boots against concrete. Farmers from around the dale had gathered at the mart. Fifty men sat round the auction ring, faces under flat caps gloomy or humorous according to type. 'Them two yows are Neville Nixon's,' murmured the ring assistant to the auctioneer, 'and the tup's from Howgill.'

Ewes were fetching fifteen to twenty pounds, pitiful prices. Big curly-horned tups went for ten times that. One huge ram bucked and bleated furiously at an assistant with an enormous tummy overhanging his trousers. The belly creased in half under the man's shirt as he bent double and bleated right back, to an appreciative burst of laughter from the ring of shepherds and farmers.

An eight-year-old boy was helping to marshal the sheep into the ring, his small hands grabbing horns, ears and noses as confidently as the big red fists of his father. 'Get down, now,' said one of the assistants to another youngster who was clambering up the railings. He gave him a gentle clap on the backside. The lad didn't care a rap, and neither did his father. It was part of the ritual, part of the generational cycle that's been going on here since sheep first grazed the slopes of Teesdale.

*

Every walker who tramps the Pennine uplands knows the Swaledales, those hardy sheep with the black faces and white noses that are as much a part of these northern hills as the rocks, peat and heather. In sunshine they scatter the distant slopes with bright white dots. In mist you come on them suddenly, their dank grey fleeces dripping water as they give you their intense, mad stare from slit eyeballs. There's no more characteristic or more stunning Pennine sight than a fellside of Swaledales swirling like a single organism as a sheepdog marshals them towards a gateway at his master's call and whistle.

Thirty years ago the idea of the Pennines without their Swaledale sheep would have been laughable. Today, with a countryside polarized between rewilding and food production, changes in subsidies to reflect the amount of land rather than the number of sheep a farmer works, and above all the uncertainty and loss of grant income that Brexit has visited on every farmer, hill and dale alike, that prospect doesn't look quite so unlikely. The sheep farmers of the Pennines work long, hard hours in all weather conditions, day and night, summer and winter. It's thanks to them and their skilful management of flocks, fields and fells that the uplands of northern England look as we all love them to look – sheep-nibbled turf and heather on the fells; grassy, stone-walled pastures in the valley bottoms. The Pennine hill farmers are the unsung, anonymous guardians and custodians of this delectable landscape, and lovers of walking and of upland countryside owe them a tremendous debt. But they don't often get their due.

Each October local farmers gather at the tup sales at Kirkby Stephen, just across the fell from Middleton-in-Teesdale. 'Come up and stay with us then and you can see how we do,' said Dorothy Metcalfe of High Greenside Farm. That's how I found myself at the farm's kitchen table one roaring, blustery autumn morning, learning how the sheep are gathered by Sam, Ben and Laddie, the High Greenside dogs, for dipping, shearing, tupping (mating) and lambing. We delved into the mysteries of 'hefting' (how a farm's sheep know which part of the fell is theirs), and of the progression of a female sheep from gimmer to hog to shearling to ewe – and the shorter, sadder one of most males from fully equipped tup (ram) lamb to castrated wether.

'John's a good farmer,' Dorothy observed of her husband. 'He's always on with the sheep, every day, out all hours and all weathers. I can lamb a ewe; I take my share of the work. But the only thing that makes it possible for us to carry on with the sheep now is John's haulage business and my B&B. The fact is, you just can't make a decent living from sheep nowadays, even with the subsidies.'

Down at Kirkby Stephen's auction mart along the back lane, the sheds, pens and auction ring were crammed with farmers down from the hills. Flat caps and old waxed jackets, shepherding sticks and saggy trousers dark with wool grease were the order of the day. Outside against the shed wall the competitors lined up in the cold October wind, each breeder grasping the curly horns of a magnificent Swaledale yearling tup, a young ram about eighteen months old. Judging was a serious business: tens of thousands of pounds were at stake. It was John Richardson of Ghyll House at Dufton whose tup was at last declared champion. Later that day the champion sold at auction for £20,000. If upland sheep breeding and farming is dead in the water, no one has dared tell the Pennine hill farmers just yet.

Leaving Middleton southwards, the Pennine Way crosses an outlier of the Whin Sill. The green spoil heaps and grey cliffs of Greengates Quarry show where dolerite was won a century ago. Now it was moorland walking once more, passing over a succession of Carboniferous rocks: Alston sandstone laid down by deltas, the thick wedges of the Great Limestone Member formed at the bottom of a shallow tropical sea, and the marginally younger Stainmore Formation of mudstone, sandstone and limestone, each deposited in its turn as the sea level rose and fell and the coastal margins advanced and retreated. The path hurdled a parallel pair of shallow east–west dales winking with large bodies of water, Lunedale with the reservoirs of Grassholme (finished in 1915) and Selset (1960), followed by Baldersdale with its trio of Hury (1892),

Blackton (1896) and Balderhead (1965). Page 78 of my *Pennine Way Companion* contained a gem of a jeremiad from Alfred Wainwright on the subject of Balderhead Reservoir, completed shortly before he wrote his guidebook: 'No banks of flowers now ... Concrete runways! No more exhilarating days of joyous spate and leaping waters ... Life has gone from it. A river has died. Been killed. By the hand of man.'

Hannah Hauxwell of Low Birk Hatt Farm knew the River Balder when it ran in its natural course. There aren't many proper old upland hay meadows left in England, but those at Low Birk Hatt are superb examples. That's thanks to the lone woman who farmed these fields in an entirely traditional way until her retirement in 1988, and also to Durham Wildlife Trust who took them on and continued the good work.

Hannah Hauxwell became a TV star in the 1970s when a documentary, *Too Long a Winter*, and a series of programmes followed her unadorned, narrow life through the seasons. A reluctant star, she never could quite understand what all the fuss was about. And the loveable image of the self-sufficient spinster scurrying about her traditional life like Mrs Tittlemouse was belied by the reality of the daily grind and hard graft, the lack of running water and electricity, her poverty (she got by on less than half the equivalent of today's minimum wage) and above all the hours of outside work in the long, snowbound Pennine winters. 'In summer I live, and in winter I exist,' was her memorable phrase. After twenty-five years of solitary life and work, life became too difficult as she grew older and more infirm. In 1988 she reluctantly sold the tumbledown farmhouse and her few beloved cattle, and left Low Birk Hatt to live in more comfortable circumstances not far away.

I stopped at the sparse little exhibition in Hannah's Barn, and then followed the Pennine Way beside the meadow, its vanilla-scented sweet vernal grass and sedges full of old hay meadow flowers such as yellow rattle, knapweed, moon daisies and blue powder puffs of devil's-bit scabious. The farmhouse of Low Birk Hatt, standing just below on the banks of Blackton Reservoir, has been carefully restored and lovingly maintained since Hannah Hauxwell lived there in absolute poverty, and it was easier to sense her spirit moving through the grasses and flowers of the meadow than among the immaculate farm buildings.

From Low Birk Hatt the squashy, puddled track of the Pennine Way led me up and out on to Cotherstone Moor. Curlews and golden plover piped plaintively, a great crowd of starlings went swooping all together, and a red grouse planed away on stubby scimitar wings. Swaledale ewes among the sedges stared incredulously in my direction, then averted their gaze like a pew full of spinsters at the sight of something unspeakably shocking: a vicar in cycling shorts, perhaps. Away to the east the flat-topped gritstone outcrop of Goldsborough stood proud of the low-rolling landscape, a miniature table mountain footed on the Yoredale Group, a Carboniferous pile-up of sandstone, mudstone and limestone rocks that underpins these bleak moors.

It's a long slog south from Baldersdale across fairly featureless moorland, only relieved by a detour into Bowes. Here on the main street through town stands the L-shaped building of the former Bowes Academy, better known (and nowadays renamed) as Dotheboys Hall. Charles Dickens visited Bowes Academy on 2 February 1838, and his creation in *Nicholas Nickleby* of the hellhole school of Dotheboys Hall under its brutal oaf of a headmaster, Wackford Squeers, was based on what he found there that winter's day.

Dickens's novel turned the spotlight on such schools with its descriptions of savage beatings handed out to starving and terrified lads. But Bowes Academy had achieved notoriety several years before, when parents of boys at the school had sued the headmaster, William Smith, for gross neglect of their children, and had won their case. The *Leeds Intelligencer* of 6 November 1823 wrote up the story under the headline 'Cheap Schools', reporting that William Smith had charged the paltry sum of twenty guineas a year to take in boys who were mostly surplus to requirement or too much trouble to their parents. For that sum the children had bread and rancid butter five days a week for their main meal, and weak tea and no food for supper. The only dietary supplement was the maggoty fat skimmings from the stockpot, which they were obliged to buy with what pocket money they had. Eye trouble from lack of proper sustenance was rampant, and two boys went stone blind. Five pupils had to share each flea-ridden bed. Everyone was filthy and suffered from the 'itch'; soap was only available on Sundays, there were two towels between three hundred boys, and they often wore only their underwear for days at a time, even in winter, while waiting to mend their ragged outer clothing.

Just the look of the building, bulky and forbidding in grim grey sandstone, was enough to put the shivers on me, and I was glad to resume the long march along stony farm paths over the Durham moors towards the Yorkshire border. In a pasture at Sleightholme Farm an enormous ram had his nose pressed to the fence. He was sniffing the emanations from the ewes in the adjacent field. As I went by I saw him flare his nostrils and pause to ponder and roll the scent around his palate like an oenophile with a particularly fine Château Margaux.

On and on across the heather and springy peat of Sleightholme Moor, recalling the long march I'd endured with my father when we first came this way in the rainless summer of 1976. 'Penance for sins' was Wainwright's pithy conclusion, but that year's drought had crisped the ground and spared us the usual slip-and-squelch conditions underfoot. Today, travelling in the opposite direction, I became aware of an aiming point on the horizon ahead, a tiny dot at first, gradually growing and resolving itself into a set of low square buildings, all alone on the wide moor. What an oasis is to a parched wanderer in the desert, the lonely Tan Hill Inn is to the Pennine Way walker. 'Never was a pint better earned,' declared Alfred Wainwright at the foot of page 91 of the *Pennine Way Companion*, and by God he was absolutely right about that.

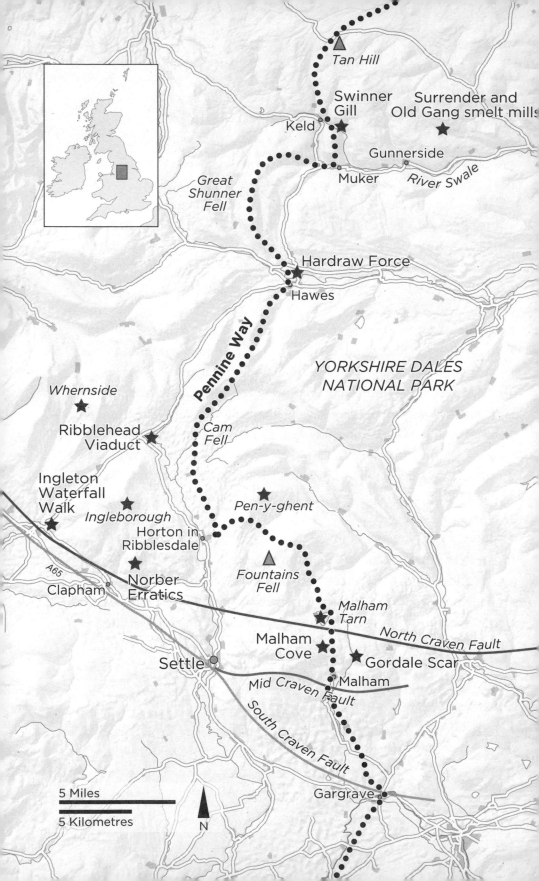

11

Scars and Faults: Tan Hill to Gargrave

THE TAN HILL INN IS the only pub where I've ever been refused a drink. Well, not refused, exactly; more like ignored to the point where I gave up and went away. Was it my posh southern tones that upset the surly black-bearded barman that rainy afternoon some twenty years ago? My peaty boots, my smell, my temerity in drawing his attention to the hearty party of walkers who'd pushed in front of me? Whatever the cause, he left me at the back of the queue, and I left him a mucky puddle of mud to clear up. Petty, eh? A twenty-year chip on my shoulder finally offloaded by a most attentive personage who pulled me a nice pint of Ewe Juice. A name of which I was sure the Sleightholme ram would have thoroughly approved.

At Tan Hill on its moorland road three counties meet: Durham to the north and east, Cumbria to the west, and North Yorkshire running on south with the Pennine Way. There has been an inn of sorts on this high saddle of ground (the Tan Hill Inn proclaims itself the highest inn in England 'at 1732 ft') for many centuries, a natural focus for packmen and other travellers across these wild hills. All around the lonely inn the heather and grass are underpinned with mudstone and sandstone, between whose layers the fallen trees and vegetation of the Carboniferous period were compressed at first into peat, then through further compression into seams of coal. Small pits were worked all round the inn; their shafts, some roughly fenced off, others still open-mouthed, make off-piste wandering a hazardous business.

A mile southwest of Tan Hill, tall black screes and scattered fragments of coal in the pale, tufty moor grass of Hoods Edge betray the site of Mould Gill Coal Level, long horizontal shafts that were connected to the Tan Hill pits in the 1840s. Coal was mined around Mould Gill since early medieval times, in shallow pits to begin with. It was poor quality 'crow coal', fit only for use in the local limekilns. The main Tan Hill seam was in a shale layer above the Upper Howgate Edge Grit, a coarse sandstone deposit. The seam was mostly only 30 inches high, a challenge to work even for an experienced hewer. But Tan Hill coal was better than the Mould Gill stuff; it burned brightly and made good coke for lead-smelting in Swaledale. The mines only finally closed in 1929; Tan Hill Colliery continued with small-scale handworking till 1938.

The rutted packhorse track of the Pennine Way runs straight among the old shafts, southwards through a landscape of long, flattish rises of sombre moorland interspersed with low hills like upturned barges. After a couple of miles it begins to descend along the brink of West Stonesdale, and there I stopped and sat down to savour the prospect ahead into the deep and beautiful valley of Swaledale.

If I close my eyes and try to picture the Yorkshire Dales, the first image that springs to mind is a green valley with a little grey village in the bottom. 'Dale' means valley, after all, and most images from TV, photographers and tourist brochures are of charming villages snuggled under limestone dale sides. It's instructive, however, to start one's imagining of the Yorkshire Dales not down in those picturesque valleys, but up on the wide moors that separate them; to think of the region as upland first and foremost, much of it over 600 metres above sea level, with the dales cutting down through this high country. That's the erosive story – the dale rivers and streams, the frost and rain, glaciers and meltwaters, landslips and fault fractures, all whittling away at rocks that were laid down over 300 million years ago.

Having said which, from my perch above West Stonesdale my eyes were irresistibly drawn down and into the bottom of the long scoop of Swaledale. Here is a classic Yorkshire dale, floored with lush green 'in-bye' pastures around the grey stone farm buildings, rising to less cultivated 'intake' beyond, and then ascending up rough olive-coloured grazing slopes to the open fell top of brown and purple. The in-bye fields lie dotted with neat stone barns and squared with stone walls, which conform to every dip of the ground as they wriggle across the intake and up over the skyline, seemingly in motion like black snakes. The dale sides bear the scars of lead mining; the land around the old workings has a velvety nap, the grey or peach-coloured spoil heaps looking naked and raw even though the mines have been silent for at least a hundred years.

In his song 'Go Little Swale' Jake Thackray, lugubrious and poetic singer and songwriter, looked with realism and love on Swaledale:

> The taciturn hill farmers patiently still
> Are pacing their hillside;
> The po-faced sheep stare as they go.
> The pinafored women go day after day
> Making their hay
> Down by the river edge where wagtails are trotting,
> By Booze, by Muker, by Gunnerside, by Crackpot.
>
> Go, little Swale: go headlong down,
> Down through your stony-faced meadows,
> Your scowling hills, your crouching towns.
> Go, little Swale, and I follow.

The little hamlet of Keld sits handsomely at the head of Swaledale. Below here Kisdon's great round lump of a hill separates the steep cleft of the River Swale on the east from the diminutive trickle of the Skeb Skeugh beck in its own narrower glaciated valley to the west, where Ice

Age debris choked off the former course of the Swale and sent it flooding down the east flank of the hill. I followed a side path high above the River Swale, past the ruin of Crackpot Hall into the narrow mouth of Swinner Gill. It was lead-mining subsidence that put an end to Crackpot Hall, and the ruins and spoil heaps of the dale's lost industry lie all around: stone-arched mine levels, a tumbledown smelt mill deep in the cleft of Swinner Gill, and the precarious trods or tracks of the lead miners.

Walking down the hill path to rejoin the Pennine Way near Muker in the valley below, I looked west across the Swale to the precipitous slope of Kisdon Side. Horizontal strips ran across the hillside, bands of alternating light-coloured limestone and darker gritty sandstone outcrops. Further down on the east side of the river, Kisdon Scar showed as a wall of limestone crags like a quarry face high over Muker and the Swale, with trees growing along its length and a lumpy apron of scree fanning out below. These rock steps and crags in the valley sides told a tale that would be repeated again and again over the next 60 miles of this journey through the bones of Britain – not just in Swaledale, but southwards through Wensleydale and Ribblesdale and on over the limestone uplands towards Airedale and the start of the great gritstone moors that cradle the industrial towns of the Yorkshire–Lancashire border.

I couldn't swerve it any longer. Down in Muker I laid out my notes, gritted my teeth and, not before time, faced up to the baffling components of the Yoredale Group.

Up to this point on the journey I had been dodging the inevitable: a proper reckoning with the Yoredale Group. These layers of Carboniferous rock form the heart and soul of the mid-Pennine landscape through which the Pennine Way runs its southward course, but what I'd read about them seemed just too damned hard to grasp. Too many references to lithofacies and diachronous bases, too much altogether of the Wales–Brabant Massif and the Serpukhovian chronostratigraphic unit. A childish and defeatist reaction, perhaps, but once again I felt baulked and belittled by the sheer technicality of geology, its thicket of thorny terms that ordinary swords such as mine couldn't hack through, that led not to a breakthrough in understanding but to more and deeper layers of geological jargon. After flailing around on the internet and in the library, and having decided to tightrope over the Yoredale Group without a downward glance, I got my breakthrough from a giveaway leaflet published by the long-defunct Countryside Agency. Entitled 'Pennine Way Profile and Geology Map', lurid purple in colour, it slid out of a moribund folder at the back of my file and in a few simple sentences dispelled the mists surrounding the subject.

The Yoredale Group takes its title from the old name for Wensleydale, where landscape features around the River Yore, now called the Ure, show off the distinctive upward series of rock layers. Bedded on a thick shelf of limestone laid down about 340 million to 330 million years ago in the stable conditions and warm sea of the early Carboniferous period, the Yoredale layers reflect a succession of changes in the climate and sea levels around the equator where Britain was stationed at that time. Limestone was deposited in tropical waters that probably never deepened beyond 30 metres. The limestone became overlain with shale (compressed mud) as light river deposits made their way into the shallowing sea, then with a layer of sandstone as a large river delta brought coarse sand and gravel, eroded from distant mountains in the north, to smother the new coastline and form sand flats. Steamy, swampy forests flourished and fell, in some places forming seams of coal. Then the sequence repeated itself as the sea levels rose and more marine limestone was deposited, then fell to allow the new limestone to become blanketed with shallow muds, then coastal sands once more. Up to a dozen of these sequences make up the Yoredale Group, each stacked on top of the preceding one, and often capped with the coarse crystalline sandstone called millstone grit, or gritstone, more resistant to erosion than the limestone, which represented the culmination of the process.

And that, basically, was it. That's why the gritstone table mountain of Goldsborough stands out as it does from the flatlands of Cotherstone

Moor, and why the sides of these Yorkshire dales, eroded by rivers and glaciers over long time, form those characteristic escarpments of horizontal bands of rock, light for limestone and dark for shale, mudstone and sandstone, that rise so evenly to the moors with their caps of gritstone hills.

However, these Yoredale cycles can't be explained simply as the result of changes in sea level due to the rise and fall of ice sheets at the poles. While the limestone was being deposited there was little deposition of sand, silt or clay, as evidenced by the abundance of corals: organisms that can't survive in muddy water. It was the deposition of the overlying mudstone that started the supply of sediment; but what actually caused that deposition of sediment to stop and start in the way it did? No one's sure; it's still up for geological debate.

Armed with a smidgeon of understanding, I took a couple of days to fossick around Swaledale and its neighbouring dales. I followed the Swale down through its stony-faced meadows to Gunnerside where the river has cut through a glacial moraine, a shelf of boulders dragged there by a glacier, now forming a steep slope down to the water. At Feetham, 3 miles downriver, there were houses backing into slopes so steep they had three storeys at the front, two at the back. They were built from the local creamy yellow-grey Hardraw limestone, the lowest of seven layers of the Yoredale Group exposed here. The characteristic sharply defined steps of the hillside limestone terraces softened as they descended into the dale, becoming curved by erosion and weathering, then smoothed out under a blanket of pliant soil and lush grass into the broad flat flood plain of the river among in-bye meadows and sheep pastures on soil of rich glacial till, further enriched by silt spread by the Swale in flood. These fertile valley farmlands exist in a different world entirely from the hardscrabble holdings of the acid peat moors above.

I decided on a whim to follow a narrow country road climbing north from Feetham into upland country. Soon it dipped to reach Surrender Bridge beside the Mill Gill, otherwise called Old Gang Beck. Here lies the ruin of Surrender Mill, a smelt mill built in 1839 at the foot of a roadless dale that curves northwest for 3 miles, rising into wild open moorland. Lead ore was dressed (sorted, crushed and washed) at Old Gang higher up the dale, then brought here to be smelted. It was roasted in an oven to remove impurities, then laid in alternate layers with fuel. This 'charge', as it was called, was then heated at about 700 degrees in an ore hearth, with bellows blowing air from an adjoining room to increase the temperature. A stone hood over the hearth was supposed to deflect the poisonous fumes away from the workers, but in those rough days the means to measure the lead content in the air around the ore hearth did not exist.

Controlling the temperature and the composition of the charge was a highly skilled job. Every twenty minutes the charge had to be pulled forward from the hearth and raked about on the work-stone to separate the slag from the molten lead, which dripped down into a channel in the work-stone and thence into a lead pot. Workers then ladled the liquid lead into moulds, to solidify into pigs. These were carted away or loaded on to packhorses for the lead market in Stockton-on-Tees.

The smelters of the Surrender Mill worked shifts of twelve or fourteen hours. It was exhausting toil, and dangerous too, given the close proximity of fire and molten lead, even without the baleful effects of the fumes. A ton of lead was produced each shift from a ton and a half of ore. Any lead remaining in the slag was crushed again and resmelted at higher temperature, and the wrung-out slag finally dumped along the banks of the adjacent Mill Gill. Nearly a century and a half has passed since Surrender Mill ceased operations, but the gill bank is still stiff with grey slabs of lead slag.

Twin stone-built flues, elegantly arched, emanate from the two hearth rooms at the back of the mill and converge to run as one amalgamated tunnel. They carried the deadly smoke, impregnated with vaporized lead, away up the fellside for half a mile to a venting point high above. However, plenty of lead vapour cooled enough during its journey up the flue to solidify on the cold stone of the flue walls. It was the unpleasant and toxic task of some wretch to crawl up the flue and recover this 'fume' by brushing it off the walls for yet another resmelting.

From Surrender Mill I set off along the old miners' track as it rose gradually beside Mill Gill under Healaugh Side. The moors above were patched brown and green, with old tough heather for grouse to hide in and young growth of tender shoots for them to feed on. A line of grouse butts foreshadowed the denouement of all this solicitous care for the birds. The softly hissing stream snaked between flattened heaps of flood stones. The track sides were a mixture of soft sandy soil and limestone boulders swept out of the banks by post-glacial floods.

Ahead, far down in the cleft of the dale, industrial chimneys rose among wedge-shaped slag banks. Ray Gill came rushing and jumping down its rock steps opposite. Greyish slabs of mine waste lay tumbled into the gill. Beyond stood a great pink-grey bank of slag as tall as a church, and the big ruined shed and tall square chimney of Old Gang Smelt Mill. A rusted frame like a medieval torture implement lay here in the ruin of the furnace house, its end plate stamped 'James Simplex Vibrator' – a device for shaking the lead ore out of its surrounding rock, though googling the name might suggest other seismic alternatives.

The broken piers of the smelt mill flue ran straight as a die up Healaugh Side. I found the bare channel of a hush not far away and followed its rocky bed steeply up to the moor above. Here a double row of stone pillars, remains of an old peat store, stood out from the fellside like an Andy Goldsworthy installation, each 6-foot monolith of gritstone rough enough to bark your knuckles on. On up to the edge of the moor and Healaugh Crag, an outcrop of millstone grit so weathered it seemed to be collapsing forward like a pile of damp sandwiches. In the crevices of the limestone band grew plants stunted by the relentless moor weather – miniature ferns with dark ribs, plump sodden mosses of emerald green, trumpet-shaped lichens and the dried skeletons of unidentifiable saxifrages. The fossils of crinoids or sea lilies 325 million years old stuck out of the limestone end-on like segments of earthworms.

Down by the river again I found the delicate white stars of the metallophyte spring sandwort trembling in wind-shaken clumps on an old slag heap, a blunt nose of consolidated ash as hard and dense as metal that squatted like a giant grey toad over a layer of river stones. Crouching there out of the wind, alone in the hidden dale, I jotted notes and watched the wagtails If I had stuck to the valley road through Swaledale, I'd never have stumbled on this extraordinary richness of geological and industrial heritage a mile away across the hill.

It's a long old 10-mile walk down the Pennine Way from Swaledale by way of Great Shunner Fell over to Wensleydale and the market town of Hawes, and an even longer 15 miles from there over Dodd Fell and Cam End to Horton-in-Ribblesdale. The Pennine Way makes it as easy as possible, with good firm going on flagstones that loop up by the flat gritstone outcrops of Great Shunner Fell, and a tremendous old green packhorse road that arrows southwest from Hawes across the sprawling limestone upland by way of another gritstone outcrop on Dodd Fell. Views from these tops are stunning, west to the distant blue fells of the Lake District, south and west to the Big Three of the Yoredale Group country – Ingleborough, Whernside and Penyghent, a trio of hills with instantly recognizable, rather leonine profiles and a layered construction that put me in mind of the sandstone monoliths of Inverpolly, 300 miles and 500 million years away.

Just before you reach Hawes, halfway over to Ribblesdale, there's a most spectacular 100-foot waterfall tumbling down past the characteristic layers of rock. You get to Hardraw Force through the teashop at the back of the Green Dragon Inn at Hardraw, like a portal in a children's story. I followed a zigzag path up through beechwoods, hearing the thump and rush of the fall long before its brown and white gush was seen among the trees. There was an unguarded lip of slippery rock where the Hardraw Beck gathered pace, jumping off the cliff in a whisky-coloured curve before gravity tugged it downwards and out of sight. Steps led down to the foot of the fall which plunged into a pool, backed by a natural amphitheatre, a great hollow in the cliff where I could clearly see the alternating layers of Yoredale rocks, the thin dark strips of muddy shale interleaved with the bands of thick pale limestone and darker reddish sandstone, all the way up to the limestone lip at the brink of the fall 100 feet overhead. I stood and wondered how long it took to build this cliff. Fifteen million years? Twenty?

It's limestone that dominates the landscape on the long descent from Cam Fell into Horton-in-Ribblesdale. Shake holes and potholes lie on both sides of the Pennine Way in the porous, easily dissolved grey rock, so that you wonder how what is there has survived all those millions of years of inundations and Pennine weather. Then you start to wonder how many supervening Yoredale layers have already eroded away in James Hutton's never-ending cycle. So you arrive in Horton-in-Ribblesdale with your head spinning once more, realizing with an inner groan how much more there is to know about the goddamn Yoredale Group and geology in general, and wondering if this is how proper geologists start on their long upward crawl towards enlightenment.

In Horton I decided I needed a couple of days off from the Pennine Way, to get my head around the Craven Faults. This rip in the Earth's crust – or rips, rather, because there are three parallel faults running west–east across the more westerly Yorkshire dales – is responsible for some truly dramatic changes in the landscape, but as yet all I knew about it was its name. Before chasing it down, however, there was Ingleborough, the jewel in the Yoredale crown, to take a proper look at. The Pennine Way passes just to the east of Horton, but everything I wanted to see lay just to the west. I did some reading, and headed to Horton-in-Ribblesdale Station and the Settle–Carlisle railway line.

The train from Horton-in-Ribblesdale to Ribblehead passed neatly between the flat-topped, crouching-lion shapes of Penyghent (694 m/2,277 ft) immediately to the east and Ingleborough (723 m/2,372 ft) a little further away to the west. The long lump of Whernside (736 m/2,415 ft), off to the northwest, was in front of the train and harder to spot. Looking out, I remembered climbing these three modestly sized but dramatically shaped mountains in the course of a long and leg-cramping day doing the Three Peaks challenge – a 24-mile circuit taking in the trio of summits in under twelve hours. I can't say their geology even entered my head during that very tiring day (except when grunting, 'Christ, another bloody step up!'), but now I could see the horizontally layered construction of Penyghent and Ingleborough, the Yoredale Group succession of sandstone and limestone, sandstone and limestone, up to their flat, or actually slightly concave, caps of gritstone.

Just before rattling out above the twenty-four arches of the Ribblehead Viaduct, the train pulled into Ribblehead Station, closed in 1970 but reopened in 1986, like the other stations on this iconic moorland line, in a rare act of retrenchment by British Rail once they realized how much business from walkers and railway enthusiasts they could have. I alighted in the company of a bunch of cheerful hikers from Leeds; they were making east for the Pennine Bridleway, and I soon lost sight of them as I joined a wide track that passed under the viaduct. This remarkable structure trumpets its year of completion, 1875, in gold numbers above the central arch. It took six years to build the Ribblehead Viaduct across the sodden waste of Batty Moss. Thousands of navvies descended on this remote moorland to live in isolation and appalling conditions, three thousand together at peak times, in crowded, self-built shanty towns named Jericho, Inkerman and Sebastopol. They worked from morning till night on a weekly intake of 18 pounds of beef. Drinking, fighting and randying when off duty, they were exploited by tommy-shops which fixed their own prices and were the labourers' sole source of supplies.

During its construction the Ribblehead Viaduct, a thing of delicate beauty in its isolated upland setting, consumed 1.5 million bricks and the lives of more than 100 men. Dozens more died excavating the Blea Moor Tunnel, whose southern portal opens a mile or so north of the viaduct. The Dales High Way crosses the railway over an aqueduct, and from this grandstand I stared up the line and into the soot-black mouth of the tunnel. Only explosives had been able to get through the succession of hard-packed Yoredale rocks – gritstone, limestone, sandstone – under Blea Moor. When tunnelling, the navvies followed where possible the softer seams of shale, but when it came to excavating the vertical ventilation shafts they had to force their way down from the moorland above through the horizontal layers, starting with the thick slather of boulder clay left behind by the glaciers. In the prolonged rainstorms so common over these western Pennine moors, cuttings could suddenly fill and sturdy embankments dissolve into gluey clay. The tunnel itself became a death-trap for three navvies who drowned there during a cloudburst in July 1870, and a fourth was only able to save himself because he was tall enough to press his mouth against the tunnel roof and gulp in air while the floodwater raced past up to his ears.

In St Leonard's Church in the hamlet of Chapel-le-Dale a wall plaque commemorates the nameless men 'who through accidents lost their lives in constructing the railway works between Settle and Dent Head'. But their true memorials are the viaduct and the tunnel, as well as the greened-over humps of spoil they dug out of the Yoredale beds that litter the sides of Blue Clay Ridge, Blea Moor and Batty Green like so many Neolithic burial mounds.

From Blea Moor Tunnel I turned back along the track of the Dales High Way, a good hour's tramp with the long shoulder of Whernside filling the western skyline. At Ellerbeck Farm I got a superb view south across Chapel-le-Dale to the bulky back of Ingleborough with its distinctive rock steps; the profile of its south-face slope looked more like a baboon than a lion from here. I recognized the path that I followed down to Chapel-le-Dale and up to the skirts of Ingleborough beyond; I'd trudged it on sore feet twenty years before, nearly 20 miles of the Three Peaks Challenge done and still 7 or 8 to go. I recalled the enormous grassy depression by the path with the intriguing name of Braithwaite Wife Hole, in former days Barefoot Wives' Hole, a shake hole caused by the collapse of a cave roof underground. The very steep northwest aspect of Ingleborough, appropriately named Black Shiver, had looked a daunting obstacle back then. But today I was going to give the climb a miss in favour of exploring the clints and grykes of Ingleborough.

The Great Scar Limestone, underlying the Yoredale layers that make up the rock pyramid of Ingleborough, was laid down in shallow seas, a bed of limestone more than 200 metres thick in places. Exposed by erosion, this ancient sea floor now forms a series of superb limestone pavements on the northern side of the mountain, from Southerscales Scars near the bottom right up to where the steep flank eases off near the summit. Rainwater has sought out joints and cracks in the flat limestone exposed by erosion and has eaten away at these weak places, cutting the level expanses into clints (cobbles) and grykes (gullies). And in the depths of the grykes, warmed and sheltered by the intervening clints as they absorb the sunshine and channel the rain, grow beautiful plants, a mass of them. On this afternoon, the mountain air scented by wild thyme, I spotted bushes of juniper with its gin-perfumed leaves and fruits, rockrose with papery yellow petals, mountain melick with petal-less flowers like grass seeds, a delicate fern like a starfish called green spleenwort, and something I suspect was baneberry with leaves divided into broad fingers, the remains of fluffy white flowers, and a hint of greenish berries turning black. I could have had a nibble to confirm the identification, but only if I didn't mind dying on Ingleborough. Baneberry, as its name suggests, is pretty toxic.

All is not rosy in the rock gardens of Ingleborough, nor in the surrounding farmlands, delectable though they look. The advent of intensive farming with its chemical fertilizers and insecticides, its draining of unproductive but water-retentive boglands, has led to a tendency to flooding and to impoverishment of the area's flora and fauna. Wild Ingleborough is an initiative launched in 2021 by an impressive partnership of environmental bodies and academia comprising the Yorkshire Wildlife Trust, Natural England, the Woodland Trust, the University of Leeds and the United Bank of Carbon charity. This conservation scheme aims to tackle 485 hectares initially, and to connect up scattered nature reserves and re-establish the natural treeline going up the mountain from broadleaf to dwarf shrub, heather moor and lichen heath. It's a bold vision, and one that has caught on locally with rewilding of the countryside being the flavour of the decade. Joining up separate areas of wild ground undoubtedly helps with increasing biodiversity and breadth of habitat. But the uncomfortable fact remains: when you introduce, or re-establish, anything in nature, you get rid of something at the same time – even if it is only a landscape pale and bare, its wildlife treasures tucked away, that touches many people.

Before resuming my southward journey down the Pennine Way from Horton-in-Ribblesdale, I wanted to get a grip on Craven Fault country. So much of the eye-catching local geography relies on those three great rips in the rocks beneath, remarkably exposed and expressive of a literally earth-shattering moment among the long-buried bones of Britain.

The Lune Valley lies out west of the Pennines. It runs north–south, separating the Pennine Hills from the Lake District, and the Craven Faults extend southeast from here. Tectonic movements under the seabed during Carboniferous times, perhaps 350 million years ago, were causing the ancient Rheic Ocean to close, and the supercontinents of Laurussia and Gondwana to approach and collide with each other. These collisions brought about a series of vertical fractures which created three great steps in the seabed.

The North Craven Fault made a tear about 30 miles long. It runs from where Kirby Lonsdale is today across to Pateley Bridge in Nidderdale, passing north of Ingleton and Clapham, crossing Ribblesdale north of Settle, then going south of Malham Tarn to cross Wharfedale just below Grassington. The layers of rock north of this fracture – Yoredale Group rocks, lying on top of older shale, slates and sandstones – were uplifted and tilted so far backwards towards the northeast that some of those underlying rocks were levered to the surface for the first time in 100 million years. At the same time the seabed south of the fracture dropped downwards. Altogether the upward and downward movements accounted for a vertical displacement of the rocks by about 200 metres, creating a submarine cliff taller than Beachy Head.

This was nothing compared to the vertical displacements along the line of the succeeding Mid and South Craven Faults. The Mid Craven Fault, about 20 miles long, diverges from the South Craven Fault just east of Settle to run through the town and on east through Malham Cove. But it's the South Craven Fault, also about 20 miles long, that was the largest disrupter of the landscape by far. This massive rip in what was the seabed runs south of Ingleton, Clapham and Settle, then curves to run just north of Hellifield and Gargrave. Where the North Craven Fault near Ingleton had a downthrow or downward drop of about 200 metres, only a stone's throw away the South Craven Fault dropped a mighty 1,200 metres. The total incremental and cumulative displacement associated with the Craven Faults amounts to not far short of a mile – a staggering disruption of the Earth's crust.

Over a couple of days I went walkabout on the local limestone, finding evidence everywhere of the Craven Faults and their reworking of the landscape. A little further south, above the unfrequented cleft of Clapdale, a path led up to a slanted upland the map named Norber. Huge boulders stood there, so-called erratics, brought from elsewhere and dumped by retreating glaciers. These truck-sized rocks were laid down as sedimentary beds around 425 million years ago. They were folded and split during the Caledonian orogeny which culminated around 400 million years ago. Erosion exposed them during Devonian times before the Carboniferous period, when they were buried beneath Carboniferous and probably younger rocks. When the North Craven Fault ripped across the seabed and tipped the buried rocks upwards and backwards, these slates and sandstones came to the surface in what's now Crummackdale. There they stayed until broken off by a glacier and trundled a couple of miles down dale before being dropped on Norber. They are 100 million years older than the limestone pavement on which they stand.

Just southwest of Ingleborough the River Doe rushes from the mouth of the Chapel-le-Dale Valley to meet the River Twiss issuing from neighbouring Kingsdale. Setting off from Ingleton to follow the Water-falls Walk along the narrow path that shadows the River Twiss, I was almost at once enclosed in the dark walls of Swilla Glen, a gorge where the river ran fast among mossy stones splashed with dipper droppings. Beside the path lay a money tree, its hide as scaly as a lizard's with tens of thousands of copper coins hammered into the boughs for luck. The trail crossed the line of the South Craven Fault, where the Great Scar Limestone was bent into an arch and thrust hundreds of feet down into the crust of the seabed. A little further up the glen I came to the site of the North Craven Fault. The tilt of the limestone was clear to see in the riverbank, and next to it an old lead-mine tunnel in a scaly bank of rock caught my eye – a squashed-up mass of shale laid down as a layer of mud in Ordovician times, 100 million years before the limestone it was pressed up against. Here was the work of the North Craven Fault, perfectly illustrated, as it slanted the Carboniferous limestone and brought up to the surface the ancient shale rocks below.

The path climbed the wall of a canyon above swirling holes where the south-going river chased round and round before escaping, sculpting semicircular hollows in the rock walls with a continuous swallow and gurgle. Its cold breath and smell of stone and earth came up to me as I crossed the gorge on lattice footbridges under which the peat-charged water sluiced as dark and frothy as a gush of porter.

Pecca Falls came crashing down a staircase of slippery limestone steps, almost directly over the site of the North Craven Fault. Beyond the cascade the trail left the trees and followed a curve of the Twiss. A wonderful view opened ahead towards Thornton Force, pride of the walk, descending a series of rapids before hurling itself in freefall into a smoking pool. Above this thunderous weight of water I followed a walled lane into the mist. Unseen and offstage, sheep bleated, a farmer whistled and a quad went puttering over an invisible field by Twisleton Hall. Imagination had to supply the scene up on Scales Moor high above the farm, the grey pavements of the Great Scar Limestone whose broken-off 'other half' lay far below the ground back down in Swilla Glen.

Below Twisleton Hall the River Doe echoed and hissed in its own steep walled canyon, leaping down towards Ingleton and its confluence with the Twiss through S-shaped channels carved through the limestone and shale by the force of water alone. I crossed above potholes boiling with toffee-coloured bubbles, and skirted backwaters where the surface lay marbled with scarcely moving patterns of foam. Below the white wall of Snow Falls the path snaked past another money tree and on through mossy old quarry workings, to emerge at the foot of the gorge with the church and houses of Ingleton lying beyond, as muted and dreamy-looking as any faded Victorian lithograph.

Back at Horton-in-Ribblesdale I packed up for the next leg of the journey south. The Pennine Way makes a curious northward loop on leaving the village in order to take in the summit of Penyghent, the third of the Three Peaks and another fine example of the geology of the Yoredale Group with its neat horizontal bands of rock. There's no doubt about it, Penyghent and Ingleborough are recumbent lions, both in shape and aura. On a windy morning of high dashing cloud Penyghent rose impressively a mile to the east as I climbed the walled track of Horton Scar Lane, looking out for two famous potholes caused by rainwater dissolving the limestone. Unlike many of their kind, this pair do not form pretty little dimples in the grey cheeks of Penyghent. Hull Pot, lying a little off the path, is an enormous wound in the flank of the mountain, a flat-bottomed hole as broad and deep as a city block. Hunt Pot, by contrast, opens right beside the Pennine Way, a tight black gash in the limestone like the entrance to a goblin cave. Boggarts were goblins that haunted the imaginations of dales dwellers in times past, and black slits in the hillsides such as this were reckoned to be their lairs.

A short sharp climb to the sandstone spine of Penyghent, an easy stroll to the summit, and at 694 metres a wind-blown and breathtaking view to all quarters, north to Dodd Fell and Great Shunner Fell over which I'd come a few days ago, southeast along the Pennine Way to the rising shoulder of Fountains Fell, further south to the far-off hulk of Pendle Hill in Lancashire, and particularly westward where Ingleborough ('looking noble, as usual', said Wainwright) and Whernside rose from their moors. Then it was cautiously off the steep southern brow of Penyghent, clambering down rock steps from gritstone to limestone and back to sandstone, till the descent and the wind both eased at the pothole depression of Churn Milk Hole and the flagstoned path ran on down towards a country road.

On Fountains Fell you need to stick to the path, because old coal mine shafts lie open all around the flat summit. Why did it feel so weird to be gazing down these unguarded holes at the top of this particular rise of ground? It must have been the name of 'Fountains' with its ecclesiastical connotations – Fountains Fell was part of the vast estates of Fountains Abbey in medieval times. If the hill had been dubbed 'Black Top Fell', I decided, I wouldn't have batted an eyelid. Gritstone interleaved with shale forms the topping of Fountains Fell, and it was the layers of shale that the miners delved for coal. The coal was reckoned noxious while burning, but it made good coking coal. A square, solid little stone oven building still stood at the summit, its arched entrance just big enough to crawl through into a dark interior. A boggart house, for sure.

Now the Pennine Way began to run over the last of the limestone by way of Malham Tarn, a lake that shouldn't be there. With shake holes and water sinks all round, how does the tarn, a body of water half a mile across, prevent its contents from draining away into the porous limestone? The answer is that the tarn is dammed by a glacial moraine and floored with near-impermeable Silurian slate, levered up from the depths as the fault lifted the rock strata and tilted them backwards.

From Malham Tarn I made a circuit of the fells above Darnbrook House Farm, where the limestone pavements held a wonderful flora more suited to spring than summer: buttery yellow mountain pansies, stout early purple orchids, spatters of mountain violets, and the intense pink flowers of bird's-eye primroses. Tucked down in the grykes or hollows of the limestone was a woodland flora, bizarrely flourishing in this open, treeless terrain – wood anemones, dog's mercury and lush ferns.

Another round walk from the beautiful little village of Malham, far down in the dale below, took me up through Gordale Scar, a fantastic twisty split in the Earth's crust 200 metres tall, sculpted by raging flood-waters during the last Ice Age. 'Going up Gordale?' asked the farmer mending his stone wall near Janet's Foss. 'That's quite a climb, mind.' I'd been offered an identical warning before doing the climb ten years ago. It could have been the same farmer. The route up Gordale Scar looked formidably steep, but it was conveniently broken into two stages, the first one up a solid block of Cove limestone with a cataract sluicing down alongside. A respite at the top among fallen boulders in a huge rock funnel where a big waterfall tumbled from a crack in the walls high above. Then on up steep scree between crags of Gordale limestone, a more fragile rock broken across into cracks and more prone to fall – hence the litter of boulders that floored this upper chamber.

Up at the top the world opened out from the claustrophobic clammi-ness of the rock funnel to gently undulating limestone uplands. I sat and caught my breath before turning back along the Pennine Way. The well-trodden track dipped downhill through the narrow dry valley of Watlowes to come out on a broad limestone pavement at the lip of Mal-ham Cove, a great sheer 100-metre cliff formed by erosion. With my nose to the limestone clints I found the white and grey remnants of corals, fragments of fan-shaped sea lilies, and end-on shards of the ribbed shells of brachiopods, or devil's toenails to give them their very evocative nickname, all residents in life and death in that warm tropical Carboniferous sea that makes such an appeal to the imagination. Ero-sion has cut the cove back nearly half a mile from its original position at the fault line. The cliff face now describes a great concave bend, prod-uct of weathering and of erosion by the giant fall of glacial meltwater that plunged over the centre of the lip some fifteen thousand years ago, the same cataract that hollowed out Gordale Scar.

From Malham the Pennine Way winds south and east beside the

River Aire and over Eshton Moor to cross the A65 road at Gargrave, and here everything changes in the course of a morning's walk.

The A65 takes advantage of the Craven Gap, the northwest to southeast trough formed by the South Craven Fault, and shadows it all the way. To anyone driving that road from Kirby Lonsdale to Skipton, the step change in scenery from one side of the road to the other is as clear as day. Along the line of the road the South Craven Fault dropped the seabed down by over 1,000 metres and, despite all the millions of years of erosion, the contrast between what stands north and what lies south of the road is remarkable. To the north it's the high, craggy ground of Newby Moss, Clapdale and Ingleborough, Feizor Thwaite and Giggleswick Scar, limestone uplands all. Products of weathering of the Yoredale Group and the Great Scar Limestone, this succession of eroded and glacier-smoothed hills and cliffs stands tall and rugged like the high waves of a turbulent sea. To the south, by contrast, the landscape sinks becalmed to the small drumlin hillocks and pastures of the Craven Basin, rising gently beyond into a modest swell of hills that roll away southward to a skyline of moors. This is the subordinate side of the fault, its eroded troughs and heights underpinned not primarily by limestone but by shale, mudstone and above all the coarse sandstone known as millstone grit or gritstone. There are curious meetings of the two rock types: at Long Preston, for example, some of the village houses are built of the light-coloured limestone to the north, others of the darker gritstone quarried south of the fault, and some local field walls are pale limestone with darker capstones of harder, more weather-resistant millstone grit.

Standing on the hill above Gargrave, my southerly view was across low green pastureland to a distant skyline of flat-capped moors. Everything in prospect heralded the change from rocks of lime to those of mud, grit and sand, from light to dark and soft to harsh. The grass of the dale still looked green, but the stones of the field walls and the barns and farmhouses were now a dull dense grey. After the long walk south through the landscape glories of the limestone dales, it looked and felt as though lower and darker country lay ahead.

The dry limestone valley of Watlowes that leads down to Malham Cove.

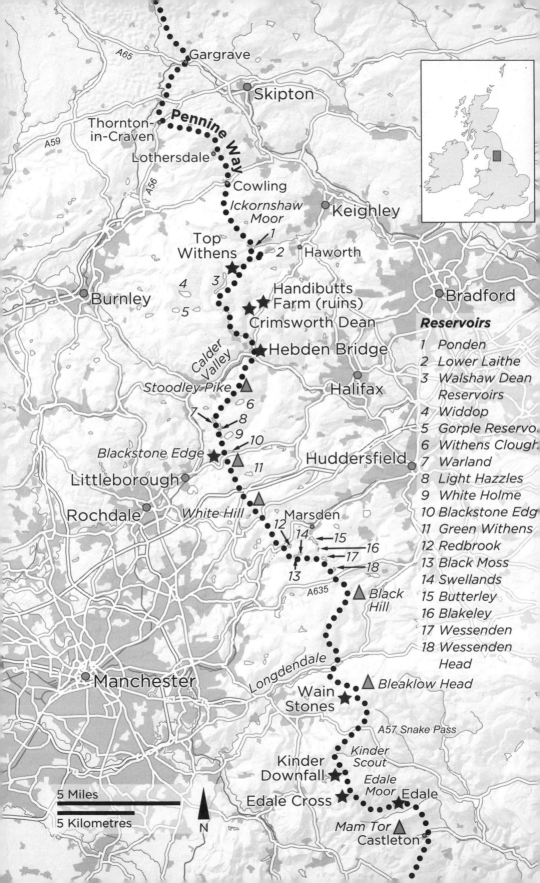

Gargrave

A65

Skipton

A59

Thornton-
in-Craven

Pennine Way

Lothersdale

A56

Cowling

*Ickornshaw
Moor*

Keighley

1

Top
Withens

2   Haworth

3

4   Handibutts
          Farm (ruins)

Burnley

5   Crimsworth Dean

          Hebden Bridge

Bradford

Calder Valley

Stoodley Pike

Halifax

6

7

8

9   10

Blackstone Edge

11

Huddersfield

Littleborough

White Hill   Marsden

Rochdale

12

14   15

16

17

13   18

A635   ▲ *Black
              Hill*

Longdendale

Manchester

Wain
Stones

▲ *Bleaklow Head*

A57 Snake Pass

*Kinder
Scout*

Kinder
Downfall

*Edale
Moor*

Edale

Edale Cross

5 Miles

5 Kilometres

N

Mam Tor ▲
Castleton

Reservoirs

1   Ponden
2   Lower Laithe
3   Walshaw Dean
    Reservoirs
4   Widdop
5   Gorple Reservo...
6   Withens Clough
7   Warland
8   Light Hazzles
9   White Holme
10 Blackstone Edg...
11 Green Withens
12 Redbrook
13 Black Moss
14 Swellands
15 Butterley
16 Blakeley
17 Wessenden
18 Wessenden
    Head

12

True Grit: Gargrave to Edale

IT'S MORE THAN LIKELY THAT my fit of the downbeat blues on looking from the limestone hills into the gritstone moorland country was subconsciously occasioned by one of my all-time heroes: walker, writer and gentle polymath, John Hillaby. 'Don't meet your heroes' is an axiom it's wise to live by, and I never did meet or walk with this fantastically fit and energetic Yorkshireman (he died in 1996), who would undoubtedly have stridden the pants off me. Hillaby wrote many books about his journeys on foot, but the one I regard as a priceless legacy was his account of walking across Britain from Land's End to John o'Groats in 1966, *Journey Through Britain*. Erudite but never preachy, humorous and serious by turns, curious about everything from pump hydraulics and local songs to botany and geology, in that book Hillaby caught the sheer joy and elation of long-distance walking through the uplands of this country in springtime. *Journey Through Britain* inspired me to become a writer, and I had a copy with me everywhere I went on my travels through the bones of Britain.

Hillaby's journey from end to end of the country was from south to north, the other way round from mine. At Gargrave he was looking northward, with the tough moorland section of the Pennine Way behind him and several days of walking through his favourite type of country to come. As he asserted:

The beauty of landscape is not something that can be reduced easily to basic geology. Preference and prejudice creep in. Mine are apparent in a love of limestone and a dislike of grit, two geological bed-fellows as unlike as chalk and cheese. [Limestone country] looks far brighter than the moors to the south. The acid squelch is replaced by wholesome springy turf. Flowers abound in the hedgerows, the bird song is clamorous, and the local inhabitants talk about something more interesting than the probability of rain.

Maybe a bit of preference and prejudice creeping in there, John. On the whole, though, it's true that gritstone country, lower-lying than the limestone uplands, sparser in its flora and darker in its rocks, vegetation and building stone, tends to feel more dour and downbeat.

Setting off from Gargrave it was mudstone underfoot, sedimentary deposits down steep undersea slopes towards the start of the Carboniferous period. On top of this, what Alfred Wainwright sums up as 'contented cows on low green hills'. These grassy hummocks – Scaleber Hill, Moorber Hill, Turnber's Hill, Langber Hill – are drumlins, conglomerations of clay and boulders dumped and then smoothed into shape by the retreating glaciers around twenty thousand years ago. The clay, rich in minerals from the minute particles of rock and fragments of the sea creatures that composed it, gives good sustenance to the grass that contented Wainwright's cows; but the drumlins that dot the low-lying pasturelands of Craven make tricky, bumpy country to farm, though easy enough to walk. Threading my way between the hillocks I followed sticky field paths, then the towpath of the Leeds & Liverpool Canal, before descending to Thornton-in-Craven and the first proper loom of the gritstone on the moor top of Pinhaw Beacon. Down again across steep fields to the deep-sunk little village of Lothersdale, footed on mudstone. And from here on, with scarcely a break until the end of the Pennine Way some 65 miles to the south, it would be a landscape of grit, and plenty of it.

Sap of the sullen moor is blood of my blood.
A whaleback ridge and whiplash of the wind
Stripping the branches in a rocking wood –
All these are of my lifestream, scoured and thinned.

From 'Northern Stone' by Phoebe Hesketh (*No Time For Cowards*, 1952)

Millstone grit covers 840 square miles of Yorkshire, most of it in the southern sector of the county. South of the Craven Gap the South Craven Fault pushed the seabed – later the land surface – down by 1,200 metres or so. North of the Craven Gap the high-perched gritstone eroded away sooner than its counterpart at the lower level, exposing the Yoredale Group series of rock layers. The Ice Age glaciers and subsequent melt-waters carved the more easily shaped limestone into dales and hills with sharp profiles, the gritstone only remaining as caps on the highest hills. Down south of the Craven Gap, though, the harder gritstone remains largely intact, its trademark tors or outcrops of coarse dark rock (locally known as 'edges') carved into fantastic shapes by wind and weather. They give tremendous character, albeit hard and austere, to the landscape.

This durable rock is made of sand, grit and pebbles of quartz with some feldspar. The original material was part of mountains formed during the Devonian period (420 mya–360 mya), when northern Britain was uplifted during the Caledonian orogeny after the Iapetus Ocean had closed and Laurentia collided with Baltica and Avalonia. The coastline advanced and receded from north to south, and a huge delta system formed along the sea margins, with masses of sandy sediments being deposited across much of the middle of England, to be consolidated as an enormous shelf of Old Red Sandstone. Erosion smoothed away a great deal of this sandy, gritty upland. The early Carboniferous period in the Craven Basin, the gentle concavity that lies low to the south of the Craven Fault, saw deposition of thick successions of mudstone with some limestone, followed later by deposition of the millstone grit. Some of this sediment derived from northern Britain, but a significant proportion came from much further north, as far away as Norway and Greenland, transported southwards by a vast river system which originated about 330 million years ago. The sediments were deposited along the river delta as sandbanks. Heavier, bulkier stones were the first to sink to the bottom, with the smaller, lighter sand and grit settling on top, and this typical river deposition process is demonstrated in most gritstone layers where coarse material with larger grains sits at the bottom of the rock, with finer, sandy stuff overlying it towards the top.

There's a fascinating feature to hunt for among the gritstone outcrops, namely cross-bedding, where sediment was deposited on the sloping side of a bed which could be anything from a dune, hump or delta bar to a mere ripple in the sand. Some of these grooves cut across others at an angle, showing where underwater dunes or hummocks of sand crossed the path of older dunes as they washed along a river bed. There are ripple marks, too, as sand and grit were deposited at river and delta mouths where water currents flowed. To see the actual motion of the sea, the slanting profile of one tiny individual ripple, caught and held in stone is to experience again, as with the Coigach lapilli and the Vindolanda pig's trotter-print, the sense of a single flake of time trapped and fixed for ever.

The BGS Geology Viewer indicates the north–south wedge of millstone grit as a great blotch of lurid lime-green. Across its moors wriggles the dotted line of the Pennine Way, flanked by the Lancastrian ex-textile towns of Colne, Rochdale, Oldham and Manchester to the west, and the Yorkshire post-industrial settlements of Keighley, Huddersfield and Sheffield on the east. The path twists this way and that as permission to pass through was negotiated with, or wrung with much ado from, landowners, grouse moor consortia and municipal water boards by Tom Stephenson and his allies from 1935 onwards. 'In several places,' noted John Hillaby in *Journey Through Britain*, 'the gap between the advancing industrial sprawl has shrunk to a little corridor of high ground on top of the gritstone edges. The marvel is that the walker can still get through.' Hillaby made that observation in 1966, a year after the Pennine Way was opened, and although the Pennine towns are all now in various stages of post-industrial decline, their urban sprawl has accelerated apace since then.

The Pennine Way crossed the Colne–Keighley road at Cowling. I looked up to see the long dark gritstone edge of Earl Crag overlooking the village on the eastern skyline. From the crag a pair of tiny pimples, widely separated, stuck up against the clouds. Man's ancient need to assert himself over a landscape had been satisfied here with two monuments, the chunky obelisk of Wainman's Pinnacle, erected to mark victory over Napoleon in 1815 at the Battle of Waterloo, and the slender castellated folly of Lund's Tower, built in celebration of Queen Victoria's Diamond Jubilee in 1897.

The Pennine Way bypassed the crag and its follies in favour of a zigzag climb to Ickornshaw Moor. Here I unwrapped a big square of home-made shortbread I'd bought at a roadside stall in Lothersdale. 'You'll have earned that if you eat it on top of t' moor,' said the woman who had made it, and she wasn't wrong. Ickornshaw Moor was wide and low-rolling, with mendicant Swaledale sheep among the rushes whose curiosity about my shortbread did not quite overcome their caution. It was a long and rainy trudge down to Ponden Reservoir, the tramp of my boots alternating between the squelch of wet peat, the rattle of duckboards and the thump of square flagstones from old mill floors. Cat Stones and Wolf Stones lay just off the path, two modest and low-lying outcrops. Their names added a smack of the wild to the surroundings; appropriately enough, because once across the valley where Ponden and Watersheddles Reservoirs lay dead and grey under the rain, the Pennine Way passed across the open moors where Emily Brontë from nearby Haworth Parsonage once wandered and let her supercharged imaginings run free.

Two of the locations in Emily's brutal battering-ram of a fable, *Wuthering Heights* (published in 1847), lie next to the Pennine Way. High on a bank over Ponden Reservoir stands Ponden Hall, a typical Pennine farmhouse, long and low among its shelter trees. This was

Emily's 'Thrushcross Grange', home of the Linton family so sadistically and remorselessly destroyed by Heathcliff and his lover and foster-sister Catherine. I stood in the lane and stared at the house, picturing Cathy and Heathcliff as naughty children, sneaking up to Thrushcross Grange by night and terrifying the gently bred Edgar and Isabella Linton by making faces at them through the window – a chink of light and laughter in the dark stormy sky of Emily's extraordinary imagination.

I followed the rutted track of the Pennine Way up past the ruins of the farmhouses of Lower Withens and Middle Withens, their bread ovens, cupboards and windows still shaped within the tumbled stones. Beyond these a hard black angle of walls jutted dramatically from the skyline, the carefully preserved ruin of Top Withens farmhouse. Did Emily Brontë actually model her lonely, fortress-like house of 'Wuthering Heights' on Top Withens? Not the individual details, say the Brontë experts. But the isolated farmhouse under the edge of the moor was well known to her, and in its harshly beautiful setting it makes by far the best candidate for Heathcliff's lair.

Emily Brontë wrote:

One may guess the power of the north wind, blowing over the edge, by the excessive slant of a few stunted firs at the end of the house; and by a range of gaunt thorns all stretching their limbs one way, as if craving alms of the sun. Happily, the architect had foresight to build it strong: the narrow windows are deeply set in the wall, and the corners defended with large jutting stones.

I sat on the hillside above the ruin, picturing the savage dogs, the surly servants, the cavernous hall and chilly bedrooms of Wuthering Heights; the gothic horror of Cathy Linton rapping on the windows in the snowstorm, crying in vain for admittance as her blood ran down the broken pane; and arching over all, the relentless will of that passionate lover and hater, Heathcliff, the embodiment of the wild spirit of the moors.

With Kate Bush's eerie screech of a song as an earworm ('Heeeethcliff, it's mee, I'm Catheee . . . ') I walked on across fox-coloured moorland and down past the three elongated reservoirs in Walshaw Dean. At the foot of the valley I turned eastward off the Pennine Way and followed a farm track and walled lane past the dark stone house of Walshaw, high above the wooded cleft of Hebden Water, and on through rough pastures to Crimsworth Dean and the Old Keighley Road, a forgotten lane from Hebden Bridge to the high moors. There's something about the gritstone farms along these high lanes, scattered communities on the edge of town but remote from it, which draws me back whenever I'm in the vicinity. Walking from Horodiddle to Coppy, Nook to Handibutts, I longed to read again Glyn Hughes's masterful novel *Where I Used to Play on the Green* (1982), which breathed life and character into the historic figure of William Darney, one of the founding fathers of Methodism, a wild Scots pedlar, cobbler and roving preacher who came striding here to save souls in the 1740s.

Darney, a friend and ally of John Wesley, appears in Hughes's book as a 'good-looking rough man, huge, with a big red beard and hair', who would roar a gathering of weavers and farmhands to their knees in an isolated barn like the one at Handibutts, and overwhelm them with threats of hellfire, glimpses of heaven and bursts of self-penned religious doggerel until they burst out with, 'I'm saved! Brothers, I'm saved!' Wildly emotional scenes of conversion were played out in many of the strong, bleak farmhouses and barns of blackened gritstone along the upland packhorse trails. The rough townees, however, didn't make life easy for William Darney. His son John described how . . .

. . . they rode him through the deepest place of the river which runs through the village, one of the mob riding upon his back with a bridle about his head. They then procured a rope, which they fastened about his waist, and then fastening the rope to each side of the river, they literally tied him in the middle of the stream. In Colne, the mob stripped him of all his clothes unto his naked skin, daubed him all over with mire, and drove him through the town streets in this condition.*

Crimsworth Dean's little Wesleyan chapel, converted to a house, still stands in a crook of the lane. But Handibutts Farm, where Darney ranted to a packed and ecstatic audience in the barn at the end of the house, is a skeletal ruin in the fields, its roof timbers naked and crooked, its long many-mullioned upper windows blank and empty. I stood here a long time, thinking of those poor weavers, spinners, cobblers and farm labourers, scourged by the impassioned preaching of William Darney until they took the proffered leap out of bondage, and of the Industrial Revolution brewing in the valleys that was about to change the gritstone country for ever and bind its people in servitude far more tightly than before.

The ironstone and the coal beneath the gritstone moors had been exploited in a small way since medieval times, when land ownership in the region was divided into large estates, many of them the property of abbeys and other absentee landlords. A state of affairs developed in which many tenants were more or less left to farm and exploit the natural resource of the country on their own account. So by the early eighteenth century and the beginning of the Industrial Revolution, the gritstone area's economy was already a mixture of cattle- and sheep-farming, mineral-mining and small-scale, cottage-style industry. The shape of the land suited the requirements for textile manufacture – rain-soaked moors on semi-impermeable rock, with steep hard-floored gullies channelling water to turn the wheels of small manufactories. There was an abundance of local wool, and the springs of the acid moors produced soft water for washing both the raw wool and the cloth the weavers made from it.

---

*   John Darney, quoted in J. W. Laycock, *Methodist Heroes in the Great Haworth Round 1734 to 1784* (1909).

As towns and cities grew across the country and the British colonies, so did their need for woollen cloth, soon augmented by cotton arriving at west-coast ports from plantations in the New World. Expansion was inevitable, as both production and the importing of raw materials increased. Local people with a head for business organized spinners and weavers to work for them, initially from home, then in small manufactories – hence the large and elongated windows one sees in the upper storeys of so many farms and houses in these parts. Gritstone quarries opened in the hills and dales. During the eighteenth century, hundreds of watermills were built, as near to their source of power as possible. By the end of the century the development of steam power led to less reliance on sources of water for power. The mills moved from the dale sides down to the valley floors and grew upwards into enormous steam-driven temples of manufacture – William Blake's 'dark satanic mills', worked by people who were employed either by the owners or by those who rented the buildings. Wool markets operated in halls grander than anything previously seen in the industrial towns. Canals began to snake through the valleys, the Rochdale Canal through the Calder Valley by way of Hebden Bridge, the Leeds & Liverpool Canal through Skipton, the Aire & Calder Canal giving Leeds access to the east-coast ports via the Humber. Better roads made their way alongside the canals, then railways too, as the textiles were sped away to consumers by barge, by wagon and by steam.

There was a great upsurge in invention, each innovation giving rise to the next. John Kay's 'flying shuttle' (1733), for example, enabled weavers to work four times as fast as before, driving up demand for yarn that was eventually satisfied by the 'spinning jenny', introduced by Thomas Highs in 1764, soon improved by James Hargreaves, and itself superseded by the spinning mule. Workers in their tens of thousands were needed; by the mid-nineteenth century there were two thousand mills operating in West Yorkshire alone, and these narrow gritstone valleys below the moors became powerhouses of ingenuity and productivity, the wonder of the world.

Along with the mills came the workers and their miseries. Cheaply built back-to-back terraces along the valley floors declined to slums, with open sewers polluting the wells. The sparkle of the newly cut gritstone soon blackened with oxidation and soot from factory and domestic chimneys. The air became foul, the valleys loud with the clatter of the looms. Living conditions in these crowded terraces were appalling. Those who were not working all day were working all night for poverty wages. The Industrial Revolution was not only a miracle, it was a Moloch, consuming the lives and health of its labourers, in particular the working children on whose narrow shoulders the great commercial edifice was built.

In his reforming book *The Factory System Illustrated: In a Series of Letters to the Right Hon. Lord Ashley* (1842), William Dodd recorded the words of 25-year-old Benjamin Gomersal from Bradford:

I commenced working in a worsted mill at nine years of age. Our hours of labour were from six in the morning to seven and eight at night, with thirty minutes off at noon for dinner. We had no time for breakfast or drinking. The overlooker beat me up to my work. I have been beaten till I was black and blue and I have had my ears torn. I found it very hard and laborious employment. We had to stoop, to bend our bodies and our legs. I found my limbs begin to fail, after I had been working about a year. It came on with great pain in my legs and knees. I had to attend at the mill after my limbs began to fail. I could not then do as well as I could before. I had one shilling a week taken off my wages.

I was a healthy and strong boy, when I first went to the mill. When I was about eight years old, I could walk from Leeds to Bradford without any pain or difficulty, and with a little fatigue; now I cannot stand without crutches! I cannot walk at all! I am very much fatigued towards the end of the day. Perhaps I might creep upstairs. I go upstairs backwards every night!

I cannot work in the mill now.

Elizabeth Bentley gave evidence to a Parliamentary Commission as early as 1815. At twenty-three years old she was living in Leeds poor-house, a cripple, in constant pain and unemployable. She began working at the age of six in Burk's flax mill as a doffer, removing full bobbins of yarn from the spinning machines and replacing them with empty ones. She walked two miles to and from the mill each day in all weathers, and worked from six in the morning to seven at night, often longer if the mill was 'thronged' (under pressure). Like Benjamin Gomersal she had no time allowed for drinking or breakfast. Treatment was brutal: 'I have seen the overlooker go to the top end of the room, where the little girls hug the can to the backminders; he has taken a strap, and a whistle in his mouth, and sometimes he has got a chain and chained them, and strapped them all down the room.'

The question-and-answer format of the interview gives an added poignancy to Elizabeth's plight.

Were you perfectly straight and healthy before you worked at a mill?
Yes, I was as straight a little girl as ever went up and down town.

Were you straight till you were 13?
Yes, I was.

Did your deformity come upon you with much pain and weariness?
Yes, I cannot express the pain all the time it was coming.

Do you know of anybody that has been similarly injured in their health?
Yes, in their health, but not many deformed as I am.

It is very common to have weak ankles and crooked knees?
Yes, very common indeed.

This is brought on by stopping the spindle?
Yes.

Where are you now?
In the poorhouse.

State what you think as to the circumstances in which you have been placed during all this time of labour, and what you have considered about it as to the hardship and cruelty of it.

The witness was too much affected to answer the question. *

The mills roared on throughout the nineteenth century, supplying the world with its clothing and cloth. But the country looked away during the First World War, and when it looked back the trade was being stolen from under its nose by cheaper imports, mainly from India and the Far East. British working practices and British prices just couldn't compete. Decline was the word all through the twentieth century, and the huge many-windowed temples of manufacture in Huddersfield and Halifax, Colne and Keighley became redundant. Small businesses crept in like mice, some to stay, some to flit out again or die. Damp, rot and vandalism took up permanent residence. The wool and cotton towns became depressed and depressing places where jobs and opportunities were hard to come by. They were bypassed by tourism and the outside world. The Pennine Way avoided them. When I, a southerner, began in the 1980s to explore the network of cobbled packhorse roads through Rawtenstall and Bacup, Todmorden and Hebden Bridge, the gritstone mill towns of Calderdale and Rossendale, it was like dropping down a rabbit hole into another world, dark but compelling, where humpbacked bridges crossed murky canals, short terraces sloped up into the hillsides, and silent mills as tall and many-windowed as cathedrals sprouted buddleia bushes from their gutters.

*

---

*   House of Commons Committee Interviews, 1832 – interviewer Michael Sadler.

The clothing factories of Hebden Bridge were still going full pelt when John Hillaby walked through the town in 1966. In *Journey Through Britain* he relates:

> I heard the old familiar roar of the looms as soon as I crossed the river ... Hebden Bridge is certainly a textile centre. The factories stand on each other's shoulders. They cling to the hillside. They are tucked away in yards and alleys; some are perched on the most improbable promontories.

Hebden Bridge, in fact, turned out to be a lucky town. Its higgledy-piggledy conglomeration of mills, factories, chapels, pubs and steep terraces along the river and canal made a picturesque impression on strangers from London and the south who began to venture here in the 1980s in search of an agreeable place to live, once the looms had ceased to roar and the chimneys to spread a pall over the town. Hebden Bridge was within commuting distance of Manchester and Leeds, cities where an arts scene was getting established. There was hilly open country on the doorstep. The buildings scrubbed up very nicely once the sombre coat of oxidation and soot had been removed to reveal the honey colour of the gritstone beneath. Over the past forty years the character of the town has undergone a metamorphosis. They are still making clothes in a room in the Melbourne Works and down the way at the Banksfield Clothing Works, but Hebden Bridge as a whole is 'sought-after' now, its houses on the pricey side, its old mills converted to bars, community centres and shops. You can eat in Il Mulino restaurant in Hebden Bridge Mill, live in an 'apartment hotel' in Croft Mill, and buy quinoa flakes ('British grown') at Valley Organics Workers' Co-op in the High Street. Quite a change from the no-nonsense manufacturing town where Hillaby enjoyed the following blunt exchange:

> 'Going far?'
> 'Gargrave.'

'Over Wadsworth?'
'Seems best.'
'It'll be mucky.'
'Could be. Know a good cobbler?'
'Try Jim Cunliffe.'
'What about some grub?'
'Alan's place. Up t'hill.'

There's a whole tangle of ways to choose from when you're walking on from Hebden Bridge. On the cobbled lane to Horsehold Farm I turned round for a last look down over Hebble End Mill and on across the tight-packed town. Then I made for the upper moors and the monument on Stoodley Pike. A pinnacle of millstone grit 120 feet high, erected (somewhat prematurely) to celebrate the overthrow of Napoleon Bonaparte after his abdication in 1814, it stands at the end of a steep edge and draws the eye for many miles around. More mendicant sheep here, their moral fibre subverted by the crisps and sandwiches of weak-willed walkers. I stupidly gave them a little of what they were after, and they followed me all the way to Withens Gate.

From the edge there were long views west along the Calder Valley to Todmorden; then, as the Pennine Way swung south, over the mills and terraces lining the deep valley to the west where the Rochdale Canal now had a new companion, the narrow River Roch. The tower blocks and chimneys of Manchester lay away to the southwest in a haze of distance as I came down to the Littleborough–Halifax road by way of a string of wind-whipped reservoirs with poetry in their names: Warland, Light Hazzles, White Holme and Blackstone Edge.

All manner of walkers, cyclists and fell runners swarmed round, into and out of the White House Inn next morning. If the Pennine Way has fallen out of fashion, no one had told this crowd. Precious few of them, probably, were tackling the whole way from bottom to top, but here they were in couples and groups, boisterous or sober, youths and greybeards, all preparing to trudge its gritty paths, stumble on its boulders and skip across its rocks, just for the enjoyment of spending a morning walking on the moor.

Across the road a broad stony track led up past the crags of Blackstone Edge Delf, meaning 'delving' or digging, an old quarry long disused. The quarry faces were green and grey with weathering and oxidation. The true colour and nature of the underlying rock was better demonstrated in the surface of the Pennine Way. A creamy, honeyed hue, the path had broken down under millions of footfalls into sand; not just sand, but a mixture of sandy particles with globules of quartz, the components of gritstone disassembled once more after 300 million years of clinging together.

A stone-walled water channel led on from the quarry, one of many that seam these moors, collecting rainwater and directing it to the reservoirs. As companionable in mist or grim weather as the irrigation *levadas* of Madeira, but rather less lethal (with one stumble you can plunge 1,000 feet to your death from a *levada*), these channels have become the walker's guide and friend on the bleak moors hereabouts.

Soon I left the drain to climb the short, cobbled slope of what has been identified as a Roman road, perhaps with its original setts replaced at some stage. On its northern edge a run of hollow stone might be a drainage channel, or perhaps the ruts worn by two millennia of iron-tyred cart wheels. It rose to a saddle of ground where Pennine Way walkers and Sunday runners were resting around the Aiggin Stone. Now sadly truncated, this medieval waymark pillar, originally taller than a man, was set up to guide benighted or mist-beguiled travellers across the moor. With nearby gritstone boulders conveniently shaped for the human posterior, and fine long views all round, it was the obvious place for a sit-down and a look-see.

From here the Pennine Way rose to Blackstone Edge, a classic gritstone ridge with cliffs jutting westward like ships' prows. Wind-distorted boulders, weathered to resemble stacks of black pancakes, stood at the edge, their rough sandy bodies studded with specks of white quartz like globules of fat in coarse salami. Tors like these are seen all over the

Yorkshire and Lancashire moors. The tiny pedestals of eroded rock that support some of these enormous boulders are so slender you marvel that they haven't snapped and sent their burden toppling.

Further back lay Robin Hood's Bed, a set of rocks cut flat by wind and weather, extensive enough to accommodate not only Robin and Maid Marian, but all the Merry Men as well. Robin once hurled a boulder from here towards the setting sun. It came down on Man Stone Edge near Rochdale, complete with the imprint of his finger and thumb. Good old Robin, with his plethora of place names all over these islands: Robin Hood's Well, Wood, Lane, Bay, Butts, Howl, Leap, Chair, Pot, Larder, Stone, Stoop, Stride, Table, Spring, Arbour, Bower and Bog. He got about a bit, did old Robin.

The narrow, stumbly path left the rocks behind and headed south down a long slope into the rushy declivity of Redmires and Slippery Moss. 'You will question your own sanity,' Alfred Wainwright warned 'the wet and weary wanderer through the ghastly mess' of Redmires. 'Standing knee deep in this filthy quagmire,' he wrote with mordant humour, 'there is a distinct urge to give up the ghost and let life ebb away.' Back in 1967 when AW was here and the Pennine Way had only just been opened, the boggy ground, the black peat slutch and pea-green sucking moss were a trial to cross and a puzzle to navigate in a relatively featureless landscape. By 1980, when I first ventured hereabouts, underfoot conditions on the Pennine Way across these gritstone moors had deteriorated to the point where the path had grown 50 feet wide as walkers attempted to outflank the bogs. In places the course of the path had vanished among peat hags and stream hollows that could be higher than one's head. In rain or mist, walking the Pennine Way here had become not only unpleasant, but dangerous. A notable photo taken at the time showed a walker on a misty day crouched in a foetal position in one of the deep peat hags, lost and exhausted, the picture of despair.

The Pennine Way was Britain's flagship long-distance path. Something had to be done. An extensive (and expensive) repair job was carried out, the Pennine Way paved with broken stone, then with flagstones salvaged from the floors of the hundreds of textile mills that were closing in dozens of little towns in the surrounding valleys. Nowadays it's a dry-shod and even-textured walk across Redmires and Slipping Moss. Yet halfway across I looked down at the sticky black slutch bordering the flagstones, and saw a boot impression 2 feet deep. Someone had gone in up to the knees. Across the peat surface a grouse had hopped, and even its miniscule weight had caused its three-toed footprints to sink an inch into the mire. The moss hasn't lost its power to smother and besmirch, and travellers still need to watch where they put their feet – the one and only piece of advice to walkers in wild country that Alfred Wainwright thought worth passing on.

From a cleft in the moor ahead came a murmur that swelled to a grinding roar. I wondered what a wanderer here in medieval times would have made of it. Nothing at all was visible until I crested the hill and looked down on the flashing highway of the M62 trans-Pennine motorway. A gracefully bowed footbridge crossed it, the four legs giving it the look of a primitive sculpture of a stretching dog – a pleasing image to carry forward.

Here begins the Pennine Way's 30-mile trudge southwards across the bleakest moors of the whole journey. These peat-blanketed wastes are the first taste of the trail for walkers heading north from Edale, as almost all do, and they can prove a very bitter starter if there's rain or hill mist about. The moors are ecologically impoverished, stripped of their trees and scrub by grazing, smothered in rain-fed peat bogs, and riven with countless slacks and cloughs, grains and groughs, all names for black peaty watercourses that wriggle about and are no help in direction-finding. What it was like for the pioneers of the 1960s, trying to find their way with 1-inch map and compass across this sodden and featureless ground, can be gauged by the comments of those two sturdy northern walkers Alfred Wainwright and John Hillaby. Wainwright reckoned Black Hill desolate, hopeless, frightening and dangerous. Hillaby on Kinder Scout found 'a silent and utterly sodden world . . .

Fed-up with the sight of peat, I took off my shoes and socks and climbed on to a crest of the soggy stuff. I didn't sink in far, but the prospect from the top was appalling. The peat extended for miles. It rose, gradually, in the direction of a pile of rocks. And it steamed, like manure.'

Hillaby found the moors extraordinarily depressing. 'I don't suppose I shall ever go there again,' he declared after struggling off Kinder Scout. And Wainwright, contemplating the wilderness of the Bleaklow plateau, remarks, 'This section is commonly considered the toughest part of the Pennine Way. It is certainly mucky, often belaboured by rain and wind, and weird and frightening in mist. But cheer up,' AW concludes. 'There is worse to follow.'

Compared with such travails, I had it easy on the moors. There was only a sprinkling of rain, and no mist worth speaking of. I had my Satmap GPS device with me, and when it wasn't performing one of its capricious tricks – switching to screens I'd never seen before, suddenly going blue and unresponsive, flashing up incomprehensible but worrying information ('Warning Radius for all OOIs!') – it was a reliable guide in the wilderness. The path through the worst of the bogs has been paved or duckboarded. And the moors are cut at regular intervals by roads crossing the hills between towns to east and west. The knowledge of civilization down there, a bus ride away, is a comfort when the legs are twanging and the heart quails at the sight of yet more peat and puddles lying in wait across the road.

White Hill looked pale enough when I got to its trig pillar, a slightly swelling grandstand for a view of pale brown moors. Down to cross the Huddersfield road, up by the flat gritstone edge of Rotcher and the gracefully shaped slopes above the Castleshaw Reservoirs. Down to cross the Manchester road and enter the Peak District National Park across an invisible border. From Redbrook Reservoir a flagged path led over the moor between the twin reservoirs of Black Moss and Swellands, where the wind-driven water lapped loudly. These water stores look more like natural lakes than reservoirs, but they were built to supply water to the

Huddersfield Narrow Canal that crosses the Pennines from Huddersfield to Ashton-under-Lyne. On 29 November 1810 at one in the morning the clay bank of the newly built Swellands Reservoir burst. The water rushed down Butterley Clough, carrying with it a 15-ton boulder which it deposited in the mill race at Marsden, 2 miles downstream – an erratic of the Anthropocene Epoch. A cottage at Bank Bottom was overwhelmed and five people drowned. The tsunami of peat-stained water gave the disaster the name by which it's remembered locally: the Black Flood.

The Pennine Way forged on east down Blakely Clough, now unflagged and properly squelchy. The black peat sucked at every step, and a red grouse hidden in the heather harshly admonished me: 'Go back! *Back, back, back!*' The sky cracked with blue and filled with whipped cream clouds, the icy wind blew at my back, and the sun came out and instantly transformed the dour moor grass into a waving sea of gold as I turned off by way of the Blakely and Butterley reservoirs for Marsden and a good night's kip.

Four reservoirs lie in the steep valley south of Marsden: Blakely and Butterley, Wessenden and Wessenden Head, strung together by the winding Wessenden Brook and built in Victorian times to supply water to Huddersfield. There's a beauty about these man-made lakes in their moorland valley that transcends their function, but there's also an implied menace. Marsden at the foot of the valley knows all about flooding; not just historic reservoir bursts, but recent inundations following catastrophic storms such as the one on Boxing Day 2015, and Storm Ciara on 9 February 2020. The town's location, in the bottom of a narrow valley under steep hills with a dozen reservoirs on the moors above, made it ideal for the operation of textile mills in times past, but also renders it very vulnerable to flooding nowadays with the advance of climate change and increased frequency of intense rainstorms.

On the steep slopes above the Wessenden reservoirs the next morning a band of volunteers was moving slowly sideways, each figure bent double. They were planting oak trees, part of a scheme of natural flood management that includes restoring rain-absorbing peat bogs and

heath, building 'leaky dams' that release rainwater gradually, and planting up steep slopes to stabilize the ground and slow the downward rush
of storm water. 'Been at this fifty year,' said the team leader as he
straightened up, grimacing. 'Gets you in the back! But gives me summat
to do on a Sat'day morning.'

On the far side of the A635 the gritstone moors stretched away, bare
and dark. Rain arrived as I slogged up the side of Black Dike to the peat
hags and sodden mire of Black Hill. Black is accurate. This is the blackest
and bleakest spot on the Pennine Way, the only sighting point being the
slender TV/radio mast, 228 metres tall, on Holme Moss a mile away. It
was a purgatorial trudge in filthy boots and trousers by Meadowgrain
Clough – no meadows, no grain – down to the reassuring sight of the
south-going cleft of Crowden Great Brook, and at last the string of elongated reservoirs that marks the cleft of Longdendale and the bus to
Glossop.

'Looked down on from above,' noted John Hillaby, 'these water-stores
remind you of moulds into which molten metal has been poured. If
Longdendale were turned upside down it would leave, you imagine, a
mountain of solid silver.' A nice fantasy to sustain me the next day
through the Pennine Way's last southerly gasp. On Bleaklow Head (bleak,
low and a long way from anywhere) a famous old couple lie by the path,
the Wain Stones, a pair of gritstone boulders with troll-like 'faces' pressed
together in a fond kiss. Just off the path a little further on an aero-engine
lay in a scatter of aluminium shards. A USAF Boeing B29 Superfortress,
an enormous bomber carrying eleven crew and two passengers, crashed
here on 3 November 1948, bursting into flames and killing everyone on
board. A modest plaque on a boulder recorded the incident. It's hard to
credit that such a large and well-equipped aeroplane could have come to
grief on these moors, low in profile and broad in outline as they are, until
you factor in the low cloud that hid the ground until the last moment.

Across the Snake Pass, a road notorious for fogs and blizzards, and a
final pull over Featherbed Moss, guided by a gappy line of stakes, to the
long gritstone cliff at the edge of Kinder Scout. I followed it down for a

couple of miles, the gritty path crunching underfoot, admiring the goblins, dogs, chef hats and shark fins carved by weathering into the black outcrops. By the waters of Kinder Downfall I sat down to peel an orange, ease my aching feet with an ice-cold dip, and look out towards the distant blue line of the hills of North Wales, with murky Manchester in the middle distance. I remembered walking here some thirty years ago with one of the city's most remarkable sons. Benny Rothman, a diminutive ball of energy and political passion, had organized and led the Kinder Scout Mass Trespass of Sunday, 24 April 1932, the most famous event in two centuries of struggle by English walkers to open all the country's moorland and mountains to everyone who wished to wander there. At eighty years of age, he'd been more than willing to climb to Kinder and show me in person just what had happened there all those years before.

The hero of 1932 proved the very best sort of walking companion – vigorous, funny, passionate in his views (still way out to the left), emanating energy, warmth and well-being like a pocket radiator at full blast. It was like walking with Moses on the mountain, or going sailing with Horatio Nelson: living history at one's elbow, the pure drop from the very horse's mouth.

'I was twenty and unemployed,' Benny told me as we tramped the moor. 'It was the time of the Great Slump, remember, and none of us had any money. We were members of the British Workers' Sports Federation, all of us young socialists or communists. There were three cheap things we could do to enjoy ourselves: cycling, camping and rambling.' Benny halted and pointed back in the direction of the distant city. 'Tens of thousands of ramblers used to flee Manchester on buses or cheap trains or just on foot at the weekends at that time. The trouble was – out of about a hundred and fifty thousand acres of moorland and high country on our doorstep in the Peak District, less than one per cent was open to the public. All the rest was owned by water companies who had reservoirs on the moors, or by landowners with gamekeepers – with sticks and dogs, and some with guns. We felt we just had to do something to gain access for all, to make our point and force ourselves to be heard.'

What the young Manchester and Sheffield BWSF members did was to organize, and publicize in advance, a mass trespass on to Kinder Scout, the high edge of moor that rises between the two industrial cities. In the quarry near Hayfield the trespassers assembled some five hundred strong to hear an impromptu speech from Benny Rothman. Then they climbed the green rocky staircase of William Clough, spreading out in a long wave across the private land on both sides of the clough. The keepers and water company bailiffs met them and tried to turn them back, but they were massively outnumbered. Up on the top at Ashop Head the trespassers stood, looking about. For a few minutes they were lords of the high moors. 'We met a party of ramblers who'd come up from Sheffield on their own trespass,' Benny told me. 'We all had a little victory meeting and some speeches. Then we all just went back down in good order – and then I got arrested.'

The newspapers had a field day, reporting hand-to-hand combat in the heather, fights between ramblers and gamekeepers, broken crowns and 'mindless violence'.

'Nonsense!' was Benny Rothman's crisp retort. 'One keeper twisted his ankle, and the rest of them just got shoved aside. But that didn't stop the police collaring me and five of my friends after the Trespass.' The six arrested ramblers received various terms of imprisonment. Benny got four months for riotous assembly and incitement. 'Oh, yes, Leicester Prison was really tough! I put on a stone in weight, and learned shorthand!'

When the jailed Manchester ramblers emerged from their cells, they found they had become heroes and figureheads for the burgeoning 'Right to Roam' movement. Ewan MacColl penned a song, 'The Manchester Rambler', to celebrate their fiery spirit. What the Manchester ramblers and their 'brothers and sisters of the boot' yearned for eventually came to pass. The Ramblers' Association was formed in 1935 to lobby for access. In 1949 Britain's National Parks came into being; the Peak District was the first one to be designated, with access to Kinder Scout and many other moors guaranteed and safeguarded. The National Trust with its open-access policies acquired much of the mountain and moorland around the industrial cities of South Yorkshire and Lancashire. And half a century later, after ceaseless lobbying, the CRoW (Countryside and Rights of Way) Act of 2000 enshrined what Benny Rothman and his colleagues fought for all those years before: the right of ramblers to roam at will over the mountains and moors of England and Wales.

At Kinder Downfall a wavering path comes in from the northeast. This was the original line of the Pennine Way, a purgatorial plodge across the peat hags, bogs, uncountable streamlets and trackless wastes of Edale Moor, now attempted only by those who relish a peat-bath. Sane south-going walkers follow the dramatic gritstone edge down to Edale Cross, a stumpy ancient wayside cross in the crook of a wall, before swinging off east for the cobbled descent down Jacob's Ladder and the easy valley walk into Edale.

Before leaving Kinder I took a long look around. It seemed as though, after a week's trudging on the gritstone, another geological step-change was imminent. Behind me ran the long black gritstone edge and the flat dark plateau of the moors. Ahead, beyond the compass of the Pennine Way, sandstone and shale were stacked in Yoredale-like bands up the flanks of Mam Tor. And further south again, curved slopes rose to a swelling green upland. There was no mistaking the shapes and the colours. Here was limestone country once more, the White Peak of Derbyshire. The BGS Geology Viewer confirmed it, the dayglo green of the gritstone where I sat giving way to the turquoise and pale blue of limestone a morning's walk away.

The Pennine Way with its volcanoes, granite and dolerite, its lead and coal mines, limestone dales, industrial decline and harsh millstone grit, trailed out and away like a dream as I put on my peat-spattered boots and headed downhill for Edale.

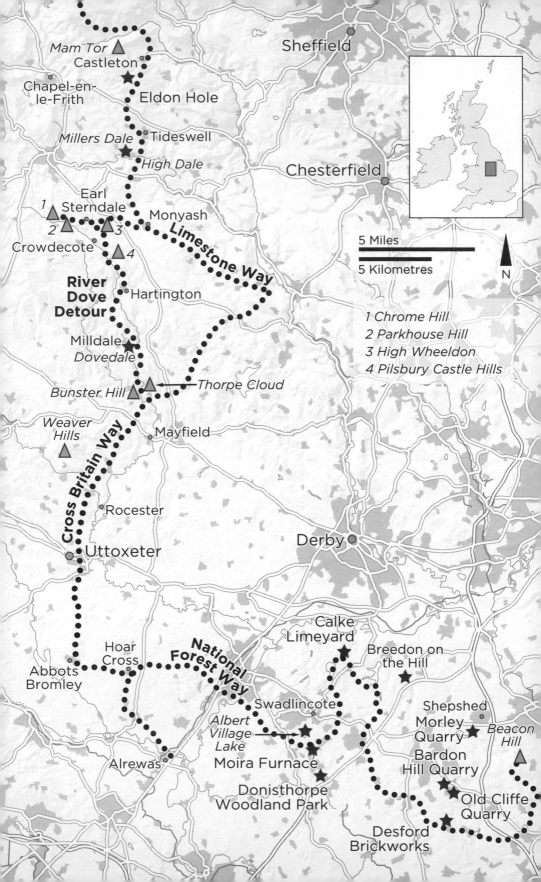

Mam Tor
Castleton
Chapel-en-
le-Frith
Eldon Hole
Sheffield
Tideswell
Millers Dale
High Dale
Chesterfield
Earl
Sterndale
Monyash
1
2
3
Crowdecote
4
**River
DOVE
Detour**
Hartington
Limestone Way
5 Miles
5 Kilometres
N
1 Chrome Hill
2 Parkhouse Hill
3 High Wheeldon
4 Pilsbury Castle Hills
Milldale
Dovedale
Bunster Hill
Thorpe Cloud
*Weaver
Hills*
Mayfield
Cross Britain Way
Rocester
Derby
Uttoxeter
Calke
Limeyard
Breedon on
the Hill
Hoar
Cross
**National
Forest Way**
Shepshed
Morley
Quarry
*Beacon
Hill*
Abbots
Bromley
Swadlincote
*Albert
Village
Lake*
Bardon
Hill Quarry
Alrewas
Moira Furnace
Old Cliffe
Quarry
Donisthorpe
Woodland Park
Desford
Brickworks

13

White Peak and 'Black to Green': Mam Tor to the National Forest

'IF YOU WANT TO SEE exactly where the gritstone and the limestone meet, I'd head for Mam Tor.' So said my godson Andy Harrison, a professional geologist who knows what he's talking about and how to put his subject over to a complete amateur. I took his advice.

Mam Tor, an upstanding ridge with peaked ends a mile south of Edale, makes a perfect grandstand for surveying the spot where gritstone hands over to limestone. On a fantastically windy morning at the start of a beautiful hot spell of weather I set off early from the summit down the paved path along the ridge, a swooping dinosaur spine rising to the peak of Lose Hill two miles away. Looking north the view was towards the moorland country through which the Pennine Way had brought me, a landscape of steps and shelves leading off and up to the sharply cut gritstone edge of Kinder Scout and the boggy moorland of Bleaklow and Black Hill, the Dark Peak encapsulated. Turning the other way, I saw below me a wide valley rising to the south towards a green plateau, gently domed. The slope that led up to the plateau, gashed by the Winnats Pass, is the northern boundary of the great limestone dome, an ancient seabed uplifted, that forms the upland known as the White Peak. A scatter of limestone quarries had hollowed the slope into neat grey concavities, as though with an ice-cream scoop.

West and east of this limestone upland, but not seen from here, are barriers of sandstone and shale, composing Shining Tor, Shutlingsloe and the Roaches to the west, Eyam How and Abney Moor to the east, with a north–south line of gritstone edges famous among climbers. South of the White Peak is shale and sandstone too, the lower lands of the Midland Plain laid down some 50 million years after the limestone. In fact, as the BGS map graphically shows, the White Peak is an island upstanding in a sea of sandstone and shale, its pale limestone lapped round on every side by the coarser and darker rocks.

The Derbyshire limestone dome, now between 200 and 300 metres above sea level, is part of the great Pennine Anticline. The White Peak (it should really be the Pale-to-dark-grey Peak) is quite small considering its landscape importance and the hold it has on walkers, climbers and holidaymakers. It's about 10 miles across from east to west and about 25 miles from north to south, a karst landscape, where soluble rocks of limestone have been uplifted, cracked and riddled with caves and sinkholes through which surface water filters into a tangle of underground passages that act as a drainage system for the area. The limestone formed at the bottom of the warm, shallow sea that covered Britain during the mid-Carboniferous period some 330 million years ago, and it's rich in the fossils of the marine life of that era – crinoids or sea lilies, shelly creatures such as bivalves, and armoured trilobites like enormous woodlice. Towards the end of the Carboniferous period, coarse sandy deposits from eroding mountains washed over the limestone, forming layers of gritstone interleaved with shale. Geothermal fluids injected mineral veins – copper, lead, barytes, fluorite – through the buried limestone. Around the same time, widespread upheavals took place as supercontinents collided to complete the formation of their most recent conglomeration, Pangaea, resulting in intense subterranean pressures that caused the region to rise in the form of a dome. The sandstone and shale capping eroded away on all sides, leaving the uplifted limestone exposed. Glaciers and their meltwater sliced deep into the karstic limestone, carving out the classic dales of the White Peak with their snaking depths and steep sides.

*

There could hardly have been a clearer demarcation between Dark and White Peaks than the view north and south from Mam Tor, unless it was down in the vale to the southeast where the tall chimney and dusty buildings of Breedon Hope Cement Works rose from the valley floor. The works is sited exactly where shale meets limestone – so precisely, in fact, that these two components of cement are dug out of vast quarries that lie cheek by jowl, separated by an isthmus of land not half a mile wide where the works buildings stand. Up to 1.5 million tons of cement a year come out of the works, Britain's largest, and it provides two hundred jobs, most of them local. It is also obliged to take extra care with its conservation and restoration schemes, given that it lies within the boundaries of the Peak District National Park. This entity was undreamed-of when the factory was founded in 1929. However, after only fifteen years of quarrying the local authorities could see for themselves the scars already inflicted, and the disastrous landscape damage that would ensue unless strong and imaginative action was taken sooner rather than later. Luck and good judgement threw up the right man for the job: landscape architect Geoffrey Jellicoe (1900–1996). In those days the brief of landscape architects tended to stop short at garden design, but Jellicoe (later knighted for services to landscape) had a much wider vision and firm ideas as to what should be done – first to minimize visual impact by disguising as much industrial operation as possible with existing landforms, and then to enhance the site with lakes, landscaping and planting. These are standard methods of quarry restoration these days, but they were bold initiatives back then.

A succession of landscape architects has kept the company to its pledges, and nowadays Breedon Hope is a showcase for site restoration that goes hand in hand with the quarrying. It includes relocating the clay and soil stripped from the still active East Shale Quarry to remodel the shape of the disused West Shale Quarry and create a mosaic of wetland, marsh, woodland and ponds where natural recolonization by plants and trees can take place. Plans for East Shale Quarry are to use its own soil and clay to re-form it as a lake with islands, shallows and slopes, rather than the usual deep lake with sheer sides, while the huge adjacent limestone quarry has already been restored in part to woodland and rough grazing. Looking at the site from the ridge I could see the chimney and the ghost-pale buildings rising from a green collar of trees in the distance, and it was good to know what was being done down there to mitigate the damage.

The wind tugged and shoved, causing walkers on the ridge to stagger and their hair to go flying. Close at hand was evidence of massive landslips, the east face of the hill dropping away at a very steep angle through a series of rock layers to a boulder field at the bottom. The narrow path dipped to Hollins Cross where an old coffin road once crossed the ridge, giving passage for burial parties from churchless Edale over to Hope Chapel. Here I dropped back down the southern flank of the hill and through the woodland around Mam Farm to come to the eastern foot of Mam Tor. At the base of the natural amphitheatre formed by the landslips lay a jumble of sandstone boulders; the shale that had fallen with them, being softer, had eroded away. From here the rock banding was clear to see as it rose to the peak of Mam Tor, a stack of rocks with a fatal weakness at its heart.

Mam Tor's fundamental instability is well reflected in its local name: the Shivering Mountain. It is a stack of turbidites, rocks made in a similar way to James Hutton's iconic greywacke cliffs at Siccar Point. Sediment was carried from eroding mountains by a river to settle in coastal shallows; then, destabilized by an earthquake or a large influx of new material, it slid on down the slope to settle in a fan on the deep ocean floor. The layers of sandstone that make up Mam Tor are interleaved with beds of more easily eroded shale, essentially compressed mud from the ocean floor, all the way down to their foundation on a bed of Edale shale. This weak and easily fractured rock has to bear the weight of several layers of stronger, heavier sandstone, a geological structure very prone to landslips. Erosion by Ice Age glaciers steepened the southeast flank of Mam Tor to a point beyond stability. Post-glacial erosion gradually increased the fragility of the slope, and the rocks were cracked as water that had infiltrated fissures repeatedly melted and froze, contracted and expanded. Some four thousand years ago the rocks began to tumble in a series of falls that still occur at intervals today. When a new road was built between Castleton and Chapel-en-le-Frith in 1819 it was routed directly below the slip area. All through its existence it needed frequent repairs as the ground beneath it shifted and slid away, particularly after heavy rain. In 1974 a big landslip took a sizeable chunk out of the road, now named the A625, and five years later the decision was taken to close it for good. History's wheel turned, and the narrow old packhorse track from Castleton through Winnats Pass to Chapel-en-le-Frith, which had been superseded by the A625, was restored once again to its pre-eminence as the only route between the towns. The fractured segments of the former main road below Mam Tor have become the playground of stunt bikers who leap the gaps for thrills.

I walked and clambered along the roller-coaster humps and hollows of the abandoned road. The central white line was still discernible in the scabby forty-year-old tarmac, eaten into holes as rainwater run-off has flensed it away from its stony foundations. Square pits in the white line showed where the cat's-eyes had been frugally removed. The tarmac had been twisted and pulled out of shape by the ground movements, in some places condensed into ripples that put me in mind of the wrinkled skin of lava flows called pahoehoe (but known to me unscientifically as elephant droppings) that I'd seen in the Canary Islands. Coming to a 6-foot drop the layers of successive road-mending were laid bare, tarmac a foot thick laid on a similar thickness of rough stones, themselves lying on layers of earlier tarmac and stones. The whole thing echoed to a remarkable degree the geological layers exposed in the landslip slope above.

The lowest layer of stones, dating back two hundred years, was salvaged from spoil heaps at the nearby Odin lead mine to build the original road. The lead veins were injected into the shales and underlying limestone beneath Mam Tor about 280 million years ago when mineral-rich hydrothermal fluids spurted up and along faults in the rock. Lead mining was good business here for centuries, as was mining for Blue John, a beautiful purple-blue mineral stone used to make decorative bowls, vases and other ornaments. This particular, much-prized banded fluorite originated at the same time as the lead from a saline fluid that intruded into limestone 2 miles below the bed of a shallow sea. Blue John is rare: it's only found under Treak Hill between Mam Tor and Winnats Pass, where it's still mined in Treak Cliff Cavern and the neighbouring Blue John Cavern.

To make sure of seeing the cream of the White Peak on these few cloudless days I decided to set course southwards from Castleton with the Limestone Way as a guide. This long-distance path snakes through the best of the White Peak's spectacular limestone landscapes, eventually burying its head in the swathe of much younger sandstone, siltstone and mudstone that rims the southern edge of the limestone dome. I had fixed up a rendezvous for the next day with Andy Harrison in Millers Dale some 10 miles to the south, and was looking forward to having my nose rubbed all over the hidden glories of this spectacular dale. First, though, a climb out of Castleton to the limestone plateau above. Chaffinches were twittering in the trees round the dark cleft of Peak Cavern, known to earthier generations as the Devil's Arse, in whose chill dark hole a whole tribe of rope spinners once squatted, their lungs and joints sacrificed to the dampness they needed to tease the fibres together. What desperate conditions our ancestors put up with, just because they had to.

The moor above was pocked with the lumpy ground, the heaps and hollows of lead mines long gone – Old Moor, Starvehouse, Clear-the-Way, evocative names for meagre pickings. Belland yard walls ran parallel with the workings: drystone barriers to keep roaming livestock away from the shafts and holes and the grazing that had been contaminated by lead. A mile's detour to the west across Access Land brought me to the brink of Eldon Hole, the largest pothole in the Peak District, with a hollow basin at its southern end where I crouched and peered

Above Old Gang lead-smelting mill and slag banks beside Mill Gill near Swaledale.

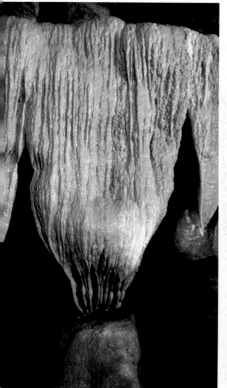

Left Calcite flow in Ingleborough Cave near Clapham.

Below Norber erratic: sandstone rock dumped by a glacier on much younger limestone.

Above The River Twiss carves its gorge down through limestone and shale near the Craven Faults.

Above Ice Age flood waters tore through the North Yorkshire limestone to form the 200-metre Gordale Scar.

Below Looking towards Malham from the rim of Malham Cove's limestone pavement.

Above Gritstone mill town: Hebden Bridge from the lane to Stoodley Pike.

Right 'Let me whisper in your ear': two gritstone trolls on Blackstone Edge.

Below The gritstone edge of Kinder Scout, looking south to Mam Tor and the limestone country of the White Peak.

Above Looking along the ridge path on Mam Tor towards the gritstone moors.

Left Landslips on Mam Tor have broken the road into 'geological' layers.

Below Coral, sea lily and shellfish fossils in a White Peak limestone wall.

Above Magma from an undersea volcanic flow bursts into the limestone at Litton Cutting.

Below Stegosaurus back of Chrome Hill, one of a pair of limestone reef knolls.

Left Unconformity at Morley Quarry, National Forest: Triassic breccia, 220 million years old, lies directly on Precambrian volcanic rock 400 million years older.

Below Stewartby brick made of 'the clay that burns', 160-million-year-old Jurassic clay.

Above Stewartby Brickworks, Marston Moretaine, Bedfordshire.

Below Tring Cutting, summit of the Grand Union Canal's passage through the Chiltern chalk.

Above Riverboat *Connaught* approaches Kew Pier in a low-level landscape.

Left Bradenham Church: geologist Allan Wheeler points out flint, limestone and sarsen.

Below Greenwich Peninsula: newly planted reed bed 'natural flood defence' looks across at the City of Oz.

Above Canvey Island's man-made flood defences, a hard sea wall 4 metres high.

Below Man works with nature on Wallasea Island's flood defences, the newly created Jubilee Marsh.

gingerly over the edge. You don't need to be much of a Freudian or a brimstone religionist to appreciate our ancestors' awe of this great bushy gash in the hillside, its rocky lips perhaps 5 metres apart, its walls clad with gleaming liverworts and mosses, chuting down into darkness.

Locals knew that anyone foolhardy enough to venture down the Hole on the end of a rope would either be scorched in the fires of Hell, or would be drawn up again a madman. 'What Nature meant', wrote Daniel Defoe in his *A Tour thro' the Whole Island of Great Britain* (1724–6), 'in leaving this window open into the infernal world, if the place lies that way, we cannot tell. But it must be said, there is something of horror upon the very imagination, when one does but look into it . . . It has no bottom, that is to say, none that can yet be heard of.' In 1771 the intrepid John Lloyd FRS made the first verifiable descent on two ropes, and estimated the depth at 62 yards (56 metres). As I knelt in the mud today and craned my head over the lip of the hole I caught a glint of metal far down in the rock walls, fastenings by which today's potholers let themselves down into the black cavern which the Devil may not yet have quite vacated.

Back on the Limestone Way I followed its rutted course down a series of steepening and deepening dales – Dam Dale and Hay Dale, Peter Dale and Monk's Dale, this last one a wooded gorge where the tops of the ash woods rustled with a wind I couldn't feel down in the shelter of the dale bottom. Scabious and harebells hazed the rough slopes with blue, and the fragments of shattered limestone pavement underfoot were treacherous with skiddy algae. It was one of those brisk afternoons when you never want to stop walking, but in the end I grew tired of having to watch my step every inch of the way and climbed up to Tide-swell and the bus to Chapel-en-le-Frith.

'The Carboniferous really was a mostly peaceful time,' said Andy Harrison with the ghost of a twinkle. 'A bit like the Great Barrier Reef is today: warm shallow seas, nice steady build-up of limestone not too far below the surface. A long smooth period from three hundred and fifty million years ago for the next fifty million. It was only at the beginning and end that volcanic intrusions disturbed the order of everything. But when they blew, they really blew.'

Geology is Andy's passion and pleasure. We don't walk together nearly as much as I'd like. But when we do, I come away marvelling at the immensity of geological events, the stately march of their processes, and the way that Andy can identify each line in the story from the minute clues left in the rocks by the edge of a bivalve shell or the ripple of an ancient estuary.

We were setting out from Millers Dale Station for our day's ramble through the Carboniferous limestone. This is prime territory to hunt for geological treasure, and as we moved off Andy set the scene for me. In those far-off days, seas covered much of Britain. The northern super-continent of Laurentia was moving south, the southern land mass of Gondwana was floating north, and the Rheic Ocean to the south of Britain was being slowly squeezed out of existence as the two contin-ents pushed inexorably nearer on their mutual collision course. Once they did finally meet, rifts opened in the hinterland where the crust was being stretched to splitting point, and the ever-waiting magma spurted up along fault lines into limestone bedding planes and out across the seabed, tumbling rocks and melting stone and shoving hitherto hori-zontal limestone beds into new positions and angles.

It was a straight climb out of Millers Dale, up through the trees on an interminable Jacob's Ladder of wooden steps. Once at the top, though, we were out in the grazing pastures so typical of these limestone uplands of the White Peak, all rolling countryside, pale stone walls and steep lit-tle dry dales. It was cool in the shade of the ash and beech trees along the rim of Millers Dale, where cowpats dropped by sheltering cattle were pockmarked with the feeding holes of *Scathophaga stercoraria*, the num-berless golden dung flies making the most of their two months of life.

Out beyond the trees the baking hot summer day struck home in a breathless shimmer. We followed a walled lane lined with verges of scabious and lady's bedstraw, where a fritillary butterfly fluttered by too quickly to be identified, a blur of orange and black. Cattle lay in the shade of the wall, their skins glossy with sunshine, too stupefied by the heat even to glance up as we went by.

At Lydgate Farm the limestone blocks of the farmhouse wall caught Andy's attention. Embedded in them were brachiopod fossils like straggly strings, slightly proud of the stone. 'That's the edge of the shell,' said Andy over my shoulder. 'It's a productid, a creature with a hinged shell from the lower Carboniferous, maybe three hundred and forty million years ago. It would bury itself in the mud of the ocean bed so that just the tip of its shell was showing, then open up a fraction, wide enough to let its feeding tendrils trail out into the water and grab any food particle that floated past.'

I ran my fingers across the productid fossil as though that action could somehow narrow the immense gap between then and now, between the haphazard sea feeder burrowed into mud that would harden to become its limestone tomb, and this man who breathed the oxygen of open air and walked upright across the outermost skin of an utterly different world.

The lane wall behind us offered surprises too. One of the top stones looked suspiciously rough and nubby. Tiny dark cylindrical protuberances stood up out of it, like the bristles of a two-day beard. A coral, or rather the feeding tubes of coralline creatures that shared the Carboniferous shallows with the productids when this land was sea and the rock beneath my feet was a slimy floor of mud, the graveyard of uncounted millions of calcareous beings, slowly piling up, slowly pressing down, slowly forming the grey dales limestone. Quite a lot of that limestone, in the form of dust, suddenly blew into our faces as five motorbikes raced by, their riders helmeted and masked into insectoid anonymity, doing wheelies and skidding to a halt in clouds of exhaust fumes and stone dust.

A dappled path led away from Lydgate Farm and the insouciant bikers, down across cattle pastures and hayfields to the mouth of High Dale, a little dry valley between tight folds of ground, an insignificant side dale off the radar of most Peak District visitors. 'An underground watercourse,' Andy said. 'You have to imagine it forming a cave or tunnel. The water went on digging down till the whole thing collapsed and the sides and roof fell in, forming the dale.' Sinkholes beside the path reinforced the image of fragile ground whose chemical interaction with water could cause such dissolution and collapse.

The slopes of High Dale were a nursery for lime-loving plants, especially the yellow spectrum – dandelion-like hawkbit, buttercups, tall frothy spikes of lady's bedstraw, kidney vetch, beloved of the larvae of the small blue butterfly; and clusters of bird's-foot trefoil in yellow and orange, recalling its two old country nicknames of 'bacon and eggs', and, more pungently, 'granny's toenails'.

A drystone wall ran along the bottom of High Dale, built of irregularly shaped chunks of limestone, now growing ragged and beginning to sag. With my eyes opened to the potential of walls like this as fossil banks, I began to notice dark shapes in the grey rock. Here a stone crammed with the C-shaped lips of productid shells, a crusty graveyard of the creatures; there the imprint of a fan, a delicate spray of coral with each individual frond disjointed. 'And what's this?' said Andy. He indicated a pale rib of rock running across the surface that reflected the sun like polished glass. 'Ah, a chunk of chert. That was a sponge-like creature or a form of marine plankton, with its skeleton transformed into silica. Like flint, but it was formed during the Carboniferous, three hundred and fifty million years ago. Flint's the same process, really, and much the same rock, but formed in the Cretaceous when all the chalk was being laid down, so only about a hundred million years ago.'

I fingered the smooth grey-green rib of chert and savoured that 'only' – 100 million years, a geologist's eye-blink of time.

The hot weather of the July afternoon was bringing out a tremendous hatch of fritillaries. In the lane above High Dale, going homeward, I spotted one keeping still on a thistle head long enough for the greenish blush of its underwing to declare it a dark green fritillary. Less beautiful, but just as fascinating, were the curious grey grubs that crawled on the fern-like leaves of Solomon's seal. Little grey cylinders with black faces, they had chomped a large hole in each leaf at exactly the same place, leaving a line of identical gaps all the way along the plant. 'The Very Hungry Caterpillar!' Andy and I exclaimed simultaneously. That informative Mr Google helped us out – they were Solomon's seal sawfly larvae, habituated to feed in large gangs and capable of stripping one of the big plants of its leaves in short order. The adult is a dull black cousin of the wasp and nothing to write home about; it's the rapacious young- sters and their symmetrical feeding that have you wondering how Solomon's seal has survived against such odds.

We descended the long steep slope of Priestcliffe Lees Nature Reserve among ranks of common spotted orchids, the pale purple flowers just going over. There was a tremendous view down into the deep gorge of Millers Dale with the grey bluff of a river cliff bulging outwards. On the quiet warm grass of Priestcliffe Lees I could willingly have slept the day away.

The ground was pitted with the hummocks and holes of lead mining. On the spoil banks flourished fragrant orchids with a subtle scent. 'Vanilla,' said my flower book, but I got a spicy whiff of pepper and cloves. Low in the grasses grew tiny five-petalled white flowers, their stamens upstanding – the metallophyte spring sandwort, tolerant of heavy metals in the ground, and last seen in Upper Teesdale among the lead-mine heaps on Cronkley Fell.

Down in Millers Dale we turned along the old railway on its hillside ledge, now the Monsal Trail. Cyclists whizzed past; walkers strolled in family groups. The navvies who hacked out the railway in the 1850s left a series of cuttings whose rough walls are fractured into squares and oblongs by bedding planes in the limestone, the southern wall stained brick red in places by iron and other minerals leaching out from on high.

All of a sudden, the brick red and ashen grey of the limestone gave way to a stretch of cutting that looked as though a giant bucket of boil- ing tar had been chucked upwards across the rock. This great sinuating splash of black was full of big round boulders of limestone. What on earth happened here?

'The Carboniferous, that's what happened,' said Andy.

At that time, about 350 million years ago, the Pennine Basin where we were standing was a shallow bowl, usually flooded. The supercontinent of Gondwana and the great craton or land conglomerate of Laurentia were pushing towards each other, and the Pennine Basin was getting severely squeezed in between these two opposing giants. Land was crumpled and pushed up and down, stretched and thinned. The waiting magma pressed up from below against weaker points of the sea floor, intruding into cracks and eventually bursting out in a volcanic flow across the seabed. The tarry spill in the cutting wall was what remained of the tongue of magma. As it made contact at several hundred degrees centigrade with the cold seawater, the surface of the lava cooled and fractured on the instant; the round blobs of rock stuck in the tar were formed when the tongue's hot interior congealed inside its hard crystalline casing. Now it licks across and up the cutting, petrified violence laid bare by the navvies' picks.

After this undersea drama was over, the limestone continued to accrete on top for many millions of years. Further on, a gap in the bushes gave a prospect north over the River Wye to a river cliff a couple of hundred feet tall, its face smeared with bluish mineral stains leached out from the cracks of bedding planes. Iron and lead were mined hereabouts, but the principal product was the limestone itself. Along the Monsal Trail, beside a viaduct over the Wye, a rank of tall limekilns, opened in 1880, stood back against a cliff above the railway. Four low arched openings pierced the monolithic face of the kilns. In the dark at the back of each drawing tunnel stood a double 'eye' or hatchway, from which the quicklime had to be drawn out. Skin burns, lung itch and eye irritation from the intensely caustic burned limestone were commonplace among the kiln workers, but quicklime was in demand for the steel industry and for agriculture, so a few burned boys were neither here nor there.

In the end, geology closed the kilns. The rock face into which the kilns were built turned out to be bedded on a tongue of that seabed lava.

The weight of the kilns and the penetration of water turned the footings to clay. The rock face became unstable, the kilns began to slip forward away from their anchoring rock. By 1944 they had closed.

Back at Millers Dale Station I bade Andy goodbye. He'd filled my head generously with rocks. Next day I turned them over, trying to spot the protuberant lips of productids and the greasy glint of chert in the limestone walls of Long Lane and Green Lane on my way down to Monyash. A man in industrial mittens was at work on the wall in Green Lane. 'Self-praise is no recommendation, I know,' he remarked wryly, with his eye on a nearby heap of stone. 'But if you were to tip a barrowload of that stuff in the road here now, I'd be able to tell you where every single piece should go in this wall. You've to choose your "throughs" carefully. They're the ones to hold it all together. Other wallers – I've noticed their poor coping and jointing. I do take a pride in my work on these walls – and in my new hips, they are a delight!'

I wondered how he was able to square up his work so precisely, and he gestured along the piece he was working on. 'If you'd put a ruler in my hand I'd not have been able to keep the line straight wi' it. But let me just put my leg against the wall' – suiting action to words – 'and get set, squint along like, and I'll keep it dead in line.'

Sitting under the wall a little further along, I laid the OS and the BGS maps side by side. There seemed an awful lot more limestone scenery of the same sort in my immediate future as the Limestone Way took a great bulge eastward across the dome. But there was an alternative. Five miles or so to the west I could see on the map the River Dove winding southward through a series of dales at the westerly edge of the plateau where the limestone met the shale and sandstone, with a string of footpaths shadowing it for 15 miles all the way to the bottom of the White Peak's showpiece cleft, Dovedale. If I turned off the Limestone Way now to follow the Dove, I'd be walking in a landscape of knolls and reefs among the narrow dales. It looked enticing. And there below Dovedale the map showed the Limestone Way cutting in again from its eastward wanderings, ready to carry me on south to its terminus near the Weaver Hills where the Carboniferous limestone of the White Peak makes its final stand.

It was an afternoon's work to cross to the Dove by way of Blackwell Lane and Hutmoor Butts. I passed the beamy old Bull i' th' Thorn inn, now all scrubbed up and more accommodatingly styled the Bull Yard – I'd always wondered how many drinkers failed to find it through inability to pronounce the former name ('All right, lad, I can see you've had enough already'). Beyond the inn I turned northwest up the High Peak Trail where the Cromford & High Peak Railway once ran a cranky operation of ropes, pulleys and hard-working horses to haul the wagons of coal and limestone up its ludicrously steep inclines. At the east end of the massive delving of Dowlow Quarry I crossed the end of a band of olivine basalt, an unbroken line of volcanic rocks forming and keeping the high ground. The BGS Viewer represented it as a very beautiful pink ribbon, rippling north to the edge of Calton Quarry, then undulating east as far as Taddington, before turning west near High Dale with a final flourish north to Priestcliffe and Millers Dale. There, only yesterday, Andy Harrison had pointed it out as a tarry smear in the railway cutting wall and brought it to life in my imagination, a tongue of boiling magma cooled and crisped in an instant as it flicked out into the tropical Carboniferous sea.

Down in Earl Sterndale I found the Quiet Woman pub. No beer, no cheer: the Quiet Woman had been closed and silenced for good. Outside still swung the fading sign of a headless woman, blazoned with the motto 'Soft words turneth away wrath'.

From here it was a meandering couple of days on or around the infant River Dove as it probed southward along the boundary of the Bee Low limestone to the east and the Bowland shale to the west, keeping just within the margin of the slightly lower shale. Leaving Earl Sterndale, the very first view I got was a breathtaker, looking west from Hitter Hill into Glutton Dale. It looked as though a pair of spiny-backed stegosauruses were walking nose to tail through the depths of the dale.

The twin limestone reef knolls of Parkhouse Hill and Chrome Hill rise dramatically from a basin of shale with the Dove sinuating at their feet. They were separated from their parent limestone plateau to the east when a fault caused the intervening shale to drop, and their abrupt northern faces follow the fault line in a steep plunge way down beneath the shale at ground level to meet the ancient seabed far below. I worked my way round by field paths to the northwest side of the knolls. From down in Glutton Dale these rugged hills had appeared formidable, but the climbs to their rocky summits were not as knee-cracking as they looked, though there were slippery bits and steep drops. The views eastward towards the craggy limestone uplands and west over smoother, more separated sandstone hills were more than worth the sweat.

Just south of the twin peaks I climbed over their rock cousin High Wheeldon, another great upstanding outcrop of reef limestone, steep and grassy, with a grand view back up the Dove to the long gritstone bar of Axe Edge where the river rises, and the broader valley around High Wheeldon through which the Dove snakes south towards the plateau and returns to a steep and deep course cut through the limestone.

Down at Crowdecote I met the river and followed its tight succession of loops and curls along the dale, a low bank of sandstone, siltstone and mudstone on my right, the steep slope of a limestone apron reef on my left, scabbed with grey crags and rising to the limestone uplands that stretched away north and east beyond sight. The path led past the crumpled hillocks where Pilsbury Castle once stood – a pair of baileys and a motte, built by the Normans shortly after the Conquest to levy tax on the traders that used the ancient packhorse route through the dale. Like the upthrusts of Parkhouse Hill, Chrome Hill and High Wheeldon, these more modest bumps were apron reef knolls, gently sloping down from the margin of the original land, now islanded in the shale that separated them from the limestone dome of their origin.

Below Pilsbury the Dove continued its compact meanders along the geological border to Hartington and on south to Beresford Dale, where it suddenly went rogue, forsook its easy bed of Bowland shale, and veered off southeast to cut Dovedale's enormous wriggle of a gorge down through the limestone. The explanation for this seemingly perverse adoption of the steep and rugged pathway lies in a process called 'superimposed drainage', in which a river will switch to a new course that mirrors an older one lying below like a subterranean shadow. In the case of the Dove at Beresford Dale, it is simply following the course it once took through the gritstone and coal measures deposited on top of the limestone shortly after its formation, which have long since eroded away.

In 1653 the old London ironmonger Izaak Walton published *The Compleat Angler*, his best-selling paean to the joys of fishing. 'Viator, the Wayfarer' was the name he gave himself in his book, and 'Piscator' was how he referred to his young friend and angling companion Charles Cotton of Beresford Hall, just up the dale. Viator and Piscator came often to fish the Dove. 'The finest river that I ever saw,' declared Walton, 'and the fullest of fish.' Gentle Walton with his sincere Christianity and the spendthrift gambler and scapegrace Cotton were an unlikely pair. But when they were together beside the Dove, plying their 6-yard hazel rods and their horsehair lines dyed with beer and soot to fool the trout, the old and the young angler were as one.

Setting off from Hartington on a cloudy morning I dropped from Beresford Dale into Wolfscote Dale, ever steepening, ever deepening between jaws of Bee Low limestone 100 metres high. Scree had been loosened by frosts and erosion and had slid down the sides, long come to rest under a grassy cover on the more gradual gradients, jumbled in naked grey heaps at the foot of steeper slopes down which it had not yet ceased to roll. Down at the junction of Wolfscote Dale and Biggin Dale I entered Dovedale proper, not much wider than either subsidiary dales but somehow feeling both deeper and darker, the path down in the bottom manoeuvring past ribs of rock and tree roots beside the river. The former milling hamlet of Milldale appeared round a bend, all in shadow and dwarfed by the gorge sides like a village in a Chinese painting. Two centuries ago lead was smelted and ochre extracted from iron ore in this shadowy hollow in the hills. The Dove ran orange, mills clattered, chimneys smoked. Standing on the packhorse bridge and surveying the silent settlement today, that all seems impossible.

Below Milldale, caves opened in the cliffs: Dove Holes with a central pillar like a uvula in its open throat; the natural arch of Reynard's Cave. Originally excavated by the Dove, they were left high and dry as the river carved its way down through the limestone to its present bed many metres lower down. The Dove swung this way and that, cutting its way out of the hills below spectacular limestone erosion features: the smooth inclined blade of Ilam Rock and the blunt bluff of Pickering Tor opposite one another; Tissington Spires and the Twelve Apostles; the bald knob of rock called Lover's Leap with its steps so polished by tourist shoe leather that the pale swirls of Carboniferous crinoids showed up as though displayed behind glass. When Dean Langton of Clougher ascended Lover's Leap on horseback in 1761 he may have leaned a little too far out from the saddle to admire the Spires or the Apostles, or perhaps the bare rock was slippery with rain. Whatever the cause, the unlucky cleric and his horse skidded over the edge and fell to their deaths. His lady companion fell as well; but her long hair became entangled in the branches of a tree and brought her up short. Someone climbed the tree and rescued her, 'much alarmed but not seriously hurt'.

Near the mouth of Dovedale a dipper was bobbing its bright white shirt-front on one of the stepping stones that crossed the river. I found the crooked path – another repository of fossils, these ones the hinged shells of brachiopods – that climbed the steep back of Thorpe Cloud above the entrance to the dale. From up there the prospect demanded a sit-down and a ponder. The cleft of Dovedale disappeared below in a fold of ground, and beyond stood the bare rock scars and tilted head of Bunster Hill. These twin guardians of Dovedale are the last of the Milldale limestone, and they mark the point where the great limestone plateau dips decisively southward. Here when all was under salt water the Carboniferous sea floor steepened its slope towards the deep muddy ocean basin known as the Widmerpool Gulf, and from here the lower lands to the south are composed of the Widmerpool Formation, a mixture of mud and limestone originally coalesced by a sticky film secreted by microorganisms before it hardened into sharp-edged rock.

All that remained of my White Peak journey was to follow the Limestone Way as it shadowed the River Dove a mile or so to the west. At Cuckoocliff Wood near Mayfield the limestone cliff of a knoll reef and a swathe of muddy Widmerpool Formation gave way to sandstone, and the horseshoe vetch and devil's-bit scabious of the limestone fields ceded to the long sombre heathland of Brown Edge, the pink sandstone of walls in the hamlet of Ousley and the conglomerate pebbles that studded the shaft of ancient Ousley Cross.

A mile to the west I came to the Weaver Hills, the last gasp of the White Peak limestone, three shapely green hummocks against a beautiful afternoon sky as blue as a thrush's egg. Looking north from the trig pillar at the summit it was all limestone heights and clefts. The southern slope of the hills, though, was a fine mishmash – the slope of an apron reef diving beneath rocks of the Triassic period (252 mya–201 mya), sandstones and siltstones deposited 100 million years later than the Carboniferous limestone seabed they rest on. My gaze roamed on south, over the great Midland Plain with its pastures, ploughlands and woods stretching away to a level skyline. The exhausted landscapes of the industrial Midlands, the search for coal and clay, iron and roadstone that had hollowed and wrecked them, and the remarkable reclamation project of the National Forest, all lay just over that horizon, a dozen miles away as the crow flew.

They were next in line.

Underpinning the gently undulating landforms of this corner of the central Midlands are coal measures, Carboniferous limestone with seams of coal. Below the coal lie thin layers of fireclay, otherwise known as 'seat-earth' (last met with along the East Lothian coast at Barns Ness), the fossil soil in which grew the Carboniferous trees, lycopods and the like, that were compressed into coal. Iron ore is here, too, in the sandstone that overlies the coal measures.

This region was thick with works buildings and with quarries, coal mines (opencast and deep), clay pits, ironworks, brickmakers. But from the 1970s onwards, old-fashioned practices and materials were overtaken by the modern age and the region slid into decline. By the late twentieth century three quarters of all Leicestershire's derelict land was concentrated in this one tiny sliver of the county, the handful of parishes round Albert Village and Moira on the south side of Swadlincote.

In 1995 the National Forest was inaugurated and the planting and remodelling began. 'Black to Green', they labelled the project. A long-distance path, the 75-mile National Forest Way, forms a great spidery 'M' with a straggly tail as it meanders across the thickly wooded landscape. Ten million young trees have so far been planted, many of them by local schoolchildren, across 200 square miles of derelict land where Derbyshire, Leicestershire and Staffordshire meet. Colliery and pottery ruins disappeared under young trees. Remnants of the old working life were rapidly cleared away. Recreation and environmental improvement descended on this most industrial of regions. Nowadays at Swadlincote, hub of the area, a ski slope slants down an old spoil heap, 'the UK's only Cresta Run!' In William Nadin Way the site of an opencast coal mine is occupied by huge storage sheds, business parks, distribution centres and a jigsaw of new housing estates. Old trades have new aspirational names – a brickmaker's at Swainspark styles itself a 'brick fabrication facility'. Houses and businesses float in a green and gold ocean of young trees.

Albert Village used to be known as Borra Nock, a place where everyone knocked on their neighbour's door to borrow a cup of this or a pinch of that. From the 1870s locals were working in Swadlincote's kilns or as clay diggers, tough lives that bred a tight-knit neighbourhood. Now Albert Village's huge clay pit is a lake in the midst of new woodland. The clay diggers have been succeeded by women with prams, runners and solitary men walking the dog. Strolling the lake shore I passed a huge bank of clay spoil, rusty brown with gorse bushes, the white clay showing in thick patches. Coot, dabchick, mallard and great crested grebe swam round an islet that swirled with nesting black-headed gulls, quarrelsome and screechy as is their way. Guelder rose, willow, ash, silver birch, dog roses and hazel were well established. Some of the trees, the silver birch in particular, had grown 40 feet tall since they were planted.

South Derbyshire fireclay was known to be the best in the country. T. G. Green's pottery in Church Gresley, next to the Albert Village clay pit, made 'Cornish' ware, whose trademark blue and white rings were known all over the world, and Moira Pottery, founded in 1922 a mile south of the pit, produced famous functional stoneware. The pottery, though, had the misfortune to be situated on top of the coal reserves of Donisthorpe Colliery. In 1972 the National Coal Board compulsorily purchased Moira Pottery and knocked it down.

Fireclay was only one of many valuable minerals hidden underground here. Ironstone, coal and limestone lay waiting to be unearthed and put to use. Moira Furnace smelted local iron in the early nineteenth century, and its hulking brick structure still stands, refurbished as a museum. But coal was the big success story, centred on Donisthorpe, just south of Moira. Donisthorpe Colliery started production in 1871, and averaged 300,000 tons of coal a year at peak production. In 1951 it won the *News of the World*'s national competition for 'Britain's best pit'. By the 1960s production had been pushed to over a million tons a year – this in an era when they were still using pit ponies. The mine did everything to maximize production, including pouring colliery waste into lagoons where the stone and clay would sink and the coal would float to the top for retrieval. But in 1990 the decision was taken to close the pit, even though the miners swore there was enough coal in the seams to continue for decades. The closure of Donisthorpe Colliery brought the coalfield to an end after 133 years.

In 1995 Leicestershire County Council bought the 36-hectare site from British Coal. Twenty hectares were planted with trees. The steep colliery slopes were regraded, drainage systems were introduced and gravel trails for walkers and riders laid down, surfaced with recycled stone from demolished colliery buildings and roadways. The lagoons were filled in, the pithead buildings levelled. It's a shame that none of the tall, gaunt and atmospheric pit architecture was retained. But the landscaping, the reclamation and tree-planting have all been beautifully done. The colliery site is now one of the greenest and leafiest in the region, a gentle wave of landscaped countryside.

My day's wandering ended in the heart of Donisthorpe's new woods with blackcap song trickling out into the evening air. On a grassy triangle of sparse grey rusty soil I found a clutch of tall and stately bee orchids with milky lilac petals, and was wondering if their ancestors had provided some miner's solace when a dog came up and dropped its rubber ball at my feet. The man who retrieved it turned out to be an ex-miner, stocky, sunburned, with tattooed arms and a strong local twang to his talk.

'I went down the pit', he told me, 'in 1982 at the age of sixteen, and worked here till it closed in 1990. I were at the coal face on one of those big circular cutters. Good work, good pay. And I loved every minute of it – *loved it*, I really did. No ill effects, though I did chop half me foot off when the roof came in one time. I were really devastated when this pit shut.'

'They were good times,' put in his wife. 'Good times. You knew everybody, a proper community. However, we'd to make the best of it afterwards.' She looked round her at the hazels and dog roses. 'One of my daughters got involved with planting these, so that helped.'

'Round here,' said the miner, 'it were all black, all pits and holes. The National Forest've done a good job, I'll give them that. But still – if I could've done just two more years, that would've been me for the full pension. Now I drive a bus three days a week to make up the money. If I'd to do that full time . . .' And he drew a forefinger expressively across his wrist.

Anywhere in Britain away from the coast, quarries are the amateur geologist's best buddy. They offer just about the only means by which to set eyes on what otherwise one can only imagine, the layers of rocks down deep below the surface. Over the next few days I fossicked around, using the zigzag course of the National Forest Way as a rough guide eastward, diverting off it to find gems of a remarkable geology exposed by a string of quarries. Red clay in great steps at Forterra Brickworks, Desford; the iridescent grey ponds, steep hummocks and canyons of Calke Limeyards, where limeburners once slaved at the kilns; the red and grey limestone quarry face at Breedon on the Hill with its wonderful Church of St Mary and St Hardulph teetering at the brink of the cliff. Inside the church were Anglo-Saxon carvings rescued from the monastery founded here some 1,300 years ago: broad-winged birds bowing and displaying; antlered deer fleeing through the thickets of Charnwood Forest whose ancient borders I had crossed.

There are not many places in England where igneous basement rocks of Precambrian age are exposed as they are here – a granite along the western edge 600 million years old or more, a central section of even older volcanic rocks, partly recrystallized under pressure from the rocks that overlie them, and in the east igneous diorites, complex rock formed from slowly cooling magma some 450 million years ago. There was little lava flow from the Charnwood chain of volcanoes; it was more a case of explosive outbursts and violent pyroclastic flows into the surrounding sea between 600 million and 500 million years ago. The hard rocks of Charnwood Forest have been in demand ever since Neolithic humans began to chip and smooth them for corn-grinding querns and stone axes. The Romans first quarried the volcanic rock for roadstone, and over the past two thousand years that demand has only grown.

It's hard to imagine that this volcanic material could contain evidence of life in the Precambrian world – so hard that in 1956, when fifteen-year-old Tina Batty told her geography teacher of a fern-like fossil she had spotted in a Precambrian rock face at Hangingstone Quarry near Woodhouse Eaves, she was told it was impossible. Shades of Sir Roderick Murchison when faced with the facts of older rocks on top of younger ones! Next year Roger Mason, another local fifteen-year-old, noticed the same fossil and took a rubbing of it, which his father showed to a geologist friend. It was the first Precambrian life form ever found, a seabed-dwelling creature which was christened *Charnia masoni* – though if everyone had their dues it should really be *Charnia battyi*.

Morley Quarry near Shepshed is a nature reserve these days; also a place of respite for local drinkers and smokers, three of whom were sharing a lazy afternoon's picnic on the quarry floor with a bottle, a baby and a beatbox. A hoarse hello issued from three throatfuls of dope smoke. I went to study the information board, though in fact the geology here was so explicit it hardly needed interpretation. The east face of the quarry was a dull grey mass of ancient Precambrian rock, fractured and lumpy through quarrying, veined across with spidery swirls of quartzy lines, orange, green and burned brown.

It was the south face and its unconformity that made me blink. Here was a smoothish wall of dark grey Precambrian, moulded during that period of volcanic activity some 600 million years ago, squeezed upward during a mountain-building period 200 million years later. Dumped on top, the joint an almost straight horizontal line, lay a bubbly, squashed-up red mass of Triassic breccia: a muddle of rock, partly fragments of the ancient rock, partly sand, deposited from an ancient desert some 220 million years ago.

The James Hutton revelation shone its light once more; the marvel of what had happened to all the layers that should lie in between, nearly 400 million years of depositions, seas, dry lands, Carboniferous forests, tropical coal measures, dinosaur fossils. How could 250-million-year-old rock be glued to 600-million-year-old rock with almost nothing to show in between? The answer, my friend, was blowing in the wind, settling in the seas and flushing away down aeons of rivers.

Three miles south of Morley Quarry, two active workings gave sensational views into the way the younger Triassic sedimentary mudstone, a startlingly vivid dark crimson, has eroded and been quarried away to reveal the ancient volcanic landscape it is wrapped around. At Bardon Hill I climbed on a pale clay track through National Forest plantations of oak and silver birch to peek over the quarry edge at the summit. The red Triassic rippled in long wavy layers, chopped into from below at 45 degrees by the Precambrian volcanic rock that appeared to be shouldering it aside and thrusting its way up to the open air. It took a moment to interpret what I was seeing: a two-dimensional section cut by the quarrymen through the mountainous shape of the ancient rock, violent in its origin and dramatically tilted, which had been cradled and smothered by the Mercian mudstone – the 'overburden', in quarrying parlance – in whose heart it had lain hidden till the quarrying brought it to light.

Down below were the tall, corrugated quarry sheds with their vaguely sinister air. A continuous shower of rubble fell off the end of a chattering conveyor on to an ever-expanding cone of stone in a haze of smoky dust. In Victorian times quarrymen would grow extra luxuriant moustaches to keep the dust out of their noses and lungs. I got the binoculars on to a group of ant-like workers round a digger 200 metres below, but couldn't make out whether masks or moustaches were their air filters of choice.

At Old Cliffe Quarry, just across the Coalville road, there was a grandstand view from a platform on the south rim. I found myself looking down into a giant circular hole half a mile wide, many hundreds of feet deep, converging towards the flat bottom in concentric terraces like a huge plughole or petrified maelstrom. On the northwest side the illusion of Bardon Hill was replicated, the ancient Precambrian rock appearing to rise in a dark grey surge like a gigantic wave, thrusting up through the overburden of rich red Triassic rock half its age to a peak at the top of Billa Barra Hill. The sense of motion, of a powerful force pushing upwards through weaker resistance, was overwhelming, and again I had to remind myself that I was looking across at an ancient igneous mountain sunk deep and static in a sedimentary blanket of mudstone and siltstone.

The National Forest Way, curling like a quarryman's moustache, gave a final northward twirl and came to rest at the summit of Beacon Hill Country Park. The title of 'Country Park' can lay a dead hand on any piece of countryside burdened with it: a smack of shaven golf course management and a suspicion of city authorities out to keep visitors off the flowerbeds. So it's especially encouraging when you find a place like Beacon Hill Country Park in the heart of the National Forest, managed by Leicestershire County Council with a light touch and a sensitive understanding of the natural environment.

On this sunny summer afternoon families were out in force on Beacon Hill, climbing the paths to perch on the ancient volcanic outcrop at the summit, the highest point in the county at 248 metres. These rough crags of squeeze-packed volcanic detritus, beautifully striped in creamy green, pink and black, were formed over 600 million years ago when the Charnwood chain of volcanoes spewed their ash into the surrounding sea. They were laid down in horizontal layers, then shoved on end some 400 million years ago. The children peeking over the edge didn't care – they just squealed and scampered all over the rocks.

I leaned on the topograph to admire the view. Quarries opened dusky red mouths in the green cover of the National Forest. Ten miles north, beyond Loughborough's sprawl, stood the waisted cooling towers at Ratcliffe-on-Soar Power Station, with the tiny clear towers of Lincoln Cathedral breaking the northeastern skyline 50 miles off. East was the way for me now, however, 15 miles along the long-distance path of the Leicestershire Round, past the quarry at Mountsorrel and the gravel pits in the Soar Valley, to reach the splendidly named village of Frisby on the Wreake by tomorrow evening. There on the Leicestershire Wolds I'd pick up the Midshires Way, penultimate link in this tremendous chain of paths, heading decisively south and east for the great clay lowlands and the chalky gateway to the River Thames and journey's end.

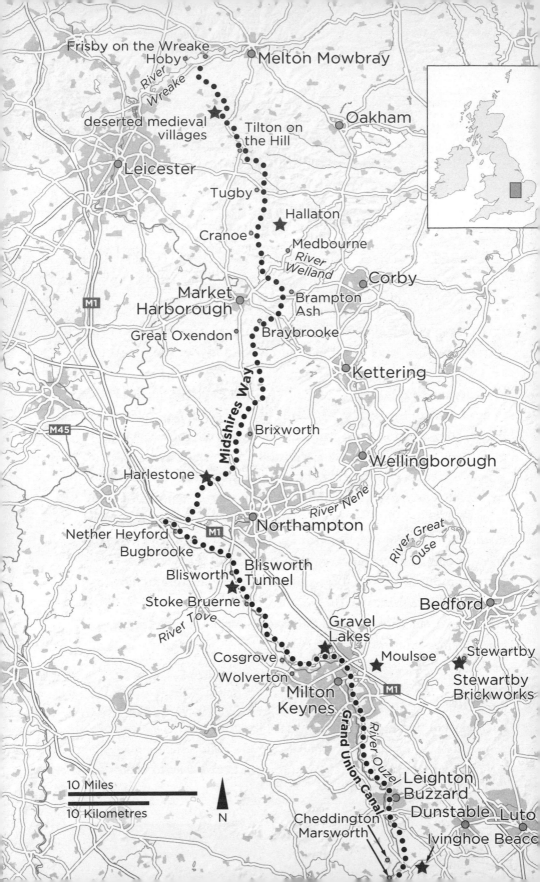

14

Bands of Bright Colours:
Leicestershire Wolds to the Chiltern Hills

FRISBY ON THE WREAKE ON a cool grey autumn morning, a winding village on a curving street. A couple outside the Bell Inn, doing the long-distance Leicestershire Round, winced as they pulled on their boots. I sat on a bench, spread out the geological map, and saw that my journey through the bones of Britain had finally reached the borders of a subterranean country whose structure it was possible to grasp at a glance.

Here they lay at last, the bands of bright colours that had transfixed me when I first stumbled on them in the *Philip's Modern School Atlas*, flowing like a band of silk scarves from northeast to southwest, succeeding one another in chronological order as the eye moved south towards the Thames Valley 100 miles away. Pink and grey Triassic (252 mya–201 mya); Jurassic (201 mya–145 mya) in mauve, orange, yellow and olive; Cretaceous (145 mya–66 mya) in turquoise and green; Palaeogene (66 mya–23 mya) in toffee brown and rose-quartz pink – geological periods that gained shape and substance over the course of that morning as I got notebooks, maps and the internet to speak to one another and to me. I pinned successive dates and changes in climate to each geological period through the enormous but comprehensible cycles of 200 million years, and saw beneath today's mild arable and pasture lands the oceans rise and fall, baking deserts give way to chalky seas, rocks harden and crumble, ice sheets advance like juggernauts and retreat like defeated enemies.

As the Carboniferous period (359 mya–299 mya) which dominates so much of the geology of northern England began to draw to a close around 305 million years ago, it witnessed an extinction event, the collapse of the Carboniferous rainforests. A mini-Ice Age, followed rapidly by a rise in global temperatures, may have been to blame, with the worldwide blanket of lycopod forests broken up into isolated 'islands' and unable to cope with the accelerated pace of climatic change. The Permian period (299 mya–252 mya) maintained the arid, desert-like conditions that followed the Carboniferous. The convergence of Gondwana and Laurussia to form the supercontinent of Pangaea had caused the great mountain-building episode to the south known as the Variscan orogeny, and for the first 30 million years of the Permian these Variscan rocks gradually eroded away. Later in the Permian the polar ice sheets melted, the sea eventually submerging most of Britain – though much of the East Midland area remained above water and submarine deposition of shale, sand, limestone and muddy marl resumed.

This shallow ocean, known as the Zechstein Sea, had evaporated by the end of the Permian, thanks to the arid conditions. A flat desert land emerged, but not one abundant in life. In fact, a catastrophic extinction event took place, starting at a time that has been quite precisely dated to 251.9 million years ago. What caused this event is uncertain. Perhaps it was a release of methane as the seas dried up, or an intolerable increase in aridity. There was a huge volcanic upheaval in Siberia at that time, too, which would have generated giant clouds of dust particles and carbon dioxide that blocked out the light and heat of the sun, sending temperatures plummeting, and also produced multiple billions of tons of sulphur dioxide, resulting in downpours of acid rain that killed land vegetation, marine plankton and the creatures that fed on them. Whatever caused it, 95 per cent of marine species and land-dwelling vertebrates disappeared, along with most insects. This event, a snap of the fingers in geological time, is known, crisply and graphically, as the Great Dying. It was followed by a new geological period, the Triassic (251 mya–201 mya), in its early stage still very hot and desert-like, with masses of reddish sand and gravel being dumped across the landscape by rivers flowing northwards across southern Britain from the eroded Variscan mountains to the south. In mid-Triassic times (247 mya–237 mya), Britain became a low desert plain dotted with shallow playa lakes, temporary bodies formed whenever water returned. Sands and muds were deposited at such times, and when the playa lakes evaporated great salt pans were left behind.

Life on Earth began to recover. There were no flowering plants, but mosses, ferns and horse-tails abounded, and early plant-eating dinosaurs like thecodontosaurus flourished. Sea life, too, surged back to health with the spread of marine reptiles such as long-beaked ichthyosaurs and lizard-shaped creatures with webbed feet called nothosaurs, as well as ammonites, bivalves and marine gastropods. Late in the Triassic, in a warm climate still tending towards low rainfall and occasional periods of intense heat, the Tethys Ocean was in the process of opening at the Eurasian and African edge of the supercontinent of Pangaea. The sea returned to deposit muds across much of Britain as a new era, the Jurassic (201 mya–145 mya), got under way with yet another extinction event. The supercontinent of Pangaea continued to break apart with an increase in volcanic activity, projecting sulphur and carbon dioxide into the atmosphere. Global temperatures rose, acid rain poured down in a repeat of the conditions that led to the Great Dying. Wildfires proliferated, plants struggled to photosynthesize as the sun's light waned, and around three-quarters of all land and sea creatures died out as the oceans acidified and food chains collapsed.

The Jurassic period saw yet another recovery of life forms. The plesiosaurs, long-necked marine reptiles, evolved warm-bloodedness and were able to evade extinction by spreading all over the world. Ichthyosaurs, though greatly reduced in numbers and variety, survived too. At sea-floor level crawled sea snails, sea urchins and crustaceans that were predated on by ammonites. Giant long-necked sauropods thumped through shallows and swamps, and the air above was stirred by the leathery wings of flying reptiles, the pterosaurs. By mid-Jurassic times ankylosaurus with its horned beaky head and the plate-backed stegosaurus were on land, and late in the period a huge meat-eating predator, megalosaurus, hunted the tropical islands of southern England.

Such were conditions in the Jurassic, and that was where in point of geological time I was joining the succession of colourful bands which stepped southwards across the miniaturized map of the *Philip's Modern School Atlas* and the rather more detailed one of the British Geological Survey. In point of place I was cutting into the succession at Frisby on the Wreake in the rolling landscape of the Leicestershire Wolds, underpinned with those late Triassic and early Jurassic mudstones and limestones. The long-distance path of the Midshires Way would carry me south from here through Leicestershire and Northamptonshire as far as the broad valley of the River Nene. There the towpath of the Grand Union Canal, pride of the Midlands, promised easier walking as it pointed a crooked finger down towards the Chiltern Hills, the chalk ridge that guarded the approach to London and journey's end out beyond the mouth of the River Thames.

From Frisby on the Wreake I followed the Leicestershire Round west along a pink earth path above the river. A flood shelf slanted down to the flat alluvial meadows around the River Wreake, whose wriggling course was marked by a line of willows. Flooded gravel pits glinted beyond. This is a country of churches on ridges, their towers and spires prominent in the landscape, and soon the stumpy spire at Hoby came into view.

Here I turned south and entrusted myself to the Midshires Way, whose 'MW' logo irresistibly brought to mind a low-slung bosom in a netted bikini top. Or indeed a brace of acorns. Whatever, these waymarks were to prove depressingly few and far between. The 230-mile path from the Chilterns to Manchester was completed in 1994 and, like many of Britain's enormous and invaluable network of public footpaths, looked today as if it could do with a bit of TLC. But the OS Explorer maps needed no scrubbing up; they were their usual reliable selves, steadily pointing out the line of the path through smothering maize, overgrown woodlands and sign-free farmyards.

Jurassic mudstone, grey and full of shells, lay beneath the Leicestershire fields. At Rotherby, All Saints Church was grey and yellow, a pleasing patchwork of this local mudstone with ironstone as crumbly as fairground honeycomb. A glance at the BGS Viewer around Brooksby with its crocketed church spire showed a lime-green streak of Blue Anchor mudstone from the end phase of the Triassic, zigzagging southwest like a cartoon flash of lightning, disappearing a few miles on, then reappearing as a tiny smear at its namesake place 150 miles away on the Somerset coast.

The path crossed the Wreake and rose between the hollows of sand and gravel diggings. To the west a rumbling elevator dribbled pale pink sand into a conical mountain, its slopes sculpted by water runnels. To the east a worked-out sandpit had been restored to nature, fringed by reeds, with mallard ducks sailing and a screen of poplars gently rustling in the wind as their leaves flicked away and spiralled down to rock on the surface of the pond. I followed the Midshires Way as it rose into fields where the path pitched over the low waves of medieval ridge-and-furrow arable farming, long gone under grass. A cream-coloured bull with shoulders humped with muscle exchanged love licks with a dappled grey cow. What a nice life they must lead on the lush grazing of the Leicestershire Wolds, and how greatly a walker with sore feet appreciates this mild country with its long, gradual descents into stream valleys and easy climbs to the following ridge.

Belts of sand and gravel are a feature of the river valleys from here on south. They were dragged and spread across the underlying mudstones, sandstones and limestones by Ice Age glaciers, as was the clay soil, thick and fertile, that blankets the uplands and grows such good grass and corn. All down the length of this journey through Britain I had been butting up against the effects on the landscape of the Ice Age, or as I now understood it, the several Ice Ages or periods of glaciation that the world has undergone in recent geological history. Ice and its concomitant meltwater gouged out the lochans in the Lewisian gneiss, ground down the mountains of Inverpolly, deepened the Great Glen, dumped the drumlins on Rannoch Moor and sculpted the Castle Rock in Edinburgh. It cut deep into the South Tyne Valley, changed the course of the River Swale, burrowed out the cleft of Gordale Scar, shifted and dropped the erratics of Norber and carved the limestone dales of Yorkshire and the White Peak. It has been a formidable landscape shaper and game changer across these islands. But where did it come from, and when, and why?

Milutin Milankovitch (1879–1958), a Serbian astronomer and climatologist, posited early in the twentieth century that small inconsistencies in the smooth orbit of the Earth around the sun might lie behind the planet's propensity to freeze and warm in long-drawn-out cycles. He pointed to the eccentricity of the orbit, which varies from circular to slightly elliptical in a cycle of around 100,000 years; to variations in the angle of its tilt over a cycle of about 41,000 years; and to the planet's slight wobble on its axis, caused by the gravitational pull and push of the sun and moon, in a cycle of about 26,000 years. Put together, Milankovitch suggested, these natural phenomena explain our climate's swings between extreme cold and great heat as the distance and angle between the Earth and the sun have altered, increased and decreased in regular cycles, sparking off and then reversing the series of Ice Ages which have gripped the Earth through half its existence. The world of science was sceptical at first, but has since fallen in line with Milankovitch's theories.

There are other causes, too. Periods of tectonic collisions and continental break-up both produce intense volcanic activity and push more rock dust into the atmosphere to block out the sun's light and heat. And when cratons collide and coalesce, the giant supercontinents they form can block the flow of warm water towards the Poles, causing ice sheets to form or grow. Unusually extended periods of low activity from the sun might have caused many of the mini-Ice Ages that the world has experienced.

Once an Ice Age starts the world becomes drier and more arid. There's less precipitation, thanks to a decline in evaporation and moisture in the air. Polar winds blow more dust into the atmosphere. The glazed white surface of the ice reflects back the radiation of the sun, increasing the big freeze. Ice sheets form at the Poles, glaciers in the mountains, as the temperature lowers.

The world has experienced at least five major Ice Ages. The Huronian (2.4–2.1 billion years ago; that is, halfway back to the origins of the Earth) was the first one we know of, and was probably linked to what is known as the Great Oxygenation, when primitive organisms spread oxygen all through the waters of the world as they began drawing energy from sunlight through photosynthesis. Before the Great Oxygenation the Earth's atmosphere was heavy with the powerful greenhouse gas methane which maintained much higher temperatures. But as the microbes that produced oxygen grew to dominate the world, the methane disappeared, the greenhouse effect diminished, and the world cooled.

The Cryogenian occurred next, between 720 million and 630 million years ago, with ice sheets advancing from the Poles to meet at the equator and create what's often referred to as 'Snowball Earth', retreating again when massive volcanic activity pierced through the ice to produce an atmosphere full of greenhouse gases and a warmer world. Next came the Andean–Saharan, 460 million to 420 million years ago, with cooling oceans aiding the spread of the ice. This glaciation probably triggered the first of Earth's five major extinction events, with something like 85 per cent of sea life being wiped out. It was followed by the Late Palaeozoic Ice Age around 360 million to 255 million years ago, probably linked to the Carboniferous period's vast increase of plant life on land that was emanating oxygen and absorbing the greenhouse gas carbon dioxide. Also, the fusion of Gondwana and Laurussia into the enormous southern continent of Pangaea blocked warm water currents and created colder summers and vast winter snowfields off which the sunlight bounced. Eventually the presence of fewer plants, killed off by aridity and the freezing conditions, reduced the oxygen content of the air and increased the levels of the greenhouse gas carbon dioxide, tipping the balance back towards a warmer world.

The Earth's climate remained more or less in equilibrium for about 250 million years until some 2.58 million years ago when the latest of our Ice Ages, the Quaternary, began with ice sheets spreading once more in the northern hemisphere. This Ice Age has been characterized by a cycle, as yet unfinished, of glaciations interspersed with interglacial periods like the one we are now in, when the planet does not revert to overheating but stays relatively cool. Ice sheets and glaciers have expanded and contracted, eroding landforms, dragging rocks miles from the place they were laid down, grinding them into clay and rubble which the ice, and the great meltwaters it has periodically released, have spread widely across the land. Rivers had their courses altered; lakes developed as the ice disrupted the subterranean drainage systems; winds picked up speed near the glaciers and spread silty dust far and wide.

What's coming next? On current models, we are six thousand years into a cooling period that will persist for another 23,000 years. But human intervention in the Quaternary glacial cycle, in the form of man-made greenhouse gas emissions from cars and ships, heating and lighting, agriculture, the cutting down of forests that absorb carbon dioxide and emit oxygen, and the pursuit of economic growth at any cost, may mean that our particular species, or indeed the Earth as a whole, never gets as far as the next glaciation.

The Leicestershire Wolds are fertile ground. Grass and crops grow in stiff dark clay. Beneath the fields lie Blue Lias and Charmouth mudstones (limestone interleaved with clay), the latter encroaching on the former and both laid down under the sea between 209 million and 183 million years ago as the Triassic period gave way to the Jurassic. In medieval Leicestershire this fertile clay grew rye and wheat, peas and beans, enough to sustain the isolated villages on the Wolds, each settlement with its three great fields. But all that changed when trade developed both nationally and internationally, and wool became a more valuable commodity than arable produce.

On the outskirts of Baggrave Park, black-faced sheep were cropping the grass. Unlike the wary sheep of the Pennine moors, these ewes were as tame as puppies. They came running to the stiles to lick my fingers and butt me with their woolly heads. Charming as they were, their medieval forerunners proved the death of arable farming on the Leicestershire Wolds. I took half a day and wandered round the adjacent estates that now occupy the land which was the property of the canons of Leicester Abbey before the Reformation.

From 1469 onwards, in the middle of a wool boom, the canons enclosed and hedged these ridges for sheep-grazing. The crop growers of Baggrave, Quenby, Cold Newton and Ingarsby were forced out, abandoning their homes and fields. Within a hundred years the wheel had turned and the abbeys themselves had been dispossessed. Rich men moved in and built themselves palaces where the beans and barley had sustained thousands of feudal lives. Pale stone Baggrave Hall stands on a knoll above its lake, the grassy park still carrying faint ridges of the vanished fields of Baggrave village. So does the parkland around the Jacobean red-brick Quenby Hall. A mile beyond, the fifteenth-century moated manor of Ingarsby Old Hall, its house and barns beautiful in rich gold and pale silver oolitic stone, presides in isolation over a field of humps and dips, seamed across with deep old trackways, all that's left of the deserted village of Ingarsby. A very poignant place to sit on the grassy hummock of someone's cottage wall as afternoon shadows threw the house platforms and lane hollows into sharp relief.

Back on the Midshires Way I climbed the escarpment to my night stop at Tilton on the Hill where St Peter's Church was built of rich honey-dark ironstone faced with pale limestone. Gargoyles leered from the church walls, in which fossil shells lay embedded among rusty orange dribbles as an evening rain shower wetted the iron-rich stone. The rocks beneath these mild East Midland landscapes lay hidden under a thick coat of glacial deposits, and from here on they would be most strikingly displayed by their use as grey mudstones, rich orange ironstones or pale and beautiful limestones in the walls of churches, cottages, barns and fields.

Next day was a skelter over the furrows and down through Tugby where a million bicycles made a tangle outside the Café Ventoux. Windy Café? I wouldn't be surprised, with the tonnage of home-made chocolate cake those Lycra-clad pelotonistas were putting away after riding God knows how many miles o'er hill and dale to congregate at this cyclists' Valhalla. The Midshires Way ran south straight as a die across the roof of the Wolds, an easy undulating track between pastures where the ewes brought their fat lambs to stare at the stranger. Hunting fences separated the fields, their upper rails smoothed by the friction of passing horse legs – a reminder that I was tramping the 'Galloping Shires', wide country under a big sky, cloud shadows and sun chasing each other across the stubbles. Wind roared in the hedge oaks, the sound of the sea here on the roof of Leicestershire in the heart of the English countryside. The hedges themselves had been carefully cut into wildlife-friendly 'A' shapes. In most parts of the country, superannuated old horse ponds have been ploughed over and absorbed into the fields, but here they lay in a string at the junctions of hedges, pools of clear water rich in weed.

Just to the east lay Hallaton and Medbourne, their houses built of Northampton Sand, a 'ferruginous sandstone' of a delicious dark honey colour. They are twin villages, with all the rivalry that implies. In the sticky fields of Hallaton each Easter Monday the youth of this 'most pagan village in Leicestershire' riot and roll with their counterparts from neighbouring Medbourne in glutinous mud and ditchwater during the ancient contest, half ritual, half riot, of the Hare Pie Scramble and Bottle-Kicking. The Hare Pie is cooked, distributed and thrown into the crowd; then a mass of young men and women strive to carry, throw or otherwise force each of three 'bottles' (gallon barrels, two full of beer and one a solid dummy) across the stream nearest either village. Strangers are 'welcome' to join in. Kicks and blows are legion, bruises par for the course, broken bones not infrequent. Beer and glory await each year's victors.

At Cranoe I dropped down off the uplands into the flat flood plain of the River Welland, with sand and gravel overspread by fans of alluvial wash from the river. A gentle, unemphatic countryside where the Midshires Way passed Market Harborough by on the east. Here the Jurassic mudstone was overlain by a tongue of till or boulder clay, a thick smear of clay, gravel, sand and boulders which was rolled along underneath a glacier and dumped in an unstratified heap around 450,000 years ago. It's a part of a great block of the stuff, shown in sky-blue on the superficial geology layer of the BGS Viewer, spattered all across the eastern Midlands in an intermittent wedge lying north–south between Melton Mowbray and Linslade, and east–west from the outskirts of Cambridge to Leicester. During the so-called Anglian Stage (between 480,000 and 423,000 years ago) of Britain's latest cycle of glaciations the ice advanced further south than before or since. When it retreated the till was left behind, unsorted as to size or weight of content, a thick coating on which fertile soil established itself.

The Midshires Way crossed the county boundary into Northamptonshire, part of a great belt of Jurassic rocks, mostly mudstone and sandstones in this northern part of the county. Up to Brampton Ash on its little knoll of Northampton ironstone; down to Braybrooke, where the tall thickset spire of All Saints' Church overlooked the rectangular earthworks, fishponds and foundations of the fortified manor house built here in troubled times around the turn of the fourteenth century. Up once more to Great Oxendon and the start of a long straight stretch of disused railway line from Market Harborough to Northampton, now the multi-user Brampton Valley Way. Two tunnels were driven through the siltstone and mudstone, each now with a door left open in the grille over its portal. The temptation to shout for an echo in the darkness proved too much to resist. I did it at the very moment a cyclist shot past me without warning. I don't know which of us was more startled, but I suspect it was him, judging by the saltiness of the language that came echoing back as he pounded on ahead.

As I neared Northampton the slender upraised finger of the National Lift Tower, 127 metres of concrete skyrocket, beckoned from the town. In a somewhat neglected grave near the tower of Northampton's St Peter's Church lies the 'Father of English Geology', William Smith (1769–1839). The bright and inquisitive son of an Oxfordshire village blacksmith, Smith overcame multiple difficulties to create and publish the world's first detailed geological map of a country. Pretty much self-taught, pushing against the glass ceiling of his lowly upbringing, in his twenties he moved to Somerset to work as a mine inspector. Visiting coal pits, canal cuttings and quarries he came to realize that, however widely separated their locations, the layers of rock always lay in the same vertical sequence, each layer could be distinguished by the particular fossils it carried, and the rocks themselves dipped over many miles.

Surveying was a sought-after skill in a Britain whose landowners were desperate to find out if their land held valuable minerals. Smith travelled and worked across the land as a mineral surveyor, building up a picture of the country's geology as he amassed a fossil collection and gathered evidence from canal and tramway cuttings, roadworks and quarry faces. In 1815 he published his pioneering geological map in ten sheets at a scale of five miles to the inch. Its title very well describes the contents: *A Delineation of the Strata of England and Wales, with part of Scotland; exhibiting the Collieries and Mines, the Marshes and Fen Lands originally overflowed by the Sea, and the Varieties of Soil according to the variations in the substrata, Illustrated by the Most Descriptive Names.* Symbols depicted roads, railways and canals along with mines for coal, copper, tin, lead, alum and salt, and geological strata were individualized in hand-painted colours.

As with James Hutton before him, 'Strata Smith' had a long struggle for recognition. His map sold only modestly. He veered into financial straits and was briefly incarcerated in a debtors' prison, had his house and possessions seized by bailiffs, and for years was obliged to pick up bits and pieces of work wherever he could find them. But in 1831 the Geological Society of London recognized his status as the 'Father of English Geology' with the award of its first Wollaston Medal. By the time of his death in Northampton, William Smith was in receipt of a pension of a hundred pounds a year (£12,000 today) and his reputation was assured.

In my backpack, along with the ripped-out geological map from the schoolroom atlas, I carried a copy of the BGS's reprint of Smith's magnificent map. I opened it out and compared it with the BGS Viewer. What a remarkable job the persistent surveyor had done. The beautiful banded colours of what he termed Blue and White Lias, Clunch Clay and Shale, Iron Sand or Carstone, Blue Marl or Oaktree Soil, Green Sand and Chalk sinuated across Smith's map almost exactly as they did on the crude schoolroom map and the BGS's online Viewer. The bold ambition and the precision of the 200-year-old map made me keen to set eyes on those promised landscapes still lying between me and journey's end.

Closer yet to Northampton the Midshires Way broke away from the old railway and veered west to bypass the town. Beyond the sandstone stable buildings of Harlestone Park – pillared, porticoed and palatial, now converted to house humans – I climbed a smooth rise of ground to Nobottle, before passing the stout little sandstone knoll of Glassthorpe Hill and descending south towards the shock and awe of the narrow bridge over the M1 motorway. A particularly old and battered van went spluttering underneath, farting out blue-grey smoke and a stink that made me hurry away down to where Nether Heyford lay in the bottom of a broad valley between the snaking River Nene and the more orderly winding of the Grand Union Canal.

It's 135 miles as the canal boat glides from the River Thames at Brentford in west London to Salford Junction in Birmingham. By the late eighteenth century Birmingham and the Black Country, with their mineral wealth and swelling population of skilled workers, were well on their way to becoming the 'workshop of the world'. But their land communication with the capital of the kingdom, its financial institutions and docks and connections to the wider world, was still by roads that were poorly maintained and subject to delays in every town and village along the way. The Act to build the Grand Junction Canal was passed in 1793, kicking off a rash of canal-building that was dubbed 'canal mania'. In fact, the system held the seeds of its own eventual decline in an over-elaborate toll structure that was different for each section, and incompatible width gauges (Midland canals were built for boats 7 feet wide, the Grand Junction for 14-footers). Also, the Grand Junction Canal took forty years to complete. By the time it was fully open, the railways were posing a threat – they were quicker, better for perishable or heavy goods, and the only realistic option for passengers. In response the canal company cut charges and began operating night and day, but it could never keep up with the competition, as illustrated by the decline in its coal-carrying business. In 1845 the canal carried 60,000 tons of coal, compared to the railway's 8,000; in 1867 the figures were under 10,000 tons by canal, and well over 3 million tons by rail.

In 1929 the Grand Junction Canal was amalgamated with several others to form the Grand Union Canal. By the Second World War the Grand Union had the biggest fleet of boats in inland Britain, carrying coal, timber, manufactured goods, gravel and ballast between London and the Midlands. But first rail, then road competition had killed it off by the 1960s, to be resurrected in new glory as a leisure waterway.

Some romantic writers of past generations idealized life on the canal, like Kenneth Grahame in *The Wind in the Willows*. But most of polite society ranked canal folk with railway navvies and drovers – aggressive primitives, rootless, drunken and foul-mouthed. Canal people were folk apart, living outside normal society. Most were illiterate and led a simple life revolving around 'the cut', as they called the canal. Nineteenth-century owner-boatmen tended to sell their boats to bigger companies and become employees. It was a precarious life – no benefits, no dole, no savings. Few 'civilian' girls would marry into a canal family and the closed world of a cramped boat cabin with no prospect of a better life. So canal society was inward-looking and much interbred, with its own vernacular, traditions and home-made entertainment – poaching, singing, playing the concertina, storytelling.

The cramped conditions, with large families living cheek by jowl in the tiny boat cabins, moved Parliament to pass the Canal Boats Act in 1877, requiring all boats used as dwellings to be registered and subject to inspection. It was not before time. In *Rob Rat: A Story of Barge Life*, written by Mark Guy Pearse and published the year after the Canal Boats Act, the author describes barge cabins as . . .

. . . the most filthy holes imaginable. What with bugs and other vermin creeping up the sides, stinking mud finding its way through the old leaky joints at the end to the bottom of the cabin, and being heated by a hot stove, stenches arise therefrom enough to make a dog sick. In the boat cabins – *hell-holes* as some of the women call them – cohabitation takes place. Father, mother, sister, brother sleep in the same bed and at the same time. In these places girls of seventeen give birth to children, the fathers of which are members of their own family.

As to that, E. Temple Thurston in his canal travelogue *Flower of Gloster* (1911) quotes his guide Eynsham Harry as saying, 'Boatmen never 'malgamate. They fructifies amongst themselves.'

What a pleasure it was to walk today with that wide, calm water as a companion and guide, to pass into the outskirts of limestone country and see the churches and cottages of canalside villages lighten in hue and atmosphere. At Bugbrooke the Church of St Michael and All Angels in cloudy light was a sullen orange in stripes of marl and ironstone; at Blisworth, just along the canal, a burst of sunshine caused the powdery limestone walls of the Church of St John the Baptist to shine with a pale intense glow. Cottages in the village were banded in polychrome, rich brown Northampton sandstone contrasting strikingly with the silvery Blisworth limestone.

When the Grand Junction Canal came through in 1805 the village was loud and rank with limestone and ironstone quarries, brickworks and limekilns. There was also the Blisworth canal tunnel, one and three-quarter miles long, excavated by Cornish miners over twelve long years through successive and progressively younger outliers of Jurassic sandstone and limestone before emerging into the older mudstone once more. The tunnel was a leaky beast, and it had another major defect – it was built without a towpath. Towing horses had to be led over Blisworth Hill to Stoke Bruerne at the southern portal, where they were hitched once more to the barges that had been 'legged' through the tunnel. What that meant in terms of effort and danger I learned that afternoon in Stoke Bruerne's excellent waterside Canal Museum.

It took half an hour for a boat to pass through the tunnel. Two 'leggers' lay on their backs, head to head, on planks in the bows. Each man put his feet against the tunnel wall facing him and 'walked' sideways, propelling the boat along. Leggers wore brass armbands to mark them out as specialists in this very demanding and dangerous occupation. You could believe the proud boast of one Blisworth legger that he'd walked twice round the world in the dark. Legging was the only way to pass through the tunnel until 1871, when steam tugs took over. As to the damage the leggers inflicted on their legs, backs and lungs in the damp dark tunnel, that went unrecorded.

Beyond Stoke Bruerne the Grand Union Canal follows the wriggling River Tove down its valley. Iron guards were wrapped round the corners of bridges to prevent abrasion by taut tow ropes passing across them; they carried deep grooves where over many years the ropes had bitten deep. Cornfields and pastures lay over the alluvial soil on either side. Ash trees in the hedges stood drooping and thinning with dieback disease. The hedges themselves were encroaching on this rather neglected stretch of towpath, and my elbows registered nettle stings and bramble scratches as I dodged between the Devil and the deep brown canal all the way to Cosgrove and its limestone buildings perched just above the wide valley of the River Great Ouse. Cosgrove's limestone is creamy and full of shells, beautifully shown off in the Moorish arch and tabernacled buttresses of Solomon's Bridge that crossed the canal just here.

The landscape descended from the Cosgrove limestone to the Quaternary river terrace deposits of sand and gravel along the valley of the Great Ouse, a winding string of flooded diggings that shone like silver as they flanked the eastward-flowing river. Sand barges went down the Grand Junction Canal to London from quarries here. Initially the canal dipped across the valley by way of nine locks, but they were always subject to flooding, and in the end the company built an aqueduct with tall approach embankments. Crossing the aqueduct, I passed from Northamptonshire into Buckinghamshire and, in a way I couldn't quite put my finger on, from the Midlands into some other regional entity – not quite yet the Home Counties, but somewhere off on its own. Canalside architecture grew less bricky, the landscape less broad, and the greetings of towpath walkers and anglers changed from 'All right, mate' to 'Hah'.

Near Wolverton Station a railway wall held a black-and-white mural of a tremendously elongated train being pulled in both directions at once by a steam locomotive at either end, with stringbag aeroplanes, balloons and cranes flying overhead. A single vandal's tag in red paint defaced this wonderful and improbable piece of art. Curse your stupidity, 'Sayerzz', you prat.

Open fields began to give way to municipal grass and trees. An early-nineteenth-century humpback bridge of red brick, seen through the square aperture of a late-twentieth-century concrete span, prompted thoughts of the passage of time since that simple and graceful arc was built, when Waterloo had yet to be fought and Napoleon was a clear and present danger. No anaesthetics, no cars, no planes, no trains. The speed of travel regulated by the speed of a horse. And thoughts, too, of how much life has accelerated hereabouts since 1974, when I bought the first series of the Ordnance Survey's 1:50,000 Landranger map, number 152. The city of Milton Keynes did not even exist on that map – only the name, 'MILTON KEYNES', stamped as a statement of intent on the blank Buckinghamshire fields, their top coating of glacial sand and gravel, and the great band of late Jurassic mudstone, full of ammonite and bivalve shells, that runs under the region and away to the southwest. There was a sizeable little town at Bletchley to the south, a line of villages from Wolverton to Newport Pagnell in the north. In the 5 miles of flat farmland in between, nothing whatsoever but a solitary reservoir and the Grand Union Canal snaking between the tiny hamlets of Willen, Woolstone and Woughton on the Green. The New Town of Milton Keynes, given the go-ahead in 1967, grew from literally nothing to fill all this space and a good deal more with its 120,000 houses and its 287,000 people, its rationale of finding agreeable living conditions for London-overspill communities, and its sky-high aspirations expressed in linear parks and woodlands, low-rise architecture, gridded layout for traffic management, proliferation of footpaths and cycleways, public sculpture and arts venues.

This morning the housing estates, business parks and shopping malls of Milton Keynes stretched away on all sides, screened by hedges and trees. For me it was all greenery and falling leaves, herons and chaffinches that occupied eyes and ears until the canal wriggled free of the city on its southern outskirts. Out into gently rolling farmland, scruffier and duller to the eye, the contrast in colour making me aware of how artificial, as well as beguiling, the well-planned new town landscape had been.

Less than 2 miles east of the Grand Union Canal as the crow flies, Nigel Richards's farm at Moulsoe is bounded on the west by the M1. Only the motorway separates the fields of Hermitage Farm from the urban mass of Milton Keynes.

'My dad came here from Suffolk in 1952,' Nigel told me. 'Just after the war there were grants for food production, so Dad put in quite a lot of field drainage and ploughed up a lot of grass for arable farming. This clay land is well suited to arable crops rather than grazing, but few people up to then had ploughed it and drained it because the available machinery was barely powerful enough to plough the land.

'I took over from my dad in 1981. Altogether Hermitage Farm is about four hundred and fifty acres. It's fairly flat land, and largely arable. The land suits autumn- and spring-sown crops, and we grow wheat, barley and oilseed rape. In the last five years the small seedling rape plants have regularly been devoured by attack from cabbage stem flea beetles, and it's a gamble as to whether you'll get a crop or not.

'The River Ouzel runs through the farm. We've got river clay along the Ouzel, where we graze the cattle. From there the land rises through a gravel terrace to a band of Oxford Clay. It's not particularly easy to work. The field shapes don't fit with the way the soil changes as the land rises, so you'll have an Oxford Clay field with patches of gravel in it, and you have to farm to the best bits of the field. The gravel tends to bake in dry conditions and to fill up with water when it's wet. The Oxford Clay becomes sticky in the wet, and in drought times it has dried into bricks. It's not a calcareous clay, either. But the land rises from that to a cap of Boulder Clay, which is better. It's got some small lumps of chalk in it, and it has better natural structure, so when rain comes it gets through the soil to the roots of the crop, which can penetrate the soil easier than in the Oxford Clay.

'What's the most important factor a farmer has to contend with? Some might think that recently it's been Brexit and Covid. They have both affected the market for our produce, and that's not to mention the global energy fallout triggered by Russia's invasion of Ukraine. With Brexit, livestock movements have got more complicated and there are more delays at ports. Feed costs, electricity, fuel, chemicals have all rocketed up in price since Ukraine was invaded. As far as Covid goes, farmers live and work quite isolated, so Covid didn't much affect us. No, the thing that most affects us is not Brexit, not Covid, not Mr Putin and what he's doing in Ukraine – it's what it always has been for farmers, the weather. Any farmer can tell you that climate change is taking place, you just can't deny it. One example: October is a lot wetter than it was; the clay soils can't drain it away; and that's when we need to get our autumn-sown crops established, which is much more difficult in the wet.

'On Hermitage Farm we've done quite a lot of eco-friendly things. While there are probably thirty per cent fewer hedges than in 1952, we've got about six miles of hedges and ditches, and we cut half of those hedges every three years instead of yearly, which helps the nesting birds.

We've planted ten acres of tree belts since Dutch Elm disease killed our hedgerow elms, and with awkward or wet field corners we've planted them with trees or flower-rich seed mixes to encourage insects. I leave buffer strips along the headlands round the fields, and we've taken part in a number of Countryside Stewardship schemes.

'However, that's all by the way now. Plans came along in 2013 to expand development over my land to help Milton Keynes hit its housing target. In view of this my son didn't want to follow me on the farm, which would disappear. It became clear that Hermitage Farm was scheduled to be part of "Milton Keynes East". In Autumn 2021 at the age of sixty-five I gave up the tenancy on the farm; I sold our cattle, machinery and crops and I retired from farming. It's a sad time for me, yes. But I look at it like this: if it's coming, it's coming. Nothing I can do about it. But I will say that it's a problem for all the southeast of England – the good arable land is here, but so's the pressure to develop it, increasing all the time. And you might have noticed that they're not making any more land.'

Farming isn't the only activity to make use of the local Lower Oxford Clay. This 160-million-year-old Jurassic clay, dark grey and sticky, forms the wide, gentle valleys of the region. Its bituminous composition gave it the nickname of 'the clay that burns' and made it cheap to fire into bricks, their rich red hue derived from the iron content of the clay. I spent a day across the county border in Bedfordshire, mooching around the old brickworks at Stewartby, five miles east of Hermitage Farm. In their 1930s heyday these works produced 500 million bricks a year for the London Brick Company. Now the grey brickfields are going back to green once more, and Stewartby's chimneys stand smokeless and gaunt over a beautiful lake where the giant clay pits once lay in all their desolation. In a green lane beside the lake I picked up a discarded brick, its surface gritty and abrasive, an intense red in colour, stamped 'Marston Bespres' – a link with a time when this quiet vale was a clattering, productive, smoke-polluted place.

*

Back on the Grand Union Canal I met a water gypsy. Bodhi – 'It means "Enlightenment", he told me – was pulling his boat *Luna* along manually with a rope. With hair knotted and plaited under an outsize cap, home-decorated trousers and plastic bags for socks, Bodhi looked exhausted, but was as cheerful and chatty as could be. 'No engine in her, so as not to pollute the planet,' he explained, indicating *Luna*, 'and my horse, well . . .' He waved in the general direction of Milton Keynes. 'I named the boat after my girlfriend,' he confided, 'and she's named after the moon.' So where was Luna the girlfriend? 'Well, she's with the horse . . . at least I think she is.' And what's the horse called? 'Oh, Narcissus, cos he's always looking in the water.'

Bodhi seemed powered entirely by romance, a true innocent. I gave him a hand to pull *Luna* through the reeds and round bushes. 'Seven days out from Harefield at this rate, I reckon. I've been on the canals since spring. Boat people will do anything for you, but in the city you don't even know your next-door neighbour, do you?' *Luna* became inextricably tangled in rushes, and I had to walk on. I looked back to see her floating free again, Bodhi still blithely hauling her on at half a mile an hour.

Through more sand and gravel at Leighton Buzzard, and then a step up the geological ladder so vividly indicated on the BGS Viewer, out of the olive band of the Jurassic period and on into the turquoise and green of the Cretaceous (145 mya–66 mya), era of crocodiles and turtles, of iguanodon 10 metres long and of baryonyx ('heavy claw') with crocodile jaws and claws 30 centimetres long, that preyed on iguanodon and on fish. In the seas swam bony fish; on the land grew the first flowering plants.

In the late Jurassic/early Cretaceous period there was an uplift in land levels and the sea withdrew from most of England, with little sediment being deposited until the sea came back around 25 million years later. I read that in my notes, and I repeated it to myself now, but my brain refused to deal with it. It was another moment like that with Peter Drake on the shores of the Coigach peninsula, when I'd tried to picture a chunk of land dropping 4 kilometres downwards. The sea withdrew from Britain for 25 million years. What? Don't just let that pass without trying to imagine what it meant. *Twenty-five million years!* How had I become so blasé about geology's gigantic numbers? Even at this late stage of the 3,000-million-year journey from the Butt of Lewis, I still found it hard to crank up my imagination in order to contend with a timescale like that so soon after contemplating the changes within my own lifetime around the farmlands where Milton Keynes now stood. It needed another massive refocusing of the mental lenses, like looking up from a microscope straight into a stargazing telescope.

The Cretaceous period ended badly for those giant dinosaurs, bony fish and flowering plants. There were falls in sea level late in the period, and volcanic activity associated with the break-up of Pangaea. The formation of the early Atlantic, Indian and Southern Oceans led to a lot of ocean-floor spreading, to volcanism and the release of greenhouse gases. Ammonites were beginning to alter their coiled shape and uncurl in an adverse response to changing marine conditions. The world was experiencing a quick rate of change, and many species couldn't cope with that. And a dramatic intervention from the sky, a bolt from the blue around 66 million years ago, put paid to most of the rest. An asteroid or comet some 7 miles wide smashed into the planet at Chicxulub in the Yucatán peninsula of southern Mexico, forming a crater 12 miles deep and over 100 miles in diameter. The huge tidal waves generated by the impact would have wiped out life in the shallows or near the shores, and the volcanic upheavals would have blocked out the sun and drenched the globe in acid rain, as often before. This mass extinction event put an end to the ammonites, the huge carnivorous marine reptiles called mososaurs up to 15 metres long and weighing 40 tonnes, and, famously, all the dinosaurs but those with the power of flight.

South of Leighton Buzzard the Grand Union Canal joined hands with the sinuous River Ouzel to run side by side through a stretch of gently undulating countryside known to the canal folk as the Fields. The hidden bones of this unemphatic landscape are all Cretaceous, laid down before that massive cataclysm at Chicxulub. Under the sandy river terrace deposits along the banks of the Ouzel lies a narrow strip of Gault Clay laid down a little over 100 million years ago, a band of blue-grey clay some 60 metres thick in places that runs diagonally across England from the shores of the Wash estuary in Norfolk to the Dorset/Devon border. Then I was into the Upper Greensand (113 mya–94 mya), familiar to me from walks in the countryside all round the rural borders of London, a softish sort of sandy, chalky rock laid down in mid-Cretaceous times, around the same time as the Gault, with southern Britain once again under the sea. It gets its name from the greenish tinge imparted to it by glauconite, a mineral which developed in the sediment that was deposited along the continental shelf of that shallow Cretaceous sea. Fruit trees love the greensand, famously those along the downs and Wealden valleys of north Kent, because of its many trace elements as well as the nourishing marine potash, iron and magnesium contained in the glauconite. Around Cheddington, in flat country with bumpy little hills, remnants of old orchards show where Aylesbury prunes or damsons were grown. I found a prune tree growing wild in the towpath hedge and picked one of the indigo fruits. The skin had the floury bloom of autumn, and the flesh was sharp and sweet. I spat out the stone and watched it plop and vanish in the canal.

On the BGS geological map the greensand faces east in the shape of a lime-green crocodile's head in the act of swallowing a wedge of pale mauve sand and clay that surrounds London and the Thames Valley. The greensand jaws do not bite directly down on the clay; separating the two is a ridge of paler, milky green, a great barrier of chalk to the north and south of the city and its river valley. From the canal bank my view ahead was filled with chalk, the dimpled wall of the Chiltern Hills facing north-west as they rose well over 100 metres from the flatlands of the Vale of Aylesbury, grass-clothed and topped with a rampart of trees. I forgot my sore feet and ankles, aching from the unforgiving pounding of the towpath. These were the first proper hills I'd met since leaving Charnwood and the National Forest, and they seemed to reach out for me.

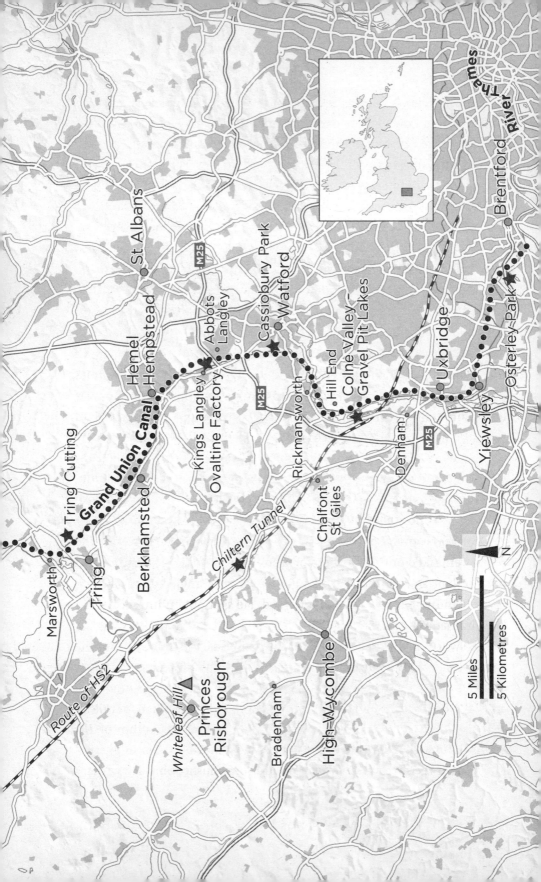

River Thames

Brentford

Osterley Park

Uxbridge

Yiewsley

Denham

M25

Colne Valley–
Gravel Pit Lakes

Hill End

Chalfont
St Giles

Rickmansworth

M25

Cassiobury Park
Watford

Abbots
Langley

St Albans

M25

Kings Langley
Ovaltine Factory

Hemel
Hempstead

Grand Union Canal

Berkhamsted

Tring Cutting

Chiltern Tunnel

High-Wycombe

Bradenham

Whiteleaf Hill

Princes
Risborough

Route of HS2

Marsworth

Tring

N

5 Miles

5 Kilometres

15

Through the Chalk:
Chiltern Hills to the River Thames

SUNDAY AFTERNOON, AND A FISHING competition. The towpath of the Grand Union Canal was obstructed by alpha anglers, every one of them male. The air tingled with testosterone. I stepped gingerly over lines and round seething trays of maggots. 'I've got tutti-frutti and vanilla,' said one, holding out a fistful of boilie bait. 'The manufacturers say fish can't resist 'em, but no one's told the fish.' At that moment his electronic bite detector went off. With maximum flourish he reeled in. Derisive cheers arose along the bank as he raised his rod tip to reveal the smallest dace in the canal dangling on his hook.

Near Marsworth I passed a pair of anglers. 'Crucian carp over by the reeds,' remarked one. 'I'll try 'em on a strawberry boilie.' His chum fired out the bait from a catapult like grapeshot. The angler cast his hook directly into the reeds, where it snagged. He spat out an imprecation. I left him hauling and cursing like a capstan hand.

Marsworth lies right at the feet of the Chilterns. Pitstone Hill and the 233-metre peak of Ivinghoe Beacon rise with dramatic steepness immediately to the east. The undulations of these chalk downs lend them an air of elasticity, as though they might be composed of some matter other than rock. Curate and naturalist Gilbert White (1720–93), author of *The Natural History of Selborne*, wondered whether chalk hills might not grow out of the surrounding clay lowlands like mushrooms. He wrote:

I never contemplate these mountains without thinking I perceive somewhat analogous to growth in their gentle swellings and smooth fungus-like protuberances, their fluted sides, and regular hollows and slopes, that carry at once the air of vegetative dilation and expansion ... Was there ever a time when these immense masses of calcareous matter were thrown into fermentation by some adventitious moisture; were raised and leavened into such shapes by some plastic power; and so made to swell and heave their broad backs into the sky so much above the less animated clay of the wild below?

Chalk is a curious substance. I'd always associated it with the classroom, the short white stick with which my teachers wrote incomprehensible chemical formulae or yawnsome Shakespeare sonnets on the blackboard. Chalk was crumbly. It could be abolished with a sweep of the blackboard duster, and it left white powder in your hair when the teacher, at the end of his tether after an afternoon's goading, threw that duster at your head. I still thought of chalk as too insubstantial to constitute a rock that a canal navvy would have to bash his way through with pick and shovel.

Chalk is very much a rock, albeit a softish one. It started to form in the late Cretaceous period, nearly 100 million years ago. Unimaginable millions of algae called coccolithophores drifted to the bottom of the warm tropical sea over a span of 25 million years, layer upon hazy layer, condensing and thickening a millimetre at a time. Looking at chalk as magnified in an electron microscope you see a bundle of grey-white plates in the shape of concentric rings joined by radiating struts, clinging together in an elaborately structured ball. The lacy beauty of those plates makes you gasp. They are coccoliths, cemented together by calcium carbonate into that ball shape. And in the middle of each sphere, armoured against mishap, resides a coccolithophore, a living organism one single cell in size.

A line of five hundred coccolith plates glued side by side would measure scarcely a millimetre in length. This is an order of minuteness that's very hard to grasp. Yet these are the building blocks that form the chalk of the Chiltern Hills. The sheer span of time it must have taken to lay down one finger's depth of such algae was beyond my powers of calculation. Yet in places under the Chilterns the chalk lies more than 300 metres thick.

The Tertiary period (65 mya–2.6 mya) followed the mass extinction that had wiped out the dinosaurs at the end of the Cretaceous period. Placental mammals, our distant mouse-like ancestors, had already established themselves before the violent end of the Cretaceous; and now, emerging from their burrows into a world where dinosaurs no longer existed, they lived hard, prospered and spread all over the world.

Between 60 million and 50 million years ago a collision between the African and Eurasian tectonic plates caused a massive upthrusting to the south and east of Britain. The seabed buckled and creased like a scrunched-up bedspread. Mountains broke the surface and grew skywards where the Alps now stand. The ripple effects from the Alpine orogeny were felt all the way from the nascent Alps to southern England. The layer of chalk that covered the southeast of England was lifted clear of the sea and bent by tremendous pressure into the shape of a syncline, a fold that dips downwards at its centre. This syncline is what forms the outer collar of chalk around London.

The youngest rock sits at the centre of the syncline. The lowest point is composed of a deep bed of gravel and successions of clays and sands some 50 million to 60 million years old, these stacked layers all showing where sea shallows and tidal estuaries came and went, advanced and receded, washing sandstone and clay to tiny fragments. On top of this lie deposits from glacial meltwaters: pebbles, rocks and flints; then a penultimate layer of alluvium, rich soil left behind after river floods had subsided. Finally, a top layer of 'made ground', the tiny covering of dirt a few inches thick that has been worked and shaped by man in the past few millennia.

At the edge of the syncline lies the oldest and highest chalk, nearly twice as old as the youngest, a lower layer of clayey stuff without any flints, and above it more solid chalk that's lined with beds of flint, the petrified remains of sponges. The erosion of the last 50 million years, seeking out and whittling away the softer rock, has left a curving rampart of steep, deeply indented downland slopes. They are covered in herb-rich calcareous grassland in which grow beautiful small flowers – sky-blue powder-puffs of scabious, rich purple autumn gentians, wild thyme and eyebright, rockrose, harebells and orchids – and topped with long-established beech woodlands or hangers. I've always loved walking in the Chilterns, particularly on slopes too steep to have ever been ploughed or up at the fringes of the beech hangers, looking out across the wide flatlands of Buckinghamshire and Bedfordshire. It felt strange this afternoon to be admiring the hills from below, 100 metres down, rather than up on a chalky path at the top of the slope where the last of the very rare Chiltern gentians might still be blooming.

From Marsworth the Grand Union Canal needs an elongated flight of seven locks to climb up to Tring Cutting, the pass dug by the canal navvies through the Chiltern chalk barrier. A bird reserve now utilizes the reservoirs the company built at the foot of the flight to keep the locks permanently supplied with water. Ospreys, the great pale fish eagles, make occasional visits, but I saw only anglers becoming morose as the sky grew damp and misty. As I was climbing past the locks it began to rain in earnest, the drops hitting the canal like bullets. A fisherman sat stolidly in the downpour, watching his float. 'You must be as mad as I am, this weather,' he growled. Not quite as mad – he continued to sit stock still, the rain pattering off his broad-brimmed hat, and I took the road to Tring.

Next morning I found the night's downpour had left a quagmire of a towpath, with bike tracks 3 inches deep. At the top of the flight the canal levelled off for 3 miles, slicing through the chalk by way of the Tring Cutting. Ten metres deep and over a mile long, it took five years for the navvies to excavate it to reach the level summit at Cow Roast. There was a profound silence in the cutting, a solemn atmosphere, almost churchy, like walking down an inverted cathedral nave. From here it's all downhill for the Grand Union Canal, a long descent from the chalk heights to the clay and sand at the heart of the great syncline, losing 100 metres in height through fifty five locks, twisting and turning down the river valleys of Bulbourne, Gade and Colne towards the distant Thames 35 miles away.

The valley sides closed in, the massed buildings of Berkhamsted and Hemel Hempsted lying ahead. The canal surface dimpled with roach and dace. I found myself herding a grey heron in front of me in short bursts of flight. It must have been a particularly good fishing stretch, because he really didn't want to be driven away from it.

There were glimpses to the left of the high ground of Berkhamsted Common, a sprawl of open ground where locals had always enjoyed the right to roam. When Lord Brownlow arbitrarily railed off a great chunk and added it to his Ashridge Estate in 1866, he thought he'd encounter little opposition. But an equally autocratic and bloody-minded grandee, Augustus Smith, took exception. Smith paid a gang of tough London navvies to come and tear down the 3 miles of railings by night, and leave them neatly rolled up for Brownlow to collect in the morning. The locals reclaimed their common land, and Lord Brownlow had to 'retire hurt'.

Down the valley at King's Langley, the River Gade and the canal squeezed together to run next to the art-deco Ovaltine factory, closed in 2002 and now a nest of luxury flats. Just the sight of the building, depicted in cheerful greens and oranges on the malted milk tins of my childhood bedtimes, brought back the thick sweet taste ('Drink Ovaltine for Health; Delicious HOT or COLD'). Ovaltine was full of wholesome goodness, just like the apple-cheeked Ovaltine Dairy Maid who beamed over factory and farm on the label like a busty angel, a sheaf of barley under her arm, a basket of eggs in the crook of her nicely tanned elbow. There was the League of Ovaltineys for children, too, with its own special song:

> We are the Ovaltineys,
> Little girls and boys . . .
> Because we all drink Ovaltine,
> We're happy girls and boys.

Innocent days! As if to point up the difference between then and now, two boys who would have been Ovaltiney age back in the 1950s were sharing a spliff of reeking skunk under the flat bridge that carried the thundering M25. We tacitly agreed to ignore each other's presence, but I could hardly help getting an involuntary lungful as I sidled by.

On through the tree-lined grounds of Cassiobury Park where the stone bridges were elaborately ornamented and the canal had been broadened and landscaped to resemble the windings of a natural river, all as a sop to the landowner, William Capell, 4th Earl of Essex, who had a seat on the board of the Grand Junction Canal. On among flooded workings where glaciers had spread a generous thickness of sand and gravel. A red and silver Central Line tube train went clacking over the canal, a sure sign that the tentacles of London were reaching out this far into the countryside. Under a slope rising steeply to the knoll of Hill End, and down to Denham among the gravel-pit lakes of the Colne Valley. Here I was edging into the centre of the syncline, a geological formation known as the Lambeth Group, mostly clay mixed up with sand, gravel and a little silt, a gritty muddy substrate laid down some 50 million years ago in shallow lagoons and on estuary margins and sea-shores as the sea flirted with the land in this southeastern corner of Britain. Just to the east on the BGS Viewer geological map a triangular segment of pale mauve, broadening towards the distant coast, lay waiting for me – a wedge of silt and mud, of beds of shells and pockets of sand, the London Clay on which my journey from the Outer Hebrides down the length of Britain would run through the capital to its conclusion in the soft muds of the Essex coast. And right at my elbow, meanwhile, the clanking of cranes and diggers, heard but only partially glimpsed behind a wall of screens just north of Denham, as the Colne

Valley Viaduct of the huge and controversial high-speed railway project, HS2, inched towards completion.

The junction of the Grand Union Canal and the River Thames at Brentford lies only a day's walk southeast of Denham. Before tackling that final stretch of towpath, however, two detours away from my Bones of Britain route would fill in, if not complete, the geological picture for me. I needed to get a view from on high across the Ice Age's tremendous field of action. And before that, here was a chance to penetrate into the heart of the chalk by way of the half-completed tunnel that would carry HS2 under the Chilterns and on towards Birmingham.

On a dull October morning I reported to the Align compound in Chalfont Lane, just outside Denham, to spend a morning on the works of one of the most controversial environmental projects of this century, the construction of the high-speed HS2 rail link between London and Birmingham. At least that was the plan. I was stymied at Chesham by HS2 itself. There was major disruption all across the region, caused by the widespread works associated with the 10-mile tunnel being bored beneath the Chilterns. The road was closed. Diversion signs and Satnav combined to land me at a works junction where I was turned back at the compound gate. Security was very tight all round the project.

'Only for heavy truck,' said the Sikh guard. 'Please go back around,' and he described an alternative way along the muddy tracks, quite self-evident to him, but hard for a stranger to interpret. I fetched up finally at the right gate, where the guard failed to find me on his list. My name eventually bubbled to the top of the stew of mangled accents and spellings, and I gained access to the compound.

My first impression was of an instant city of prefabs, singly and piled into low-rise units. Miles of wire fence; hundreds of cars. Massive banks of grey and white soil. Heaps of preshaped concrete and steel mesh. The roadways gleamed with chalk and water. Workers strode or sauntered about in the universal PPE (Personal Protection Equipment) of orange suits and white hard hats. Everyone had a purpose, none fathomable to a newcomer. At the Tunnelling Office I met geotechnical engineer Guilhem de Langlais, a Breton in a striped jersey (*naturellement*), and his French colleague, visitor coordinator Marie-Amélie Auvinet. Two smart, educated, professional members of the Align team – a bland, corporate name deliberately selected to avoid any hint of allegiance to any particular nationality. Like much of the works culture around HS2, the name was a manifestation of the project organizers' mission to head off national rivalry or misunderstandings among the hundred-plus nationalities who had come together, two thousand strong, to push this tunnel through the Chilterns by technological knowhow, massive investment, mind-spinning feats of organization and machine dexterity, along with plenty of old-fashioned muscle and sweat. It was a potential Tower of Babel, exacerbated by the post-Brexit difficulty of securing foreign workers with good English.

I thought back to Yoredale Group country and the rough lives of the thousands of navvies in the 1870s, camped out in the wilds of the Yorkshire moors as they drove the Settle & Carlisle Railway's tunnel through the gritstone and limestone under Blea Moor. Antagonism between the English and the Irish navvies had been intense there. No hint of that among the well-fed, well-housed and well-educated workers of the multinational Align team. Interesting, though, that out of all those nationalities working on HS2, the Irish were still considered top dogs as skilled tunnellers.

'When new people arrive on this project,' said Guilhem, 'they are put through cultural-awareness courses. We have to be able to understand the way others express themselves, what they really mean. A Frenchman, for example, just says what he thinks. French don't praise a job. If it's going well, that's normal. If it's bad, we'll say so. Whereas the English don't tend to criticize. They will praise things that are just going normally well. "Oh, good job, well done!"'

It was mostly men in the office, hunched pale and yawning at computer screens, and almost all men on the works outside, crunching and sloshing in yellow site boots, talking in groups of four or five, driving diggers and manipulating cranes on the sloppy banks of the South

Portal entrance. The occasional woman, passing through the office or striding the tracks in PPE, stood out by her rarity.

On the office wall was a wonderful map in greens and browns, a cross section of the tunnel in its entire length slicing through the geological layers below the Chilterns. Intervening desks prevented me from getting closer, and my request to photograph the map was turned down. They really were tight on security here.

Guilhem sat me down in an office for a quick spin through the geology. Contemplating the infinitely leisurely, yet sporadically violent processes that have been unrolling over the past 3,000 million years, the shifts of climate, and the migratory journey of these islands across the globe to where we now lie, the undertaking of HS2 seems laughable in its infinite physical smallness, while at the same time admirable in its ambition.

Faults and cracks were shown on Guilhem's charts, cutting and wrenching the surrounding chalk, trapping soft pockets, inserting hard planes and oddly angled seams. Faults are the devil to tunnel through. Each fault, each rock and each hard place of the Chiltern tunnel's route had been mapped and charted, pored over, assessed and accounted for. High tech was absolutely the name of the game. Yet the driving of the tunnel was not silky smooth, but halting and jerky. Long sections of the bore lay below the water table, meaning that it crossed several 'source protection zones' – wells, boreholes and springs that must not be contaminated by the tunnelling process. The ornery geology continually broke equipment, clogged machines, threw up surprises. And yet it was only chalk, and no other type of rock, that the tunnellers had to deal with.

The Chiltern tunnel is 10 miles long. It's approached from the east by a viaduct 2 miles in length, at that moment being constructed, partially through ancient woodland, over the Colne Valley north of Denham. The steel cages that would form the viaduct lay heaped in segments on the ground, already brown with rust. Two TBMs (Tunnel Boring Machines), each with a rotary cutterhead shaped like a circular fan and

10 metres wide, were inching their way northwest through the chalk, forming a pair of parallel tunnels, up (London-bound) and down (to Birmingham). One of them had already advanced more than a mile into the hill; the other was a little way behind.

Seven curved segments, fitted together to form a cylinder, shaped each section of the tubular casing that lined the tunnel. There was a huge warehouse-style factory for making the tunnel casing, and others for producing the concrete and the steel cages of which it is composed. The same went for the Colne Valley Viaduct that led to the tunnel, its 2 miles needing eight hundred sections of casing.

I stood in a conference room with big windows, looking out over the South Portal. The black tunnel mouths were close below. Bundled lines of rusty pipes carried the spoil slurry out of each tunnel. Twin big silver ventilation pipes clung to the tunnel roofs.

In my skull cinema an image flashed up from a 1950s comic, maybe the *Eagle*: a boy and his eccentric millionaire uncle in their fabulous rocket-powered digging jet, its corkscrew nose cone whizzing round, its armour cladding glowing as it bored through the splashy metal whiteness at the Earth's core ('Iron and nickel at *a thousand degrees*, children!') and on down to pop out of a hole in Australia. I imagined each Tunnel Boring Machine busting along horizontally, scrabbling out its tunnel, spraying a torrent of chalk and flints behind it like a terrier in a rabbit burrow.

'How fast are they digging?' I asked Guilhem.

'Oh,' he said, 'about fifteen metres each twenty-four hours.' Really? Why so slow? 'Well, you have to step back and think of the difficulties of this kind of tunnelling. The new sections to reinforce the tunnel have to be brought forward and put in place, and then grouted – a lot of infill between the outer casing and the chalk cavity the tunnel's running through. The slurry pipes and the ventilation pipes must be extended all the time as the excavation goes forward. The TBMs themselves need lots of maintenance because they work round the clock. And the chalk is not just a soft crumbly thing; it can have very hard brick-like blocks, and flints that can break a blade on the cutterhead. So we go slow.'

Ironically, considering that so few women work at the (literally) cutting edge of the tunnelling venture, the two TBMs were named 'Florence' and 'Cecilia'. The names were chosen by local schoolchildren and students to honour pioneering women of science with local connections: Florence Nightingale (1820–1910), who wrote many of her groundbreaking books on nursing at Claydon House, the Buckinghamshire residence of her brother-in-law; and Cecilia Payne-Gaposchkin (1900–1979), a brilliant Buckinghamshire-born astronomer and astrophysicist.

TBM 'Florence' began to tunnel in May 2021, her counterpart 'Cecilia' the following month. Each TBM was 170 metres long, nearly twice the length of a Second World War destroyer. In fact, they were like superclean, super-neat submarines, slender tubes full of gangways, railings and pipework, each with a top-of-the-range control cabin where live screens showed the driver and his colleagues what was going on at the chalkface. Seventeen operators crewed each TBM in rotating shifts, and these extraordinary machines would have run without stopping for well over three years by the time they reached the North Portal a little east of Great Missenden, 10 miles away from where they started.

I donned PPE: orange suit, hard hat (blue to mark me out as a visitor), gloves, goggles and big clumping boots with reinforced toes. A minibus under Guilhem's command took me out of the site and down roads slick with chalk slurry and peppered with 'closed' and 'diversion' warnings. Between the tunnel and the approach viaduct we passed a big old extraction hole, Pynesfield Quarry, where treated 'arisings' from the excavations were spread – 3 million cubic metres of chalk spoil, enough to fill the old quarry to the brim.

We crossed the M25 London orbital motorway where the tunnel passed 10 metres below ground, and the Misbourne Valley where the trains will run 15 metres below the river. In the hills around Missenden the chalk ridges showed dimples and hollows – depressions like swallet holes where rainwater had infiltrated underground fissures, causing them to collapse. The valley is orientated north, but the River Misbourne runs northwest to southeast, following the course of a fault associated with the Alpine orogeny or mountain-building episode 55 million years ago.

The geology consists of soil over a deposition layer of flints and hard chalk – what's left of the upper layer of the hundreds of metres' thickness of chalk after weathering, frosts and so on. Four vent shafts were being excavated down through this to a depth of 100 metres, in order to provide ventilation and emergency access to the tunnels. Big fans suck air and push it down the shaft into the tunnel. If a train is on fire, the air will blow smoke, fire and fumes along the tunnel, and the passengers will be directed to escape in the opposite direction. That's the theory, anyhow.

The first shaft was near Chalfont St Giles, the second near Little Missenden, slippery and smeary sites with stacked Portakabins for offices. Bulldozers and cranes with grabs pushed and pulled at sloppy speckled chalk and clay. It all seemed like a big boys' playground, the outsize diggers and drills in bright yellow, scarlet and buff colours. All the men looked young and fit.

What was very striking was the minute planning necessary to get several large and heavy machines, some as tall as a tree, within the tight circle of the shaft head, precisely positioned so they did not get in each other's way – cranes and gantry arms, cable reels, the massive drilling arm inching downwards as its screw drill spiralled up, shuddering and grinding to a halt at successive obstructions. A grab in the very centre of the tangle of machines was sinking with its operator into the ground, digging out the chalk, some stained red through weathering of its iron content. The spoil was sloshed into trucks that took it away to join the tunnel arisings in Pynesfield Quarry.

Later that day I looked up the technology available to the canal navvies when they excavated the cuttings and tunnels of the Grand Union Canal at the turn of the nineteenth century. With pick, shovel and wooden wheelbarrow, over twelve years they forged the route of 135 miles from London to Birmingham, a four-day journey by horse-drawn boat. The cost of the project was some £1,600,000, the equivalent of about £150 million today. The HS2 workforce, with all the modern technology available to them, are taking about fifteen years to build a line whose trains will reach Birmingham from London in less than an hour. The cost is estimated at somewhere north of £45 billion. If the line ever ramifies from Birmingham to Manchester and Leeds, as per the original plan, that figure could be, well, plucked out of any blue sky.

The stated benefits of HS2 – faster travel, upgraded infrastructure, job creation, cutting carbon emissions by transporting more goods and people by rail, 'levelling up' of midland and northern regions in comparison with London – are aspirational, but only putative. The 'rewilding' schemes, creation of new woodland, using tunnel spoil for landscaping, planting of embankments with wildflower mixes and other restoration schemes, welcome as they are, provide only sketchy mitigation of the damage and disruption to the environment, to wildlife and ancient woodland, watercourses, archaeology, people's houses on the direct line of the railway and the villages and churches in close proximity. These losses are incalculable, and protesters have been on the case of the company and of successive governments since the first digger scooped the ground. After seeing with my own eyes the massive scale of the works and the investment, however, allied to the kinetic energy around the project, only one thing seemed absolutely certain: the 'Stop HS2' slogan of the objectors was never going to prevent that tunnel being driven through the Chiltern chalk.

For all its ease of walking and direction-finding, and the directness of its progress across the successive geological layers, the towpath of the Grand Union Canal could only afford me a boatman's-eye view of the landscape. Before leaving the great chalk ridge of the Chilterns completely behind, I wanted to get up high on the crest and look out over the Buckinghamshire plain, to try to get a wider perspective on what happened here when the mile-high wall of ice came creeping from the north half a million years ago. I knew I hadn't a hope of picking out the clues by myself, so I was delighted when local geologist Allan Wheeler agreed to walk with me and point things out.

'Let's meet on the village green at Bradenham,' was his suggestion. 'There's a few things there that'll help tell the story. Then we can go up on Whiteleaf Hill and get a proper view.'

Bradenham lies at the outer rim of the Chilterns, 15 miles northwest of Denham as the crow flies. Allan was waiting there, a modest and quietly spoken expert in his field, very active in local geological societies as a speaker and expedition organizer. At the edge of the village green reposed a line of huge sarsen stones. There's debate about the provenance of the name, but it probably derives from 'Saracens', a term used in western medieval culture to describe the Arab invaders during the Islamic occupation of southern Europe between the eighth and fifteenth centuries. 'Sarsens' or 'saracens' referred to the foreign look of these blocks of composite stone that turn up in geological territory where they don't belong, where they seem to have arrived like strangers. Allan passed his hand across one of the sarsens, a great ragged block of Hertfordshire puddingstone the size of a hay bale, composed of nuggety flints cemented together with silcrete, a glue-like compound of gravelly sandstone and silica that sets as hard as concrete.

'You're looking back somewhere between five million and two point five million years ago, the end of what we call the Neogene period (23 mya–2.5 mya) and the start of the Pleistocene, what's called the Ice Age (2.5 mya–11,500 years ago). There were several glacials and interglacials, freezing and thawing following each other. Whenever the climate warmed there were multiple rivers depositing a mixture of sand and clay on top of the chalk – it weathers into a mottled red and yellow; you've probably seen it in various places across the Chilterns. Those rivers would flood, forcing the flint pebbles out of their banks. Silica dissolved out of the sand, and then cemented with the clay into the sheet of hard silcrete that binds these flints together. The silcrete sheets were broken up into individual sarsens when Britain was a tundra with permafrost during the Ice Age.'

Allan patted the stone. 'This sarsen came here from Naphill Common a mile away and a hundred and twenty metres higher up, gradually moving down the slope as water and weathering, and gravity of course, shifted it bit by bit. This next sarsen here looks a bit redder and smoother, doesn't it? That's because the sand in it was infused with iron oxide, and it was deposited and smoothed off by a more sluggish river. Sarsens vary, as you see. But local builders used whatever they could find to hand, and they were very grateful if they found sarsens on their doorstep.'

We walked up the grass towards St Botolph's Church at the top of the green. The church is nine hundred years old, built of the local flint, some in small round cobbles, some of shapeless fragments, some roughly knapped by medieval masons. Allan pointed to the rough grey stones that formed the inner corners of the fifteenth-century tower. 'Sarsen, that is, from Denner Hill, a couple of miles north of Naphill. A good building sandstone which they also used in building Windsor Castle.'

The outer corners of the tower, by contrast, were Portland stone, probably from the top of Quainton Hill away northwest in the Buckinghamshire plain. Around 155 million years ago in the late Jurassic, thick deposits of Kimmeridge Clay were laid down in the deep seas that covered Britain. When the sea receded and became shallower, lime from tiny shelly creatures began to pile up and form the so-called Portland stone (named after Dorset's Isle of Portland where it was massively quarried), of which only a thin layer remains in the Chilterns, capping some of the hills. I rubbed my hand across the rough, abrasive surface of the church stones, layer upon layer of minute shells, tight-packed and cemented together to create this block of matter hard enough to keep a church tower standing tall through half a millennium.

Down the centuries local stone has been used to make and mend around St Botolph's. Flints were ground to a gritty powder to stiffen the mortar for tower repairs. In the churchyard the gravestone of Peter David Stubbings (d. 2020) had been fashioned out of sarsen stone. The arched south door, tall and slim in Saxon style, was coarse local limestone, in contrast to the porch, built a thousand years later of smooth yellow Bath limestone, but still incorporating local flints roughly knapped. It's all a testament to the ingenious practicality of builders, scouring their immediate surroundings for anything at hand, on the ground or under it, that might prove useful.

From Bradenham I headed north in Allan's company, climbing up the Chiltern slope to the curving promontory of Whiteleaf Hill. I couldn't help exclaiming as we emerged from the beechwoods at the brow of the hill, for here was a grandstand view out west from the great chalk escarpment. Now the whole Ice Age field of action was revealed below, a life-size relief map of the low country to the west. 'The ice sheets came this far south during what we call the Anglian Stage of the Ice Age, that's between 478,000 and 423,000 years ago.' Allan was pointing out towards the flat claylands of Buckinghamshire, well to the west of where I'd been following the Grand Union Canal. 'The ice would have come from over there, and probably stopped there finally. Might have been half a mile thick.'

This ice behemoth at its most southerly extremity covered the Vale of Aylesbury and butted up against the Chiltern escarpment. It blocked the course of the Thames at Watford, causing it to flow south and a huge lake to form. The large Bradenham Valley was excavated by seasonal meltwater roaring off the ice in the Vale of Aylesbury. 'It would have been a wide, shallow valley, scoured clean by this massive meltwater river flowing across the surface of the permafrost; probably a braided river like the Test today, but several hundred feet wide and tremendously fast.'

The meltwaters came out on the charge in a twisting string of courses that joined and diverged and rejoined, as powerful as an inland tsunami. They tore open the Bradenham Valley, sluiced and scooped out the sides and bottom. 'See the promontory down there in the southwest, Wain Hill? That's the mouth of the valley. Nowadays it dips northwest into the plain, but back then it sloped the other way with the meltwaters rushing away southeast towards the Thames across a bare and empty surface of permafrost.'

The noise, the crack and crumble of ice, the roaring of the water and rumble of boulders, flints and pebbles bowling along must have been stupendous, but half a million years ago there was no human ear to hear it. Our snub-nosed, deep-browed ancestors, *Homo heidelbergensis*, had been driven out southward by the intense cold and unworkable conditions.

'An ice wall half a mile thick.' Allan Wheeler's words had dropped casually enough, but at the crest of Whiteleaf Hill I rolled them round like gold nuggets. Half a vertical mile of ice – what was that? – getting on for 1,000 metres, two and a half times as tall as the Shard, five times the height of the chalk cliff at Beachy Head. I sat and looked out, imagining the ice face like the Wall in *Game of Thrones*, or more probably a great sloping arrowhead rising to a wrinkled upland of ice, smeared grey and brown with dust and squashed-up rock, cracked with fissures, no plants, no trees, no hill peaks tall enough to break its surface and rise higher into the freezing air.

Next morning my feet went on strike, demanding respite from long days of towpath walking. I told them to shut up, and promised them a nice easy time of it once we'd done those last 15 miles from Denham to the Thames at Brentford. Sulkily, they agreed, and in fact I hardly registered the passing miles, the increase in traffic noise and smell, the encroaching buildings, the rumble of the motorways, the roar of Heathrow Airport. It was nose to the path and mind elsewhere.

Gravel diggers, house builders and road makers have unearthed a mass of prehistoric evidence under the streets, parks and riverbanks along this urban stretch of the Grand Union Canal. Casual grubbers in the earth have happened upon animal bones and tools of flint, bronze and iron. Archaeologists have minutely investigated and interpreted these finds. And it's clear that our ancestors have come and gone and come again along the shallow valleys of the Colne, the Crane and the Brent, rivers that course through the gravels, peat layers and silt on their way south to join the Thames.

First the Palaeolithic or Early Stone Age hunter-gatherers and fishers, nomadic dwellers in caves and makeshift huts, users of stone and bone tools. They only very gradually developed their skills as they morphed from the *Homo antecessor* hominoids that reached Britain 900,000 years ago through many developmental stages into *Homo sapiens*, our own species, some 40,000 years ago. Mesolithic or Middle Stone Age hunters began to settle along river valleys after the last glaciation some 12,000 years ago, crafters of delicate stone blades, polished and sharply pointed, that they attached to handles of bone, wood and deer antler to give better purchase and power. The Neolithic or Late Stone Age farmers followed around 6,300 years ago, still reliant on stone tools for cutting and chopping, ploughing and tilling, but capable of domesticating animals and growing cereals, able to plan with reasonable confidence for the next year and the year after that as they established a non-nomadic way of life in the fertile plains and developed skills that embraced pottery, sewing and weaving. Then the huge jump up in culture and understanding represented by the people of the Bronze Age (4,200–2,750 years ago) with their large settlements of round houses, their wheeled vehicles and ox-drawn ploughs, and their minds creative enough to solve the equation that said copper + tin = bronze, an alloy far tougher than either of its components for making weapons and tools. Finally, the folk of the Iron Age who learned to forge iron and smelt steel, to fashion multi-roomed houses, and to build hill forts for defence and temples for pantheistic worship. Their way of life was still flourishing some 750 years later when the Romans invaded Britain and gave the native culture a radical shake-up.

Mesolithic flint blades have been discovered in fields at Denham, along with Neolithic flints and scrapers, all round a mess of burned flints indicating a camp fire. A couple of miles down the canal at Uxbridge, round another hearth where the gravel had been cracked by the heat of a fire, were flints worked on a thousand years or so after the last Ice Age, scattered among bones of red deer, beaver and otter that had already recolonized the Colne Valley. At Three Ways Wharf nearby, archaeologists found five separate scatters of flints along with bones of wild horse and reindeer, and with them were delicate flint blades over 20 centimetres in length, the best example of such 'long blades' found in these islands, made with great skill as, or just after, the last ice withdrew from Britain.

A great leap back in the story of human development was found at Yiewsley, another mile or so along, where gravel extraction yielded troves of early Palaeolithic hand tools, some used by our ancient ancestors 300,000 years ago and others by Neanderthals as late as 70,000 years ago. And a mile or so on around Osterley Park, at the highest point of the plain that slopes down to the Thames, there's masses of evidence of a very broad spectrum of occupation. Here above the River Brent, Palaeolithic nomads had left hand axes and flaked flint tools from as early as 440,000 years ago. In interglacial periods people here had hunted lions and rhinos, and left their bone fragments scattered. Neolithic people with flint axes were the first to prepare and farm the fertile land near the river; Bronze Age inhabitants built ditches, banks and mounds; Iron Age farmers constructed drainage moats that circled their round houses.

From Osterley down to the Thames the River Brent runs closely intertwined with the Grand Union Canal. Around this area flint tools from the Early to the Late Stone Age have been unearthed, along with the remains of animal bones: Ice Age mammals such as mammoth and woolly rhinoceros that roamed a landscape of tundra and permafrost; and a perfect menagerie of animals that were hunted in the warmer interglacial periods – hyena, hippo, ox, red deer, bison, straight-tusked elephant, and the magnificent giant deer that stood 2 metres high at the shoulder and carried massive antlers, more than 3 metres from tip to tip and 40 kilos in weight. Thoughts of these, and of the snub-nosed, low-browed, strong and active people who shared the land intermittently with them for the best part of a million years, carried my battered feet and me on down the towpath beside the amalgamated River Brent and Grand Union Canal, until their united waters gave a final wriggle and passed through Brentford Lock to join the slate-grey Thames.

On the terrace of the cutely named One Over the Ait pub (it overlooks Brentford Ait, or elongated islet, just upriver of Kew Bridge) I sat at a table, boots off, a pint of London Pride standing ready next to an already emptied glass. On the Thames below, Brentford Ait was bushy with willows. The Three Swans inn, which flourished on the island in the late eighteenth century, was once a notorious place of resort. It was in a room at the Three Swans in 1780 that actress Mary 'Perdita' Robinson succumbed to the seductive charms of George, Prince of Wales. Robert Hunter, whose house on Kew Green was close by, later wrote in complaint to the City of London:

> Brentford Ait is a great nuisance to this parish and the neighbourhood on both sides of the river . . . The house of entertainment has long been a harbour of the men and women of the worst description, where riotous and indecent scenes are often exhibited during the summer months and on Sundays.

Hunter got rid of the nuisance eventually; in 1812 he bought the whole ait, and promptly closed down the Three Swans.

All kinds of artefacts have been cast up on the foreshore where the Thames bends between Brentford and Chiswick. They were the focus of obsessive interest to Thomas Layton (1819–1911), a coal merchant and local resident. Layton was a very private man who jealously guarded his collection of antiquities, many retrieved from the riverside hereabouts, which grew to fill his house and overspilled into thirty sheds in his garden in a state of increasing chaos. After decades of confusion and wrangling following his death, the Layton Collection was transferred to the Museum of London. Two items, enigmatically styled as 'Late Iron Age', yet made of bronze, caused me to linger the very first time I saw them. One was a horn cap, shaped like a shallow bowl on a pedestal, with intricate swirls and curls of Celtic design; it was some form of accessory for a chariot, though experts seem unsure whether it was a handle to help the charioteer mount his vehicle, or a kind of prestigious hub cap, the equivalent of today's flashy alloy car wheels.

The other artefact from the Layton Collection that stayed in my memory was a great tankard, a mighty half-gallon quaffing mug fashioned from staves of oak and sheathed in bronze. Daydreaming on the pub terrace above the shore where this heroic drinking vessel was found, I saw a warrior of mighty thews and sinews raise the bronze mug, heirloom from another age now wrapped in myth, to the rafters of some drinking hall.

The sounds of the modern age came to burst my bubble. 'Get it down you, go on, all in one!' shouted some lads behind me. Sounds of gulping, beer spilling, a roar of laughter. Some things don't change.

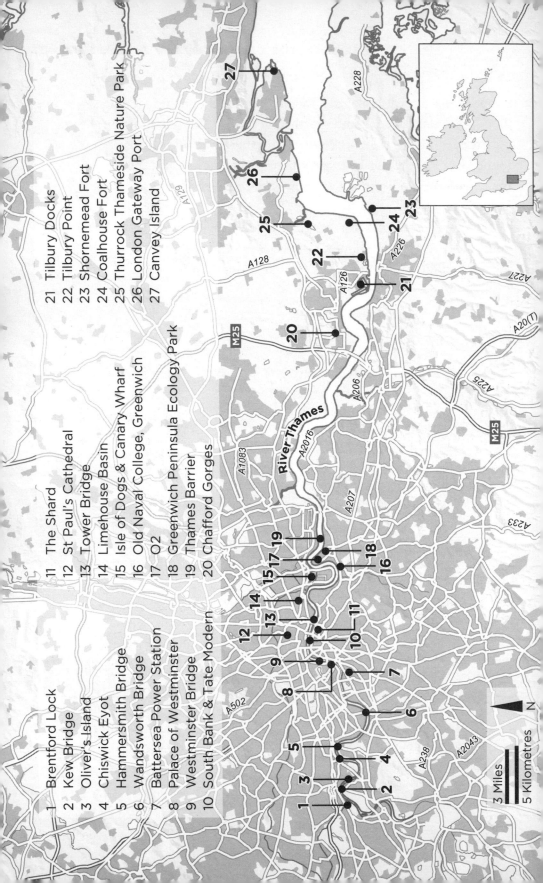

1   Brentford Lock
2   Kew Bridge
3   Oliver's Island
4   Chiswick Eyot
5   Hammersmith Bridge
6   Wandsworth Bridge
7   Battersea Power Station
8   Palace of Westminster
9   Westminster Bridge
10   South Bank & Tate Modern

11   The Shard
12   St Paul's Cathedral
13   Tower Bridge
14   Limehouse Basin
15   Isle of Dogs & Canary Wharf
16   Old Naval College, Greenwich
17   O2
18   Greenwich Peninsula Ecology Park
19   Thames Barrier
20   Chafford Gorges

21   Tilbury Docks
22   Tilbury Point
23   Shornemead Fort
24   Coalhouse Fort
25   Thurrock Thameside Nature Park
26   London Gateway Port
27   Canvey Island

River Thames

N

3 Miles

5 Kilometres

Concrete and Clay:
River Thames to Canvey Island

A BRISK LATE AUTUMN AFTERNOON at Kew Pier where I waited for my boat to London. The River Thames at low water occupied only the central strip of its bed of grey stony shillets. Ash and willow hung limply over the banks, their spear-blade leaves gently showering the pebbles with green and gold. Black-headed gulls in white winter hoods screamed in their harsh accusatory voices as they harried a herring gull weighed down by a sandwich stuck sideways in its beak. Low walls of patchy brick and concrete kept the river from the red and white houses and pubs along the waterfront, though it was hard to imagine this rippling trough of slack olive-green water, a few feet deep at best, ever rising with strength enough to overtop its banks. Hard to believe, too, that a few miles downriver from these spacious surroundings of mellow brick and leafy banks, of old pubs and houses whose meaning you could understand, a solid jam of glass boxes with dark polarized faces as blank as Hollywood aliens would be cramming to the brink of the river and jostling for space halfway up the sky.

Old barges and neatly done-up houseboats lay moored on the thin coat of mud under the wall. A rower in a red vest worked a long white shell under the central arch of Kew Bridge. Above her a metal cylinder came tinkling down the stonework on the end of a scarlet rope. It splashed into the water, tilted, filled and righted itself, and began its slow journey up again to a young woman in high-vis who leaned over the parapet and reeled it in. A couple of minutes later she was down on the pier beside me, repeating the exercise. 'Just testing for ammonia levels; Thames Water need to know whether there's sewage pollution.' She paused to correct herself. 'Well, there's always sewage pollution at some level, but whether it's unacceptably high.' She dipped a probe into the container of water, connected the lead to her laptop and consulted the screen. 'And . . . here, today . . . it isn't.'

Kew lies at the western end of a 40-mile stretch of the Thames that in 1957 was declared unable to sustain animal life. The river had become so grossly polluted with emergency sewage discharges and heavy metals from road run-off and industrial waste dumping that there was a complete absence of fish between Kew and Gravesend on the Thames estuary. In the 20-mile section of water through the city itself there was no measurable dissolved oxygen at all. The Thames, the 'dirty old river' of Ray Davies's song 'Waterloo Sunset', was biologically dead.

Something had to be done. From 1976 all raw sewage was banned from entering the river. More water-treatment works were opened. After the water companies were privatized in 1989 the National Rivers Authority was established to protect rivers, though the responsibility for water supply and, crucially, sewerage passed to the newly created privatized companies. Over the ensuing decades, Thames Water has continued to dump millions of tons of sewage into the river each year in the form of emergency overflows. Every privatized water company has done the same in its own region. England's rivers are still among the dirtiest in Europe. But London's river, in many ways the country's showpiece waterway, did benefit from its initial clean-up. Salmon, flatfish and eels returned to the Thames. Water quality in the river began to be properly monitored. And today oxygen levels are boosted by special boats, charmingly styled 'Thames bubblers', which cruise up and down injecting oxygen into the water. A fleet of flat barges sucks up the plastic and other pollutants. As usual, modern man has come up with a technological fix at the very last moment. And reflecting on that familiar human cycle of accumulate, degrade and renew, accumulate, degrade and renew, I detected a curiously geological, Huttonian feel to it.

Here on the Thames my journey was all but over. I'd walked up the great geological mountain of Britain, up through the aeons from a baseline of lifeless rocks measured in thousands of millions of years, all the way up the slopes to the uppermost clays and gravels laid down by the retreating glaciers, with man the hunter and gatherer in their train. Epochs had come and gone under my boots, epochs with names and dates ever-changing as geological science continually refines its understanding. Hats off to any amateur who can keep up with the subdivisions of the Palaeocene, the Eocene and the Oligocene, the Miocene, the Pliocene and the Pleistocene, names that reminded me of the modelling clay which my children shaped into bulbous dinosaurs. At last I had my head in the clouds and was stepping out on to the summit plateau, the post-Ice Age, still-unrolling epoch called the Holocene. And now I was there, I realized that there was one more tiny nebulous layer to contend with, a stratum still in active creation, a geological eye-blink of time during which humans have first shared, then dominated, their surroundings. The Anthropocene epoch, some call it, though that term is a source of contention. The National Geographic Society defines the Anthropocene as 'an unofficial unit of geologic time, used to describe the most recent period in Earth's history when human activity started to have a significant impact on the planet's climate and ecosystems'. Some think it is self-excoriating bunkum, a stick with which western liberals can beat themselves up about climate change. Others condemn it as unscientific, or too open-ended. When should the Anthropocene epoch be reckoned from, for example? From the so-called 'Great Acceleration' in population growth, human consumption of natural resources and degradation of the biosphere after the Second World War? From 1950 and the first appearance of plastic particles in sediments, or from the detonation of the first atomic bomb in 1945? From 1712 when Thomas Newcomen first effectively harnessed the power of steam? With the start of intensive agriculture along the River Nile around

8,000 BC? Or way back some 46,000 years ago, when humans hunted Australia's megafauna to extinction?

Trying to reconcile what I could grasp of scientific understanding with what I had experienced for myself along the way, I'd already come to appreciate that geology is not just about a pile of rocks. Geology is entangled, right down to its roots, with soils and plants, with wind and water, sand and silt, rain and fire, the breathing of a salmon and the crawling of a snail; in short, with what I understood as 'nature', everything about the world that's not to do with man. If the Anthropocene is a measurable geological epoch, then it's a quantum leap from all the other epochs. Up to this wafer-thin and putatively non-existent layer, the natural processes of the planet have ruled the roost. But in the Anthropocene the human factor becomes the rogue element, exerting a grotesquely disproportionate influence, poisoning the air, fouling the soil and water, uprooting and unbalancing, crashing about like a bull in a china shop, but without the bull's excuse of lack of consciousness. The deleterious things that man does, he does in spite of, rather than in the absence of, knowledge and understanding. But it's this very knowledge and understanding that also enables him to box his way out of the corners in which he traps himself and everything that shares the place with him.

Watching the Thames Water technician at work, I felt very acutely aware of that. Genuine geological epoch or handy metaphor, the Anthropocene was where I'd landed today.

In terms of formally recognized geology, though, I had left the chalk behind when I'd joined the Grand Union Canal among the flooded gravel pits of the Colne Valley. The chalk was still there, though, far underground, part of a geological club sandwich eight layers thick. Climbing upwards from the deepest and oldest rocks that have yet been reached by boreholes under London, these layers are of Silurian mudstones and sandstones (444 mya–419 mya), covered by Old Red Sandstone from the Devonian period (419 mya–359 mya). Above those there's a gap in the geological record, where depositions of New Red Sandstone from the Triassic have been eroded away. Next comes a layer of thick blue Gault Clay about 112 million years old; above it, a layer of Upper Greensand, laid down a little under 100 million years ago; above that again, a great chalk basin up to 200 metres thick. Infilling the basin, a belt of blue-grey London Clay almost as thick and solid as the chalk, full of fossils, formed from marine deposits around 60 million years ago when Britain was under warm tropical seas. On top of the London Clay, laid down in shallow water some 10 million years afterwards, a layer of Thanet Sand, fine-grained and warm coloured; and approaching the land surface, beds of sand and gravel deposited by the Thames during the Ice Ages. As water became locked up in ice and the sea levels fell, the river piled up these sand and gravel terraces along its margins as it shifted its location south to its current bed. It also deposited brick-earth, a mixture of sand and clay that has proved perfect for London's brickmakers over the past two thousand years as the settlement by the Thames grew from nothing to one of the world's great cities. And that growth has itself laid down another layer over which geologists of the future will pore, a layer of bricks and clay pipes, of iron rods and concrete slabs, of glass and chemicals and plastics – the layer on which London stands, to which it adds daily, and from which it subtracts 7 million tonnes of waste each year for recycling, for incineration to generate heat and power, or for landfill – the Anthropocene layer, many metres deep in the oldest quarters, that underpins the city.

A neat little pleasure boat, *Connaught*, was slowly approaching Kew Pier and the handful of passengers waiting there. I'd been expecting a big fat bully of a boat, the sort that rushes about the Thames in London itself, but here was this trim Edwardian craft to ease us away downriver. I crossed the wobbly gangplank – literally a plank – and got up into the bows. As we moved into midstream and swung around under Kew Bridge, a flight of Canada geese landed in the shallows with maximum flap and fuss, lifting their heads with the white chinstraps as one to watch us depart. Beyond the geese a heron stood by a muddy sluice, head cocked forward, dagger beak poised, yellow eyes concentrating and alert for any movement in the shallows. It was the first of many I saw along these semi-urban miles of the Thames, hunched grey opportunists who wouldn't have been wasting their time there had the river still been devoid of fish.

Connaught slipped smoothly along the narrow low-tide waterway past a succession of eyots or aits, modest little islands shaped long and slim by the constant whittling of the river. Tree-hung Oliver's Island, its humpy back consolidated with old brickwork and sandbags, was overlooked by the white-faced Bull's Head inn behind the north river wall. Legend said that in November 1642 Oliver Cromwell, trapped in the inn during the Battle of Brentford, escaped through a secret tunnel under the river to hide on the island. But if every secret tunnel under the Thames was a reality, there wouldn't be enough river bed left to hold any water.

Flat shillets of pebbles had smothered the steps below a boathouse and half choked a barred opening beyond, evidence of the dynamism of the river, always shifting its mud and stones, uncovering centuries-old cobbled hards on very low tides such as today's, sweeping away landing stages and willows from the eyots at flood times. Near the three cantilever spans of Barnes railway bridge, sluice gates and open concrete pipes had laid lines of bright green algae across the mud as they trickled who-knew-what into the river. I wondered what findings the Thames Water technician's laptop would show if she were to deploy it here today.

By Chiswick Eyot a line of cormorants held their wings out to dry like ragged black washing. A luxuriant growth of willows and reeds behind a hand-woven flood fence obscured the island, once prized for the rich grass fodder that grew in its silty soil. A party of wellington-shod schoolchildren grasping worksheets plodged through mud and pebbles along the north bank, while from the opposite shore a fisherman patiently watched his float from the tideline – like the herons, a sign of a river restored to better health.

Hammersmith suspension bridge spanned the river in Victorian wrought-iron elaboration. Two workmen clung to the top of its northern portal. The weight and numbers of modern traffic have cracked the old structure, opened in 1887 when the heaviest thing it might be expected to carry was the occasional traction engine. Beyond the bridge a fleet of rowing shells propelled by schoolgirls was passing the pink Edwardian terracotta palace of Harrods Furniture Depository, pursued by a coach in a motorboat. 'Amelia – into the bank a little! Heather – upper body!' *Connaught* soon left them behind. At Barnes on the south bank the massed trees and bushes of the London Wetland Centre, a green oasis reclaimed from a clutch of disused reservoirs, gave the impression that we were cruising through the countryside. Bankside buildings were still low-rise, carefully painted and preserved to enhance the river frontage. But reality entered the picture in the shape of a flat barge, a PDC or 'passive debris collector', cheerfully labelled 'I Eat Rubbish', which held a mess of plastic and other flotsam it had sifted from the Thames, far more than you'd expect from any country stretch of river. And at Wandsworth Bridge, still a good 5 miles from the centre of London, the change from suburban to city architecture came up like a smack in the face.

Suddenly the Thames, the dominant feature in the landscape up till this point, was overshadowed by buildings that sprouted one above another like toadstools after rain. The first high-rise appeared on the south bank, and then a mass of twenty-five-storey blocks, close-packed, followed by giant flights of buildings stepped like Italian hill villages grown monstrous, dwarfing the low spans of Wandsworth and Battersea bridges, crowned with scarlet cranes and filling the forward horizon as though blocking off the river. Scale was given to this mountain range of buildings by the fat Rhine barges moored four deep along the river wall below, each twice the size of *Connaught*, but toy-like under the stacks of luxury flats. The designers had aimed to differentiate each development from its neighbours with curves, swirls and 'venetian blind' effects. But any character redolent of this particular city and its unique charisma seemed absent from such prosperous 'could-be-anywhere' architecture.

As for the rubbish generated within these gilded courts, that was travelling away at a brisk lick in barge-loads of yellow containers pulled by tugs, headed downriver.

Albert Bridge and Chelsea Bridge, Grosvenor Bridge, and then the brick bulk of Battersea Power Station heaving up on the right, its white fluted chimneys soaring. I looked instinctively but in vain for a Pink Floyd pig high in the sky between the chimneys. It was bizarre to realize that the eye accepted the enormous art-deco power station as familiar and comforting in scale compared with the multi-storey glass pencil boxes that reared up on end alongside. On under the flat crimson-and-gold arches of Vauxhall Bridge, then the pink spans of Lambeth Bridge, with a whisper of the Thames of olden days emanating from the wall of lion's head mooring rings along the south bank. Those rings were there when Arthur Conan Doyle came down to the insalubrious river front-age of Lambeth in the late 1880s in search of colour for his epic Sherlock Holmes tale, *The Sign of Four*. The Albert Embankment had been built along Lambeth's waterfront by that time to carry one of the great sewers that revolutionized public health in Victorian London; but it was here among the remaining wooden jetties and river stairs, ramshackle houses and low pubs that Doyle sited the wharf of Mordecai Smith, owner of the steam launch *Aurora*. How I had thrilled to that yarn, reading in bed as a boy in Gloucestershire, following Sherlock Holmes and Dr Watson through this strange place called London, as with the help of Toby the mongrel they nosed out the *Aurora* and pursued her down the Thames with her exotic cargo – the wooden-legged villain Jonathan Small, his sidekick Tonga the Andaman Islander with a blow-pipe loaded with poison darts, and a box full of Indian treasure! No trace today of wooden wharves, tumbledown dwellings or steam launches among the towering flats and offices of Lambeth's river front, but there was still a whiff of romance for me as little *Connaught* slid beneath Westminster Bridge. Right in front of Big Ben and the palatial Parliament building, the stone steps of Speaker's Stairs led down to the water – 'Westminster Stairs' in *The Sign of Four*, where Holmes and Watson, pistols in pockets, embarked in a fast police launch on a sunlit evening for the breathless chase of the *Aurora*.

At Westminster Pier those who intend to continue down the Thames and don't have a fast police launch on call must change boats. I disembarked and watched *Connaught* burble off upriver. Larger ferries surged under the bridge, barges swung to their moorings and restaurant boats jostled by the bank. Opposite the giant bicycle wheel of the London Eye lay another rubbish gobbling PDC vessel, this one proclaiming 'I Clean the Thames'. Soon a lumbering sightseeing boat, *Thomas Doggett*, scooped me off the jetty and proceeded downriver. Here were the big bullying boats I had expected, dubbed 'Thames clippers', the fluvial equivalent of those fat gangster cars with smoked windows, zooming along at 20 knots with grimy bow-waves and deeply furrowed wakes that rocked anything smaller crossing their sterns. Some carried the Uber name, though they were too big to be at the beck and call of individual commuters. They charged up and down, spewing out water and fumes, already obsolete in design, soon to be superseded by a new breed of Thames clipper with hybrid electric engines that recharge from the excess power they themselves generate.

Thomas Doggett's skipper was a master of the cheeky-chappie commentary. 'Just behind Cleopatra's Needle, layz-'n'-gemmun, the Savoy Hotel,' he offered as we emerged from under Charing Cross railway bridge. 'I was in the American Bar there last month, twenny-five quid for a gin and tonic. You can take the glass home with you, mind . . .' Pause. '. . . but just don't tell them you're doing that, or they get a little bit upset.'

The classic landmarks of the south bank's post-war river front slipped by. The Royal Festival Hall of 1951, concrete with a light touch; the Royal National Theatre, blocky and brutalist; Tate Modern, housed in the squat but enormous 'cathedral of industry' that was once Bankside Power Station. Opposite, beyond the north bank, the dome of St Paul's Cathedral, finally dwarfed out of its long-guarded prominence. 'Until 1967,' said our skipper, 'nothing in London was taller than the cross on that dome.' Ahead loomed the twenty-first century's take on vast glass architecture for the City of London: gherkins, mobile phones, cheese graters, prawns. And of course the Shard, literally scraping the sky over the south bank, a cloud streaming from its fractured summit 1,000 feet above the ground.

Now the river had reached full tide, as high as the ebb had been low at Kew. I'd hoped to catch sight around Bankside of a posse of mudlarks, those dedicated collectors of abandoned treasures from the sand, mud and rubble of the Thames foreshore, but the rise of the tide had already driven them up the stairs and away. People have been scavenging the tidal banks of the Thames since time out of mind, but the practice reached a peak with those who did it for survival in savage old Victorian London. At low tide these mudlarks, most of them young children, would scavenge the foreshore around barges, wharves and riverside pubs for anything to sell – bottles, old iron and copper, bones, nails, discarded tools. Threepence a day was the best they could hope to realize. In 1851, the year that the Great Exhibition showed off Britain as the wonder of the civilized world, Henry Mayhew published his book *London Labour and the London Poor*. In it he wrote of the Thames mudlarks: 'These poor creatures are certainly about the most deplorable in their appearance of any I have met with . . . Their bodies are grimed with the foul soil of the river, and their torn garments stiffened up like boards with dirt of every possible description.'

The Thames that Mayhew's mudlarks worked was a totally filthy river. Since the turn of the nineteenth century the population of London had more than doubled to 2.6 million, but the only methods of sewage dispersal were medieval ones – cesspits, waterways or the gutter, with the solids being carted off between dusk and daybreak by the euphemistically named 'night soil men', and dumped way downriver as farm fertilizer. A cholera outbreak in 1848–9 flagged up the problem, but Parliament's mind was not focused on it until the 'Great Stink' of 1858 when the river smelt so bad that sittings at Westminster had to be suspended. Father Thames appeared in cartoons as a revolting shit-smeared anti-Poseidon, with his offspring Disease and Disgust. Parliament hurried to pass laws for a new London sewage system. Civil engineer Sir Joseph Bazalgette was appointed to plan it, and he came up with a design that collected the city's sewage, run-off and domestic waste, and fed it in pipes downriver until they discharged in the lonely marshes east of the city where no one wanted to go. The new sewers ran in embankments on either side of the Thames – Victoria and Chelsea Embankments to the north, Albert Embankment across the river to the south.

Bazalgette designed his sewerage system for a city of 4 million, but in the ensuing century London's population more than doubled. Second World War bombing destroyed a lot of the system. There was no more Great Stink, but the dirty old river of 'Waterloo Sunset' was nothing for the city to be proud about.

As for today's environmental problems along the London Thames, rising temperatures associated with climate change continue to deoxygenate the river. Although recession-driven factory closures and tighter anti-dumping laws have seen fewer heavy metals enter the river, the level of this pernicious form of pollution remains high as tiny fragments stick tenaciously to the London Clay at the bottom of the river. Plastics and their minute particles deal death to fish. We still don't know enough about the pharmaceuticals we keep inventing – anti-depressants, diabetes and contraceptive medicines, for example – which can slip through the processing at water-treatment plants. Run-off from the roads still charges the river with microplastics from tyre and brake dust, as well as the oils, heavy metals and other pollutants from exhausts. And raw sewage can still be discharged into the Thames if sudden enormous rainfall, increasingly part of our climate-change experience, overwhelms the system.

The hardening of the banks on both sides of the Thames as it flows through London means there's no room for the most natural and inexpensive flood defences – gradually shelving salt marshes, river meanders, Lammas meadows, trees and vegetation – to slow down the rate of flow and absorb the flood water. Slick, impermeable man-made surfaces like tarmac and concrete do the opposite. They are designed to shed water away from themselves and the development they are part of as quickly as possible. As soon as rain hits them it washes away downhill and into the Thames, piling up and racing along the hard walls and banks, looking for somewhere to break out. Planners and developers have been thinking and acting in ways that run counter to nature. So you can't help wondering whether all their Thames bubblers and electric ferries, their planted reed beds and rubbish patrols amount to anything more than futile greenwashing as we go on kicking the environmental can down the road.

Under the bridges went *Thomas Doggett* – plain and ugly London Bridge, ornate and wonderful Tower Bridge. Here the scene and the story changed, from pure glass-hearted exhibitionism to the massive warehouses of a working river, from stocks and shares to shipbuilding, cinnamon and coal. Yet, lamented our skipper, these waterfront warehouses once filled with the world's produce were now luxury flats, the formerly pulsating docks all gone to eateries and drinkeries, every old shipyard filled with brand-new apartments, and only a tiny glimpse of cranes and gantries behind St Saviour's Dock to remind him of how lively the Port of London had been when he was a lad just starting work in the 1970s.

In sketching out the vibrant dockland economy of his youth, the skipper of the *Thomas Doggett* was donning a pretty outsize pair of rose-tinted spectacles. I had memories of those days, too. I recalled setting off to explore the north bank of the river from Tower Bridge to Limehouse on a summer's day in 1973. There had been a buzz of Dickensian thrill to the weedy walls and river stairs, the rotten old jetties and the whiff of spices around the dark lowering warehouses on Cinnamon Street. But those warehouses were empty, and some of them were already coming down. There was dust in the air, and the crunch of wrecking balls. What a transformation since then. And what a revolution in the docklands of London, forever moving downriver, away from the city, away from the valuable real estate they occupied, east and east again to reclaimed marshland that nobody wanted, each time growing, expanding, modernizing and declining, doing the Huttonian dance, each new development drawing nearer and nearer to the mouth of the Thames.

Central London had had its wharves since Roman times, and they ceded trade to wharves that lined the Pool of London, downstream of the Tower. The Pool became too jammed with shipping to cope once the commerce of the British Empire began flooding into London. From then on London's docks were developed further and further downriver – Wapping Docks, Limehouse Basin, the West and East India docks on the Isle of Dogs, Surrey Docks at Rotherhithe, the Royal Docks way down at Silvertown. They and their communities withstood Hitler's bombs, industrial strikes and changing tastes in goods. But none had been able to handle the vast new container ships that arrived from the 1970s onwards. Those stopped at Tilbury on the north shore of the Thames, 25 miles from London, a port cut from the Essex marshes in Victorian times and furnished with deep-water docks that opened in 1978 as all the other docks were falling into obsolescence and closing.

'Wapping Old Stairs on your left,' announced the skipper, 'and that pub at the top is the Town of Ramsgate. See the big "E" painted on that building just along on the right? That marks the site of Execution Dock, and that's where they hung Captain Kidd the pirate. That's Rotherhithe on the right with the Mayflower pub, nice little boozer; the Pilgrim Fathers set off for America in the *Mayflower* from there in 1620, though I doubt they had a pint before they set sail. And on the left again, all that building going on just in front of the King Edward Memorial Park – that, layz-'n'-gemmun, is where they are building the Thames Tideway Tunnel to keep our river nice and clean.'

Beneath London, where the Anthropocene layer gets down and dirty, miracles of inventive architecture are being put in place. Just by the Memorial Park a shaft plunges down through the London Clay, connecting a notoriously leaky sewer with the Thames Tideway Tunnel some 60 metres below. This giant sewage pipe, 16 miles long, runs from Acton storm tanks in the west to Abbey Mills pumping station in the east, most of it following the line of the Thames deep underground. The Tideway Tunnel connects up over thirty of the worst sites in London for sewage overflow into the river, Blackfriars' foreshore being one of these, and hurries the stuff away eastwards from Abbey Mills to Beckton Sewage Treatment Works some 15 miles downriver. It's the latest skirmish in the never-ending battle to prevent London's ever-swelling, ever-more-demanding population from poisoning its river, and itself.

'And what do we think about *that*?' enquired our skipper rhetorically as we came in sight of the Isle of Dogs and its shopfront, Canary Wharf. 'Once those docks were done for in the 1970s, over there on the Isle of Dogs, it became the sort of place you wouldn't wanna go for a nice day out. Full of dereliction and unemployment. Something had to be done. So they did that.'

A banking city of mirrored towers pressed against the sky and crowded in upon one another along the prestigious river frontage. Another upright cheese grater, another mobile phone (this one a 'brick' from the 1990s), another pencil box seventy-five storeys high, semaphoring back the sun from one window after another. Some aeroplane propellers; some flights of steps descending from the heavens. And behind this showboating shopfront, a darker forest of oblong monoliths clustered round the slate-grey pyramid roof of One Canada Square, for twenty years the tallest building in the land at 235 metres until the Shard overtopped it by a third as much again. Plain and functional in comparison with the Shard's showy stylishness, the pride of Canary Wharf once overshadowed all around it, but now stands back in the shadows itself. A boxy structure with tiny windows in strict ranks like the honeycomb cells of a hive, its worker bees, once fled, have proved reluctant to return. 'Only half the workers have gone back into those office buildings after the Covid pandemic,' said the skipper. 'The upside is, now you can buy a nice flat in any of 'em with a view of the river for a couple of million quid.'

Now the Thames made a great southward loop around the Isle of Dogs peninsula, cutting through clay and silt and a tongue of Thanet Sand. The silvered City of Oz fell behind and the skyline ahead dropped abruptly to a relatable scale once more. At the bottom of the loop *Thomas Doggett* passed the masts and spars of the beautiful old tea clipper *Cutty Sark*, and swung across the tide towards Greenwich Pier. Here in grand symmetry stood the pavilions and cupolas of Sir Christopher Wren's waterfront masterpiece in pale Portland limestone, a home for disabled sailors that became the Royal Naval College. 'Built in 1694, and a very fine sight to end our little cruise with,' declared the skipper, his crumpled pug face alight with humour as he left the wheelhouse to bring round a top hat for our coins. 'After he designed St Paul's, Sir Christopher had his reputation to keep up, didn't he? If he'd had all that glass on the Isle of Dogs to work with, and a better slide rule, I reckon this place would've been another Canary Wharf. Lucky he didn't, eh!'

I disembarked, leaving the skipper of *Thomas Doggett* jingling his hat and his wisecracks, and followed the Thames Path eastward along the frontage of the Royal Naval College. As I was admiring the fluted columns, something cold and wet slapped my ankles and poured over my boot tops. The Thames, unnoticed by me, had risen to the top of the tide and was spilling over across the walkway. 'Does it all the time, mate,' said the man coming the other way. He performed a sort of sashay, skipping adroitly in and out of the shallow waves, and I followed suit to reach drier land.

Here commenced a hybrid river, grand apartment developments interspersed with uncompromising industrial premises, the only segment of working waterfront along this stretch of the Thames: a boat-breaking yard, a rubble dump, a grimy vessel groaning and clanking at a wharf as she spewed gravel on to a conveyor that fed Hanson's huge ready-mixed concrete works. Tugs dragged rubbish barges downriver, their wash flooding the path a minute or more after they had passed. A reed bed planted at Harrison's Wall swayed and thrashed as a Thames clipper went by, the ripples of its wash running through the stems like fingers through hair, plucking out some, bending others. The disused industrial Alcatel Jetty sprouted mosses and stonecrop, planted to entice birds. Coarse, vigorous sprigs of red valerian and buddleia planted by nature pushed their way through railings around abandoned yards not yet developed into apartment buildings. The Thames Path was roughly screened from this gritty hinterland with lengths of hardboard, sprayed with the only graffiti I saw on the journey – 'Alcee Youts', 'Fuck The Tories'.

The City of Oz reappeared across the river, an otherworldly jangle of rocket profiles and silvered glass. On my side rose the flattish dome of the O2 Arena, bristling with yellow antennae like a creature in a horror B-movie breaking out from deep beneath the Earth's crust to wreak havoc on the human race. These images, jotted in my notebook in an even wilder scrawl than usual, might have been occasioned by an empty belly and an overstimulated brain. Once past the O2 a vast development of tall apartment blocks and futuristic sculpture, the Tide, hid everything near and far. At the feet of the Tide another sparse little strip of reeds bowed and whispered over the greenish mud they had been planted in. The Greenwich Peninsula Ecology Park promised refuge among its reed beds, waterways, bird hides and wildflower meadows, but a notice fixed to the chained gate lamented a boardwalk in disrepair, funds that hadn't materialized, contractors that couldn't be contacted. When would the hides, ponds and meadows be open once more? The notice shrugged its shoulders: no date in sight. In my inner ear I heard the spectral voice of David Cameron, swerving from urging 'Vote Blue, Go Green' in 2006 to telling his officials in 2013, as Prime Minister, to 'get rid of all the green crap' from the government's environmental policies in order to cut energy bills.

More secret fences; more machines beeping out of sight. A gravel boat unloading, a clatter from a scaffolding store, a quiet hum from the British Awning Company's works. Five hundred starlings screaming as they perched on a crane, with the peregrine that had spooked them screeching from its perch on a rusty gantry opposite. The Anchor & Hope pub, with three men joking round a winkle stall outside. Here were signs of life beyond the enormous silent blocks of flats with their empty ground-floor restaurants and key-twirling security guards. And just beyond the pub, as my feet told me that I'd done quite enough for the day, the Thames Barrier straddled the river with a line of steel towers shaped like the helmets of drowned warriors rising to defend the city.

I blinked away the feverish simile and tried to focus. Those ten dully gleaming silver towers looked, in fact, more like cross-sectional slices of fish with the scales left on. The Barrier was completed in 1982, at the time one of the engineering wonders of the world. The gates towards either bank stand close above the river, ready to be lowered if needed, and their presence only a few feet above the water makes both margins of the Thames unnavigable. But ships need to be able to pass up and down in normal conditions, so the six 'rising sector' gates in the centre lie flat on the river bed, allowing traffic to go by. Hydraulic cylinders in the towers rotate to drag each gate up to its vertical closed position with a pair of arms like yellow praying mantis legs. When fully raised the Barrier presents a flat face upstream, and a convex one downriver to mitigate the force of the surge. The designer, Charles Draper, based the hydraulic cylinders on the taps of the gas cooker in his mother's house. That was somehow nice to know – like Barnes Wallis conceiving the design of the Dambusters' bouncing bomb while skipping stones with his children.

The Sea Level Projection Tool produced by the USA's National Aeronautics and Space Administration (NASA) is a map that predicts rises in sea level from a baseline of 1995–2014 if nothing is done to mitigate the threat. It shows about one-third of central London under water by 2030. The Thames Barrier, ingenious and beautiful as it is, can't deal with that. Back in the 1980s it was closed four or five times a year in response to flood threats; now that number has trebled. In the flood season of 2013–14, the same year as David Cameron's 'green crap' outburst, there was a quantum leap to forty-one closures. Altogether there have been over two hundred closures since the inauguration of this flood barrier, which protects 50 square miles of central London from flooding. That amounts to 640,000 properties and 1.5 million people, as well as hospitals, schools, tube trains, shops and offices.

It seems inevitable that we are going to see a very great increase in water from both sky and sea in the approaching decades. 'Even if we reach the government's required target of net zero by 2050, there is likely to be 59% more winter rainfall, and by the end of the 21st century, once-a-century sea level events are expected to become annual events.' This forecast is not from the mouth of some breast-beating catastrophist, but from the Environment Agency's sober and sensible policy paper 'Thames Estuary 2100: 10-Year Review', updated in September 2022. But aren't things bound to be better downriver where the estuary with its flatter and wider banks, less built up, offers so many opportunities for natural flood mitigation through the retention and creation of salt marshes, winding creeks, reed beds, flood meadows and mudbanks? 'The outer estuary', says the policy paper, 'is recognised by government as a prime growth opportunity area, driven by the creation of the Thames Estuary Growth Board in 2019.' And that Board's website posits 'green growth' along an estuary flanked for 60 miles by business parks, 'multi-modal logistics super-hubs' and tens of thousands of new houses. 'This growth must be resilient to flooding,' warns the Environment Agency. We know that flooding is coming. But can flood resilience really be compatible with the continued hardening of the estuary banks and flood plains with infrastructure, and the introduction of hundreds of thousands of people who weren't there before? There were proposals in 2005 to build an enormous barrier 10 miles long between Sheerness and Southend. But how high and how hard, how long and how wide would any man-made barrier have to be to shut out a self-inflicted, existential threat of the magnitude we appear to be facing, here at the apex, and perhaps also at the end, of the Anthropocene epoch, the one tight corner from which our species of marvellous and endless invention might finally prove unable to magic itself away?

An elderly woman and a little boy were bending and stooping along the tideline below the Anchor & Hope. At first I thought they were mudlarking, and then I saw the bin liner they were filling with discarded tins and plastic bottles. That tilted the picture a bit. I got a pint and leaned on the rail and stared at them and at the reflections of the setting sun on the water and the silver towers until I felt better. Couldn't face the winkles, though.

It's hard to say precisely where a river becomes an estuary. The lower Thames doesn't really widen out until it has wriggled its way to Canvey Island, 30 miles east of the Thames Barrier. But the estuarine feeling, a fresher east wind with a hint of sea salt and the more definite surge of the tide, starts to make itself known downstream of the Barrier; and hereabouts the fish on an incoming tide tend to be flounder, smelt, bass and mullet, sea fish that can cope with a bit of water in their salt. A mile downriver the Woolwich Ferry chugged me across the Thames, crossing a tongue of chalk deposited a little less than 100 million years ago. A hop and a step along the north bank crosses the invisible boundary that separates London from Essex, and from here a series of walks brought me eastward along the river as it veered across the strata from London Clay to Thanet Sand, but generally close to or actually on that chalk with its comforting echo of a warm, shallow sea. I was sliding away, for the moment, from the questionable Anthropocene to rejoin the epoch of the Cenozoic, the 'period of new life' between 66 million years ago and the present day, to the geology of chalk and sandstone, gravel and clay through which the River Thames cuts its way eastward to its dissolution in the cold and murky North Sea.

My head was still reeling with the ominous dazzle of the City of Oz. I needed to get my feet back on solid geological ground, to visualize this buried landscape of the lower Thames through some practical illustration. Fifteen miles downriver, beyond the great arc of the Queen Elizabeth II Bridge, the South Essex dormitory towns of Thurrock and Chafford Hundred stand around a string of abandoned chalk quarries named Chafford Gorges, canyons choked with vegetation that open unexpectedly off side-roads and at the back of suburban gardens. In spite of the best efforts of Essex Wildlife Trust, money and manpower are in short supply. The information centre is only unpredictably manned, picnic tables are rotting, and wild growth is creeping or, more accurately, racing in its green vegetable way to take back its native territory. The quarry sides and tramway cuttings, formerly clear-cut, are losing the contest. But before they do, it's possible to discern – as I did gradually, with a little illicit scrambling and a lot of head-scratching – the story of what happened when the chalk grew and the sand drifted, and then again after the polar oceans froze, the sea level fell and the ice began its stately dance across the northern hemisphere – advance, retreat, advance, retreat; repeat, repeat, repeat.

In the former quarry pit of Warden Gorge, the chalk laid down 65 million years ago stood tall as a stadium wall, topped with a darker cap of Thanet Sand 10 feet thick or thereabouts, deposited around 10 million years after the chalk. The top of the older rock undulated where erosion had bitten down into it, and the younger sand had drifted to fill in all the crenellations before continuing to build in a solid layer above.

Sandmartin Cliff stood a little further down the slope towards the Thames. Here the iron gate was padlocked, the fencing resisted entry, and a thicket of spiny municipal shrubs offered further deterrence. But I pushed in among the thorns like the prince in *Sleeping Beauty*, and got my reward with an excellent view of a smooth upright face of very fine, apricot-coloured Thanet Sand, at least 10 metres high. Among the undergrowth at the base of the cliff the sand was footed on a bullhead bed, a crumbled, clotted mass of chalk, thick with nodules of flint known to quarrymen in times past as 'devil's dung'. They were left behind in a heap when the chalk layer they were part of dissolved completely away. Up at the top of the cliff the sand lay under a carpet of gravel laid down some 400,000 years ago in an interglacial period when the Thames flowed at this height, 30 metres above its present level. Chalk under sand under gravel, a succession my understanding could get a grip on.

In the south of the quarrying area another zigzag path led down to Lion Pit. This cleft lies even lower down the slope of the land, less than a mile north of Fiddler's Reach on the Thames. It was dug as a cutting for a tramway that took chalk from the quarries to a jetty on the north shore of the river. Curtains of convolvulus, old man's beard and scrub trees hid most of the cutting sides, but intrepid geologists had battered out an informal network of pathways through the tangle, and there I fossicked about until the puzzle sorted itself out.

The Thames had occupied this site on four separate occasions, said the information board – 400,000, 300,000 and 200,000 years ago in the intervals between glaciations, edging ever southward, digging ever deeper into the underlying Thanet Sand and the chalk below that. The chalk lay clear to see, a porridgy mess in the bottom of the cutting sides down at tramway level. Fractured chalk chunks the size of hens' eggs lay where they had been kicked out by the hind legs of burrowing rabbits. The Thanet Sand, weathered to a greenish ochre, lay above, deposited some 57 million years ago in very shallow seas, probably in a subtropical climate of hot summers and mild winters. On top of the sand, spread by the Thames as it moulded its wide old bed into terraces during interglacial periods, lay a thick cap of orange clay seeded with pebbles and chunks of rock, the Thames gravel that construction businesses are still digging out along the river on either side of London. And cleaving like a pitted skin to all this ancient musculature, the measly few inches of soil that have formed from fragments of rock and vegetation, animal bones and human ones, rags and tags and bits of breakfast eggshell since the glaciers finally went away.

Now it's a hard shore as the Thames narrows to take a southward bend around Tilbury. At Tilbury Point they left the body of Captain Kidd the pirate to dangle for three years as a warning after he was hanged at Execution Dock in 1701. Here on 21 June 1948 the liner *Empire Windrush* disembarked 1,027 passengers, mainly West Indians, pioneer immigrants to help Britain back on her feet after the war. And here, 25 miles from the city, are the docks of London that finally stuck, the massive deep-water berths that could handle the big container vessels whose size and capacity killed off all the other docks upriver. A wedge of concrete islands shapes this corner of the river, lined with container stacks, bristling with cranes and glinting with enclosed water. Those containers of London's rubbish that I saw going downriver make landfall here, and 15 per cent of Tilbury's exports each year are of the city's waste to destinations abroad for recycling or landfill, out of sight and mostly out of mind. Thirty-two million tonnes of cargo go through Tilbury every year, and these docks, recently expanded, are due for more of the same in the plans of the Thames Estuary Growth Board.

From chalk to Thanet Sand and silt and back to London Clay, the Thames makes its final eastward meander. From Tilbury Docks I set out to walk the shore, a strange journey of incongruities. Separated from the bulk and clangour of the docks only by a moat stood a diminutive and beautiful star-shaped Tudor fort. Here on 9 August 1588 Queen Elizabeth I addressed her forces, assembled to oppose the expected invasion by troops from the Spanish Armada, with the splendidly defiant declaration, 'I have the body of a weak, feeble woman; but I have the heart and stomach of a king, and of a King of England too.' Immediately beyond the fort: the wreckage of Tilbury Power Station, demolished in 2019, the site forlorn, plans for Britain's biggest gas-fired power station on hold. A stretch of bleak marshland, leading to a northward turn of the river where Coalhouse Fort lay low and ominous under a grey, rain-speckled sky. Any damned Frenchie coming up the Thames to burn London would be blown out of the water before he'd even spotted the bastion at the bend of the river; that was what General Gordon of Khartoum surmised when he built the fort in the 1860s out here in the marshy wastelands. The French never came, though.

There is a haunting beauty about this Thames landscape where the river is about to broaden into its estuary proper. Oystercatchers piped on the tideline and the water slid gently seaward with the last of the ebb. Across the river rose the ghostly outline of Shornemead Fort, where Charles Dickens sent poor little Pip in *Great Expectations* on a foggy Christmas morning with stolen 'wittles' and a file for escaped convict Magwitch.

This is all moody country, looking downriver over bird-haunted marshes and mudflats. The flat riverside land hereabouts was for centuries the dumping ground for London's rubbish, and the biggest landfill tip of all – destination for decades of those barge-loads of yellow containers – was on the appropriately named Mucking Marshes. The methane-rich gas produced by the decomposing organic matter underground is piped off for generating electricity, a neat piece of recycling in itself. The site closed in 2012, and a phoenix has arisen from the ashes there in the shape of Thurrock Thameside Nature Park, a big reserve of reed beds and grasslands, woods and lakes. From the roof of the Visitor Centre I had a wonderful view over the sullen grey river, the green hills of the landfill, the towers and spires of London floating dreamlike on the western skyline, and to the east a line of massive cranes dipping and rising like skeletal giraffes at a waterhole.

Tilbury, eat your heart out. Here on your doorstep looms an even bigger port, an even harder shore. London Gateway Port, financed by DP World of the United Arab Emirates, was opened in 2013 at the point where the Thames leaves its last meander and widens to meet the sea. It's twice the size of the City of London, extends 400 metres out from the original shoreline and occupies two miles of the Essex bank in one continuous wedge. Thirty million tonnes of silt were dredged to form deep-water berths that can take the biggest container ships in the world, monsters 400 metres long carrying a quarter of a million tonnes each. The cranes ranked along the quay are 138 metres tall, three times as high as anything else in the landscape. London Gateway Port already has links to over fifty countries and aims to handle three times the tonnage of Tilbury Docks. Lorry carriageways feed from the port into the overcrowded roads around London, rail links ramify to huge freight parks and distribution centres in other parts of the country. How much diesel is expended in fuel and fumes, sailing the goods to the port and driving them away, is incalculable, but then the enormous Shell Haven Oil Refinery that was there before London Gateway was no slouch at emissions itself. Ditto the adjacent Coryton Refinery, closed in 2012 and now being transfigured into what the Thames Estuary

Growth Board styles 'a vibrant manufacturing, food, energy and multi-modal logistics super-hub'.

Immediately downriver of the port, separated from the super-hub by the muddy channel of Hole Haven Creek, lies Canvey Island. This dead-flat wedge of land, shaped like a horse's head, exemplifies all that's contradictory about the Thames Estuary Growth Board's bullish oxymoron, 'good green growth'. The western half of Canvey is all green freshwater and salt marshes, one of the most diverse bird reserves in Britain, with marsh harriers over the reed beds, lapwings in the fields, curlews on the muddy foreshore and cuckoos among the scrub bushes in spring. The eastern half of the island is tight-packed with housing and edged with gas storage silos. The dividing line is drawn at an oil refinery that never was, a plan that never came to pass, its memorial a great black pipeline that rises from the green marshes and hurdles the shore on skeleton legs to a terminus of empty air above the river.

'Stand and watch the towers burning at the break of day,' growled Lee Brilleaux, gravel-voiced singer with Canvey Island's 'greatest local band in the world', Dr Feelgood. The song, a familiar earworm, was 'All Through The City', the year 1974 when the Feelgoods took off like an R&B rocket, and the words were those of the band's guitarist Wilko Johnson. The 'Bard of Canvey', who died in November 2022, was a thoughtful songwriter, interested in exploring what he termed a 'submarine consciousness' that was rooted in the strange atmosphere of his birthplace, an island that lies wholly below sea level and relies for its existence on a ring of hard concrete sea walls. I was first drawn to Canvey by the power and pungency of Johnson's short, sharp songs. Under their spell I went to stand and watch the refinery towers of Coryton burning at the break of day. Later I walked the sea walls and saw gulls in their thousands wheeling in the distance over the rubbish hills of Mucking Marshes.

Over the years of walking the walls of Canvey, something has never ceased to peck at my own submarine consciousness: the spectre of the east coast flood disaster of January 1953 when the sea rose to overtop the defences and flood the island. 'I looked out of the back window at the fields,' Wilko Johnson told writer Zoe Howe, 'but there were no fields, it was the sea, there were waves, waves coming in.' Fifty-nine Canvey Islanders drowned that night, and the memory still haunts the island. It clung to the subconscious of Wilko Johnson, Canvey's erudite R & B chronicler, till the end of his days.

They raised the sea walls after the Canvey flood disaster. Now the island is girdled in concrete that reaches 4 metres higher than the sea at high tide. A 5-metre surge, as yet unlikely but not unthinkable, would top those walls. I went along the walkway inside the walls, past the fuel silos and amusement park, above the sub-sea-level streets of neat houses and on past the graffito that reads 'Canvey Is England's Lourdes', as the grey Thames widened to meet the open sea. Out at Canvey Point a fisherman slumbered peacefully in the shadow of the wall. I sat on the concrete step near him, watching the incoming tide invade the strip of salt marsh, infiltrating itself among the sea purslane and sea lavender. The sleeping fisherman mumbled something under his breath. It sounded like, 'Could be us one day,' and I wondered if his thoughts and mine were mingling.

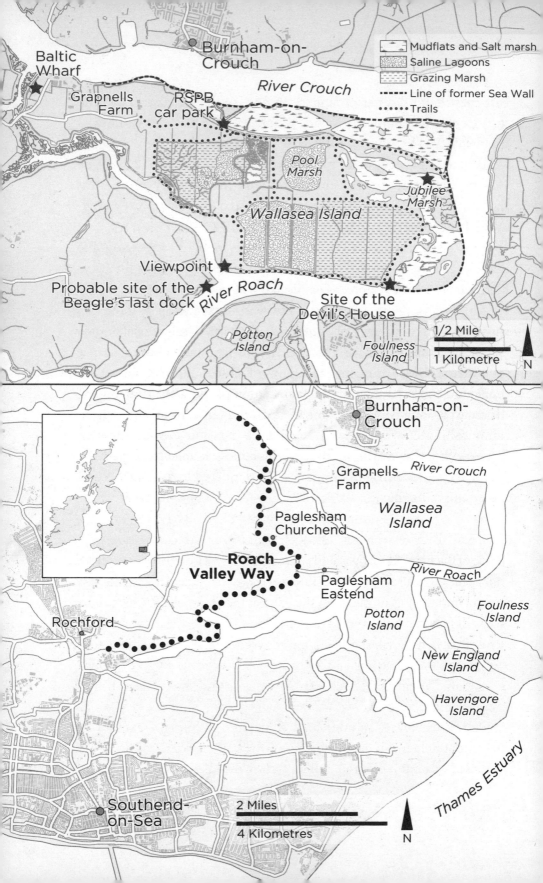

17

Breaking the Wall: Wallasea Island

A COLD MORNING AT THE turn of the year in the remote flatlands of
southeast Essex. Ice skinned the puddles and a keen north wind cut out
of a streaky sky. Through flat farming country as lonely as any in
England I followed the muddy path of the Roach Valley Way towards
the coast among enormous prairie fields, the uniform green of chem-
ically fertilized winter wheat. No hedges, few trees. Everything lies
tremendously low in this level clay landscape reclaimed from the sea, a
bleak place, productive but somehow barren.

On this isolated coast six dead-flat islands make up the Essex Archipelago: Wallasea, Potton, Rushley, New England, Havengore and
Foulness. They tessellate neatly, their snaking shores conforming with
those of the neighbouring islands, separated one from another by muddy
tidal creeks. Wallasea (in Old English *walas ey*, 'outsider's island') fits
snugly into the jigsaw puzzle at the edge of the archipelago, its southern
sea wall facing the flat shores of Potton and Foulness Islands across the
River Roach, its north shore looking across the River Crouch to the snug
red-roofed waterside town of Burnham-on-Crouch.

At Baltic Wharf, where cranes and pallet stacks and piles of timber
stood against the sky, I crossed a bridge from the mainland on to Wallasea Island. There was an immediate contrast in colour and texture with
the smooth chemical greens through which I'd approached the island.
Wallasea was dun-coloured, tufted, a place untidied by nature that had

been left to its own devices. At the RSPB car park the wind waved reed fronds and trembled dried teasel heads in the ditches under the sea wall. There was a great clamour from birds I couldn't see, hidden over the far side. I climbed up to the grassy grandstand of the wall and the view went leaping out to the north. The red and white waterfront of Burnham-on-Crouch was suddenly in view. A sailing boat went pitching seaward down the river. It wasn't these that made me draw in my breath, however, but the changes wrought on Wallasea Island since I was last here some fifteen years ago. The landscape I remembered had vanished, and another island had materialized where that former Wallasea used to lie.

Four miles long, less than half that wide, Wallasea Island lies mostly below sea level. It is bedded on London Clay, the muds and silts that underpin London and the Thames Valley. Above this London Clay the island is coated with a thick layer of silty clay deposited by the tides after the last Ice Age. In medieval times Wallasea consisted of five separate tidal islands used for grazing sheep. Then in Tudor times Dutch experts built walls round Wallasea to consolidate and reclaim it as one piece of ground.

The Essex Archipelago was notorious for its fogs and fevers, chiefly 'marsh ague' or malaria, which the locals treated with liquor and opium. No landowner lived on Wallasea; only the poorer tenants resided here. In the early twentieth century there were several scattered farms on the island. By the 1950s only one farmer remained, working some two hundred small, irregularly shaped fields. Then came the disastrous floods of January 1953, the same floods that drove up the Thames to pour over Canvey Island's sea walls and drown fifty-nine of its residents. No one died out here on Wallasea, but the east and west sea walls were swept away and three-quarters of the island was submerged, the salt water impregnating and poisoning the soil. It took ten years for the farmland to recover. New sea walls were built, new drains dug, and the fields were reshaped into a handful of large north–south strips of intensively farmed cereal crops and peas. That's how Wallasea Island was when I first ventured here in the 1980s, a dead-flat, dead green crop factory.

The one public footpath petered out halfway round the crumbling sea wall of the island, and there was no particular welcome for wildlife or the walker.

Early in the twenty-first century, concern was beginning to flicker in political and environmental circles over the projected rise of a metre in sea levels by the end of the new century. There was also a push to create a wildlife-friendly wetland habitat to replace that which had gone under concrete to form the new docks at Felixstowe, a few miles up the coast. The sea walls of Wallasea Island, constantly in need of repair, would cost up to £3 million a year to maintain. That was too much. Lonely Wallasea, with its 700 hectares of flat land, poorly maintained sea defences and proximity to the tidal River Crouch, was chosen for the innovative experiment of 'managed realignment' – in other words, breaking the walls to let in the sea, an attempt to work with natural forces rather than continue with the futile policy of trying to outface them. Many farmers and landowners thought it was madness, a show of defeatism, to give in to nature in this way and surrender good growing ground that had been hard won from the sea. But DEFRA, the Department for Environment, Food & Rural Affairs, was behind the project, and in 2006 it went ahead. The RSPB bought Wallasea Island for a wildlife reserve in 2009, and have very meticulously managed and monitored it ever since.

The last time I set foot on Wallasea was back in 2007, just after 300 metres of the northern sea wall had been deliberately breached to admit the sea. Back then most of the island still consisted of straight-edged fields under intensively farmed corn. The marsh called Allfleets at the northeast tip of the island, newly breached and subject to the tide, was no more than a few hummocks of glutinous mud scattered with a light green haze of infant glasswort, a pioneer plant in vegetating salt marshes. But sightings of avocet and godwit, wigeon and dark-bellied brent geese, peregrine and marsh harrier had already risen significantly, and overall the wildlife experiment looked like being a success. As for managing the effects of global warming, it was too early to say.

Looking from the sea wall today, I really could not believe it was the same place. The position of the wall under my boots was disorientating me; it had been moved inland, and the old one, breached and broken, ran like a gappy set of teeth along the river 100 metres away. More than that, though: where were the wall-to-wall fields of corn and beans? Staring south across the island, I saw the whole regimented prairie had gone. In its place lay pools and lagoons, coarse grasslands and marshes, as far as I could see. And on this bitter winter day the island was alive with the movements and noises of birds. Curlews bubbled; oystercatchers piped, *pik-pik!* Redshank and shelduck patrolled the shore. Brent geese with white rumps and black heads came in to land on the grass with hoarse hound-like yelping, their wings held stiffly up and behind them at the moment of landing.

Having taken some of this in, I turned south along the embankment between the stippled sheet of Pool Marsh's brackish lagoons and the slimy mudbanks and ever-encroaching saltings of Jubilee Marsh. Embankment, marsh and lagoons, all were new. And all bore witness to how different sectors with different aims, unlikely bedfellows, could work together if they chose.

Between 2007 and 2015 Wallasea Island saw the arrival of 3 million tons of 'arisings' or spoil, excavated during the building of tunnels and ventilation shafts for the Crossrail rapid transit railway across London. The tunnels were bored through the same stratum of London Clay that underlies Wallasea, so there was little geological conflict. The spoil was taken to Barking, most of it by rail, to be transferred to barges for the 50-mile journey down the Thames and north round the coast to Wallasea. Conveyors laid along walls newly built across the island carried the spoil to stacking points, where it was loaded into dumper trucks and carted off to form the islets, banks and foundations of Jubilee Marsh. This coastal portion of the island, the lowest lying, was raised above sea level, creating 115 hectares of intertidal salt marsh, mudflats and islands. The tide itself, flowing in and out, would manage the marsh naturally with its deposition of sediment, seeds, invertebrates, fish and plant fragments.

The spoil was also taken to the interior of Wallasea to reinvent the cornfields as saline lagoons with their own creeks and grazing marshes. Sluices control the water levels in the lagoons for different bird species; cattle grazing keeps the grass at a level to suit short-eared owls and kestrels. And intertidal areas around the island have been designed to soften the impact of climate change. Long slopes gradually descend from the new sea walls to the tideline, gently shelving ground on which salt-marsh plants can establish themselves to diffuse the power of the incoming tide as sea levels rise. There are multiple benefits from these accelerated geological processes, facilitated by the ability of our remarkable species to cooperate and think creatively.

From the new cross-island wall I marvelled again at the birds, their vigour and variety. A host of lapwing took off from a lagoon shore, circled, and settled again to their muddy browsing. All of a sudden the sky flickered with long lines of the black-and-white birds, more jumping up from the marsh to join them, like marching men at a tattoo, different groups converging and dividing. On the muds a congregation of dunlin faced the wind, heads bowed like little old chapel-goers in grey cloaks. Out over the distant sea a great cloud of knot a thousand strong swerved this way and that like iron filings drawn across paper by a magnet.

From the flat land beneath the flying mass of birds rose skeleton towers, communication masts and observation posts, a shock to the eyes in these elemental surroundings. It took a moment to reorientate myself and remember that there was land, not sea, to the east of Wallasea, the wedge-shaped island of Foulness that lies at the outer edge of the Essex Archipelago. Foulness is used by a Ministry of Defence contractor for testing 'defence technology', and is closed to the outside world. At the time of my first visit to Wallasea I'd talked my way on to Foulness for a fascinating day among experimental gun batteries, cast-off Sea Harrier jets riddled with bullets, mysterious rings of scorched earth, targets dangling from scaffolding towers, and 'atomic and environmental establishments'. I'd also sat by a red-hot stove in a tiny wooden-walled pub and heard the farmhands of Foulness talking with a thick and twangy accent not heard outside this isolated outpost for many decades.

Beside the Jubilee Trail a warning had been sprayed in angry letters, now fading: 'Vaccine = Bioweapon. Wake up! Warn others!' Nearby lay the rusty cutterhead of 'Victoria', a TBM or tunnel boring machine from the Crossrail excavations. Victoria had been a hefty girl, 150 metres long, 1,000 tons in weight. I patted her disembodied head and thought of her slightly bigger cousins Florence and Cecilia hard at work back in the Chiltern Hills, boring out the tunnel for HS2.

Just along the wall a wooden post stood out of the marsh some 30 metres away. A magnificent peregrine sat there unafraid and stared me down as I went by. Its handsome black hood and stripy back were clearly outlined against the steel-grey waters of Jubilee Creek. On the path nearby, in a burst of feathers, I found the remains of a knot. The little wader had been the peregrine's meal, its breastbone, as light as paper, pecked into sharp notches by the hunter's beak.

Suddenly a tremendous commotion arose over marshes and lagoons. Shelduck, brent geese and avocet rose in panic. Dunlin flew up, knot raced away low to the ground and climbed to coalesce into a defensive ball. Ten thousand voices shrieked, piped and whistled their alarm. And through it all, serene as could be, sailed the dark shape of a female hen harrier. With wing fingers uptilted she cruised across Wallasea with utter insouciance, making no move to chase or grab any prey, seeming just to be teasing and enjoying her power to spark this circling and clamouring of every bird in sight. On the Wallasea of old, a couple of goldfinches might have panicked among the ditch thistles, or one or two brown hares might have galloped away along the plough furrows. Never this explosion of birds into the air.

The far point of Wallasea overlooked a wind-blown wilderness. A wooden hide offered a viewpoint over fresh and salt marshes and towards Foulness and Potton Island across the wind-ruffled waters of the River Roach. Here at this outer corner of Wallasea once stood the haunted farm of Tilebarn, or Denval's House, or the Devil's House, a landmark until the great flood of 1953. At some time in the past it was the abode of Mother Redcap, a local witch. When Mother Redcap wasn't flying across Wallasea on a hurdle, she would sit peeling potatoes and chanting, 'Holly holly, brolly brolly, redcap! Bonny, bonny!' Cows would turn mad around the farm, and anyone bold enough to enter the house might encounter a terrifying black shape with glowing yellow eyes, or feel an impulse to do away with themselves at the urging of a voice that cried, 'Do it! Do it!'

Tilebarn Farm was reputed to have been bombed to ruins during the Second World War, and when the great flood of 1953 receded from the island it dragged the remnants of the Devil's House away down the river and out to sea.

From this haunted ground I turned along the southern shore of Wallasea above the River Roach. The map marks this section of tideway as the Devil's Reach, so old stories linger long. Sun and cloud shadows chased each other across the freshwater marshes and saline lagoons, the russet and dun grasses under the sea wall. The wind rippled the river where grey seals were going seaward with the ebb, casually floating and watching me.

Opposite the low black shore of Paglesham I stopped and sat down. After the agitation of the hen harrier everything was very still and quiet. I stared through my binoculars across the Devil's Reach, looking for something I'd hoped to spot there, a notch or inlet in the opposite bank. There was nothing to see but a rampart of mud rising to a ragged sea wall, but somewhere in that blank grey slope lies the outline of the rudimentary dock where HMS *Beagle* finally came to rest.

Beagle was originally commissioned as a single-masted, 10-gun Royal Navy sloop, but in 1825 she was refitted as a two-masted barque and detailed for survey work. Ninety feet (27.4 metres) in length, she carried a crew of seventy in cramped conditions. Her shallow bottom enabled her to operate in coastal waters, but made her liable to roll in heavy weather.

In 1831 the young biologist and geologist Charles Darwin, twenty-two years old and newly graduated from the University of Cambridge, joined *Beagle* as ship's naturalist for a voyage to chart and explore the Southern

Hemisphere. Captain Robert Fitzroy nearly turned Darwin down because he didn't care for the shape of the young man's nose. It was lucky for the world that he relented. Darwin was often seasick over the five years that the voyage lasted. But as he visited and explored Cape Verde, Brazil, Patagonia, the Falklands, the Galapagos Islands and Australia, he took meticulous notes on the fauna, flora and fossil specimens he collected, the geological phenomena he witnessed, and the variations he observed among related but geographically separated species. When he returned it was the start of a long struggle over twenty-five years, through frequent bouts of debilitating illness, agonizing over the theories he'd begun to formulate on board the *Beagle* about how the seas and lands of the world made and remade themselves, and about the necessity for species, including humans, to adapt to their circumstances and to pass those adaptations on down the genetic line, or face extinction.

On the Origin of Species was published in 1859. The key sentence:

> As many more individuals of each species are born than can possibly survive; and as, consequently, there is a frequently recurring struggle for existence, it follows that any being, if it vary however slightly in any manner profitable to itself, under the complex and sometimes varying conditions of life, will have a better chance of surviving, and thus be *naturally selected*. From the strong principle of inheritance, any selected variety will tend to propagate its new and modified form.

By the time he died in 1882 Darwin had earned worldwide fame, religious opprobrium, the opposition of traditionalists opposed to the idea of natural selection, and after the publication in 1871 of *The Descent of Man* a general acknowledgement of his related theory that man, a mammal like any other, was descended from the apes. His summary, beautifully expressed:

> Man with all his noble qualities, with sympathy which feels for the most debased, with benevolence which extends not only to other

men but to the humblest living creature, with his god-like intellect which has penetrated into the movements and constitution of the solar system – with all these exalted powers – Man still bears in his bodily frame the indelible stamp of his lowly origin.

After her return from Darwin's voyage in 1836, *Beagle* sailed the following year for Australia and did not return to England until 1843. Two years later she was assigned to coastguard duties, and made her last short voyage to anchor in the Devil's Reach at the heart of the Essex smuggling country. With its mazy miles of creeks and tideways, out-of-the-way farms and many impoverished families in its lonely villages, this corner of the coast was up to its elbows in smuggling. Paglesham was a hotspot, particularly at the turn of the nineteenth century when the village grocer William Blyth, a.k.a. 'Hard Apple', controlled the trade. Hard Apple cut a fantastic figure, keeping hold over his ruffian gang by deeds of prowess such as wrestling bulls, munching wine glasses and drinking a keg of brandy at a sitting. His cutter *Big Jane* ferried contraband, outwitting and outsailing the Revenue. In his capacity as Paglesham's churchwarden, Hard Apple used the tower of St Peter's Church as a hiding place for smuggled goods. Occasionally apprehended, always wriggling through the net, he died in 1830 at the age of seventy-six in the odour of sanctity, uttering his final words: 'I'm ready for the launch.'

Hard Apple had long ago slipped away to heaven, or the other place, by the time *Beagle* and her crew were stationed here, but smuggling was still rife, and the old barque had a long, lonely duty ahead of her under the name *Watch Vessel 7*. She spent her last years moored on the Paglesham bank, with a maximum allowance of five pounds a year for maintenance. Barnacled, weed-bottomed and rotting to pieces, she finally fell victim to Royal Navy cuts. No one tried to save the rotten old boat. No one mourned her, or honoured her as the little barque that once sailed round the world with Charles Darwin on board. In May 1870 the hulk of *Watch Vessel 7* was sold to two local men. They broke her up in situ, just as her now elderly and celebrated naturalist was preparing *The Descent of Man* for publication and worldwide acclaim.

Over on Foulness someone had lit a monstrous fuel fire. Black smoke rose from a red glowing point and billowed into the sky. The volcanic look of the fire sent me back through the bones of Britain, back up the long road past the glittering city and the gently settling chalk, the Great Dying and the miracle of life on land and in the sea, past gritstone and limestone, coal and shale and the metamorphic mountains, all the way back to the fire and fury at the beginning of the world. There was a tingle in my feet and my head seemed full to bursting. What a hell of a journey I'd been on.

Somewhere out there beyond the island the sea was turning. Soon it would be whispering in along the coast, shifting the curlews and dunlin inland, bringing particles through the broken walls of Wallasea, taking them away, depositing, building, eroding in the old Huttonian waltz.

The Devil's House is long gone out to sea. So are the particles of the *Beagle*, washed to nothing in the tides. But as for the young naturalist who sparked a new enlightenment as he lay seasick in his cabin, and as for the old Scots geologist who couldn't get a hearing for his theory of the operations of the Earth, their voices are always dinning in our ears, admonishing and encouraging, more powerful than storms; louder, as yet, than the beating of waves on walls.

Acknowledgements

It's a pleasure to thank Peter Drake of Achiltibuie for leading me around the Coigach Geotrail, Andrew Bateman of Scot Mountain Holidays for the Ben Nevis walk and the Cairngorms snowholing adventure, Kath Norgrove for our walk in the White Peak, and Allan Wheeler, Field Meetings Secretary of the Harrow & Hillingdon Geological Society, for explaining the remarkable effects of ice on chalk. A special thank you to my godson, geologist Andy Harrison, for his patience and enthusiasm in making plain the crooked ways of geology.

It was delightful to have the company of members and officers of the Forth & Clyde Canal Society (forthandclyde.org.uk) – Stewart Procter, Chairman; Christine Hammell, Secretary; Robert Irvine, Treasurer; also Dave Connell, James Pert, Bill Knox and Willie Boyle – for my boat trip in the Clyde puffer *Maryhill* along the Forth & Clyde Canal.

My very grateful thanks are also due to Dave Richardson for putting me up and putting up with my mandolin-mangling in Edinburgh; Dr Neil Clark, Curator of Palaeontology at the Hunterian, University of Glasgow, and David Webster of the Geological Society of Glasgow for showing me round the Fossil Grove in Victoria Park, Glasgow; Dr Mike Browne, Edinburgh Geological Society, for supplying some very helpful trail leaflets; and Dr Jonathan Ford, Chief Geologist England, British Geological Survey, and Dr Kathryn Goodenough, Principal Geologist, British Geological Survey, for giving their valuable time to helping me set up my journey.

Thanks also to Ronan Clancy, Edward Tattersall and Rob Hutchinson for arranging my visit to the HS2 tunnelling works in the Chiltern

Hills, and to Guilhem de Langlais and Marie-Amélie Auvinet for hosting me there.

For all sorts of helpful advice I'm very grateful to these kind people:

Liz Chiu, Online and Meetings Secretary of the Harrow &
     Hillingdon Geological Society
Dr Jill Eyers of the Bucks Geology Group
Pete Harrison, geologist at the North West Highlands Geopark
Paul Hildreth, Secretary, and Mike Bowman, Membership Secretary
     of the Yorkshire Geological Society
Neil Mackenzie, President, Edinburgh Geological Society
Rosalind Mercer of the Essex Rock & Mineral Society for her
     suggestions of geological trails in Essex
Sarah Nice, British Geological Survey
Michael Palmer, Keeper of Natural History & Geology at Bucks
     County Museum
Sue Plumb of the Manchester Geological Association
Dr Nick Riley, President of the Yorkshire Geological Association
Professor Alison Rust, Head of School, Earth Sciences, University of
     Bristol
Walter Semple, Hon. Secretary of the Geological Society of Glasgow
Peter del Strother, Lancashire Group of the Geologists' Association
Derek Turner, Secretary of the Bedfordshire Geology Group

For checking my manuscript and saving me from any number of geo-illogicalities and misunderstandings, I'm indebted to Dr Dave Millward, field geologist, formerly with the British Geological Survey in Edinburgh; palaeontologist Dr Katie Strang; and geologist Mathilde Braddock, founder of Steps in Stone geology walks (mathildebraddock. wixsite.com/stepsinstone).

Thanks to Nigel Richards for a good chat about farming on gravel and clay soils at Moulsoe on the outskirts of Milton Keynes. As ever, my sincere thanks to Penny Grigg for her speedy and skilful typing. And it's a great pleasure to thank Carry Akroyd (carryakroyd.co.uk) for creating another wonderful cover illustration.

Finally, I'd like once again to thank Jane Somerville for being my constant companion and inspiration in walking, exploring, learning and enjoying.

Reading and Viewing

BOOKS

Reading the Rocks: The Autobiography of the Earth by Marcia Bjornerud (Basic Books, 2006)

The Famous Highland Drove Walk by Irvine Butterfield (Grey Stone Books, 1996)

London Clay: Journeys in the Deep City by Tom Chivers (Doubleday, 2021)

The Making of the British Landscape: From the Ice Age to the Present by Nicholas Crane (Weidenfeld & Nicholson, 2017)

On the Origin of Species by Charles Darwin (annotated version, Harvard University Press, 2011)

Voyage of the Beagle by Charles Darwin (Penguin Classics, 2003)

The Drove Roads of Scotland by A. R. B. Haldane (Birlinn, 2019)

Journey Through Britain by John Hillaby (Constable, 1995)

Earth Environments: Past, Present and Future by David Huddart and Tim Stott (Wiley-Blackwell, 2020)

The Geological History of the British Isles by Arlene Hunter and Glynda Easterbrook (Open University, 2004)

Why Geology Matters by Doug Macdougall (University of California Press, 2011)

Geology for Walkers: For the outdoor enthusiast curious about the geological story around them by Steve Peacock (self-published, 2021)

Hutton's Arse: 3 Billion Years of Extraordinary Geology in Scotland's Northern Highlands by M. H. Rider (Rider-French Consulting, 2005)

Stone Will Answer by Beatrice Searle (Harvill Secker, 2023)

The Geology of Britain: An Introduction by Peter Toghill (Airlife, 2007)

The Map that Changed the World by Simon Winchester (HarperCollins, 2009) – the story of the 'Father of English Geology', William Smith

MAPS

The BGS (British Geological Survey) 1:625,000 maps of the bedrock geology of the UK in two parts, North and South (online mapping, see below)

Ordnance Survey's Maps App is a handy location tool

INTRODUCTION TO GEOLOGY

British Geological Survey (BGS) – bgs.ac.uk – has many geological excursion guides, and other helpful information

The BGS's zoomable Geology Viewer online map (geologyviewer.bgs.ac.uk) shows the UK's bedrock and superficial geology

Two TV series explain the geology of Britain in easy-to-grasp terms – Iain Stewart's *Earth: The Power of the Planet* (BBC) and Tony Robinson's *Birth of Britain* (National Geographic Television), both available on YouTube

The BBC's *In Our Time* programme 'The Geological Formation of Britain' (https://www.bbc.co.uk/programmes/b00n8t48) is a helpful overview

There's an excellent introduction to Scotland's Five Terranes in a short online course, 'The Geology of Scotland' by Dr Alex G. Neches and contributors (youtube.com/watch?v=zNAnSLSNKxo)

National Stone Centre, Porter Lane, Wirksworth, Matlock DE4 4LS (01629 824833, nationalstonecentre.org.uk) has a visitor centre and a geological trail

HELPFUL WEBSITES

dalesrocks.org.uk – easy to follow geology of the Yorkshire Dales

geograph.co.uk – photographs and map sections of features in every square kilometre of Britain and Ireland

geological-digressions.com – discursive site, simple explanations

geologistsassociation.org.uk/geotrails-building-stones-walks – a superb collection of geological walks in Scotland, England and Wales

geologytrails.wordpress.com

geolsoc.org.uk – the Geological Society

nationalgeographic.com/science/article/devonian – 'Life on Earth' series (substitute desired geological period at end of link)

nature.scot – 'A Landscape Fashioned by Geology' series – wonderful
explanations of Scottish geology, written in layman's terms and easy to
understand
scottishgeology.com – Scottish Geology Trust publications and leaflets

Walking the Route

walkhighlands.co.uk has details of hundreds of trails across Scotland, including several on the Bones of Britain route

geowalks.co.uk offers geological walks in Scotland, run by Angus Miller

The Long Distance Walkers Association (ldwa.org.uk) details long-distance paths throughout Britain

Interactive map at www.christophersomerville.co.uk gives walks on or around the Bones of Britain route

SELECTION OF GEOLOGICAL SITES AND TRAIL GUIDES

Asterisks denote geological sites

Geoparks along the Bones of Britain route

UNESCO North West Highlands Geopark – nwhgeopark.com

Lochaber Geopark (Ben Nevis, Glencoe) – discoverglencoe.scot/lochaber-geopark-glencoe

UNESCO North Pennines Geopark – northpennines.org.uk/unesco-global-geopark

1. *Geology Bombshell: Isle of Lewis*

*Dail Mòr and Garraidh beaches, Isle of Lewis – 3,000-million-year-old gneiss

*Hushinish jetty, Isle of Harris – oldest rocks in the Outer Hebrides

Walking on Harris and Lewis by Richard Barrett (Cicerone, 2010) – helpful for the West Side Coastal Path on Lewis

2. *Purple, Red and Black: Isle of Skye*

*The Quiraing, Isle of Skye (walkhighlands.co.uk) – basalt

*Broadford marble quarries, Isle of Skye (walkhighlands.co.uk) – Marble
  Line trail
*Loch Coire Lagan, Black Cuillin, Isle of Skye (walkhighlands.co.uk)
  – gabbro
Walking the Skye Trail by Helen and Paul Webster (Cicerone, 2016)

3. *Strikes and Thrusts: Coigach Geotrail and Knockan Crag*

*Coigach Geotrail – cornucopia of Precambrian geology
*Knockan Crag (nwhgeopark.com/knockan-crag) – Moine Thrust explained
*Inverpolly NNR (inverpolly.com; walkhighlands.co.uk) – 800-million-year-
  old sandstone mountains
Coigach Geotrail – visitcoigach.com

4. *The Famous Highland Drove: Isle of Skye to the Great Glen*

*By Loch a' Choire Bheithe, and several other places along the road from
  Kinloch Hourn to the Great Glen – exposures of metamorphic rock:
  psammite, quartzite and pelite
*Wester and Easter Glen Quoich – mountains of psammite and pelite with
  microdiorite dykes
*Along Loch Quoich – hummocky glacial till
The Famous Highland Drove Walk by Irvine Butterfield (Grey Stone Books,
  1996); video of the walk in 3 parts available on YouTube

5. *Plutons, Volcanoes and Military Roads: Great Glen to Rannoch Moor*

*Ben Nevis – a pluton or intrusive tongue of igneous rock
*Glencoe – from the A82 Pass of Glencoe, view the volcanic rocks (sills, lava,
  tuff, volcanic glass) of the Three Sisters of Glencoe (Aonach Dubh, Meall
  Dearg, Geàrr Aonach) and the Aonach Eagach ridge
  – from the Devil's Staircase and Stob Mhic Mhartuin (OS ref NN209574),
  view the volcanic rocks of the Big and Little Shepherds of Etive (Buachaille
  Etive Mòr and Buachaille Etive Beag)
*William Caulfeild's old military road from Inveroran Hotel, PA36 4AQ, to
  the West Highland Way at the Black Mount (NN282469) – blanket bog
  over granite; glacier-carved mountains
Great Glen Way by Jim Manthorpe (Trailblazer Guides, 2017)

West Highland Way: The Official Guide by Bob Aitken and Roger Smith
(Birlinn, 2018)

6. Whinstone and Wheel: Kirkintilloch to the Kelpies

*Fossil Grove, Victoria Park, Glasgow G14 9QR (fossilgroveglasgow.org) –
fossilized stumps of 325-million-year-old lycopods (primitive trees)

*Auchinstarry Quarry, Croy, Kilsyth, Glasgow G65 9SG – fine exposure of
the dolerite Whin Sill

*Five Sisters Bing, West Calder EH55 8PN – spectacular oil-shale spoil heaps

John Muir Way by Sandra Bardwell and Jacquetta Megarry (Rucksack
Readers, 2018)

A Geological Guide to the Fossil Grove, Glasgow by Iain Allison and David
Webster (ringwoodpublishing.com)

Forth & Clyde Canal Society (forthandclyde.org.uk)

7. Volcanic Lumps and Lightbulb Moments: Edinburgh to the Border

*Salisbury Crags and Arthur's Seat, Holyrood Park, Edinburgh – volcanic
landscape that inspired James Hutton

*Barns Ness Geotrail, near Dunbar – succession of geological wonders

*Hutton's Unconformity, Siccar Point, between Dunbar and Eyemouth – the
place where James Hutton proved his theory of Earth's great age and
cyclical processes

John Muir Way by Sandra Bardwell and Jacquetta Megarry (Rucksack
Readers, 2018)

Pamphlet guides to Holyrood Park, Barns Ness and Hutton's Unconformity
at Siccar Point by the Edinburgh Geological Society (edinburghgeolsoc.
org)

Leaflet guide to Berwickshire Coastal Path by Scottish Borders Council
(scotborders.gov.uk)

8. Saint, Dragon and Devil: Holy Island to Hadrian's Wall

*Farne Islands and Beblowe Crag on Lindisfarne, Northumberland coast –
Whin Sill and dolerite dyke exposed

*Doddington Moor, near Wooler (NU018327 approx.) – cup-and-ring
marked rocks

*Housey Crags and Long Crags, Harthope Valley, near Wooler – remnants of a volcanic roof pendant

St Cuthbert's Way: The Official Guide by Ron Shaw and Roger Smith (Birlinn, 2018)

Pennine Way Companion by Alfred Wainwright, revised by Chris Jesty (Frances Lincoln, 2012)

Pennine Way 2019: Edale to Kirk Yetholm: Route Guide by Stuart Greig and Henry Stedman (Trailblazer Guides, 2019)

9. *Mighty Wall, Black Nazis and Leaden Ore: Hadrian's Wall to Garrigill*

*Hadrian's Wall – dramatic Whin Sill outcrops all the way from Sewing Shields to Thirlwall Castle

*Banks of River South Tyne between Garrigill and Low Redwing, Cumbria – metallophytes (plants tolerant of heavy metals, e.g. spring sandwort and alpine pennycress) growing in lead-mine spoil. See also Moss Shop near Birkdale Farm (NY796723), and White Moss mine ruin on Cronkley Fell, Upper Teesdale (NY853280)

Hadrian's Wall Path by Henry Stedman and Daniel McCrohan (Trailblazer Guides, 2017)

10. *Hill Farmers and Teesdale Jewels: Garrigill to Tan Hill*

*Old corpse road between Garrigill and Cross Fell – lead-mine ruins

*Dunfell Hush, Great Dun Fell – man-made gully to reveal lead seams

*High Cup (NY746261), on Pennine Way between Dufton and Cauldron Snout – viewpoint over England's most perfectly formed U-shaped glacial valley

*Cauldron Snout and High Force, Upper Teesdale – more dolerite drama as the River Tees tumbles over two spectacular steps in the Whin Sill

*Thistle Green, Cronkley Fell, Upper Teesdale (NY843283) – exclosures (areas fenced off from sheep, rabbits, etc.) where relict arctic-alpine flora flourish: spring gentians, bird's-eye primroses

11. *Scars and Faults: Tan Hill to Gargrave*

*Swinner Gill near Keld, N. Yorks (NY909005) – lead-mine ruins; view across River Swale to Yoredale Group strata in Kisdon Side

*Surrender Mill (SD991998) and Old Gang Mill (NY975005), near Feetham,
   Swaledale – ruins of lead-smelting mills, flues, chimney and slag banks
*Hardraw Force, behind the Green Dragon Inn, Hardraw, near Hawes – a
   100-ft waterfall in a hollow where Yoredale Group layers are exposed
*Penyghent, Ingleborough and Whernside – three hills with classic exposures
   of Yoredale Group sandstone, limestone and gritstone. View from the
   summits of each
*Ingleborough, northwest flank – superb flora in extensive pavement of
   Great Scar Limestone
*Norber (SD764698), near Clapham, N. Yorks – limestone upland with
   slate-sandstone erratics brought by a glacier from Crummackdale 2 miles
   to the north
*Waterfalls Trail, Ingleton LA6 3ET (ingletonwaterfallstrail.co.uk) –
   spectacular, strenuous circular walk across exposed rocks of South and
   North Craven Faults; Carboniferous limestone and Ordovician shale
*Ingleborough Cave (ingleboroughcave.co.uk), near Clapham – show cave in
   limestone; stalagmites, stalactites, weird formations
*Gordale Scar, Malham (SD915640) – scramble up two-stage limestone gorge
   sculpted by Ice Age floodwaters
*Malham Cove (SD897641) – 100-metre limestone cliff formed by erosion;
   fine limestone pavement on top with fossils

12. *True Grit: Gargrave to Edale*

*Hebden Bridge, W. Yorks – gritstone mill town of high-perched mills,
   factories, terraces, canal, river, humpback packhorse bridges, cobbled
   lanes
*Blackstone Edge, near Littleborough, Lancs/Yorks (SD973163) – classic edge
   or cliff of millstone grit with weathered tors
*Wain Stones, Bleaklow Head, Derbs (SK092959) – pair of gritstone 'kissing
   trolls'
*Kinder Edge, near Hayfield, Derbs (SK083889 approx.) – another classic
   gritstone edge

13. *White Peak and 'Black to Green': Mam Tor to the National Forest*

*Mam Tor, near Castleton, Derbs (SK129835) – wonderful viewpoint over
   meeting place of Dark Peak gritstone (north) and White Peak limestone

(south). Landslip exposure of bands of turbidites (sandstone and shale) on east face, with former road twisted and broken at the foot

*Treak Cliff Cavern (bluejohnstone.com) and Blue John Cavern (bluejohn-cavern.co.uk), near Castleton – show caves: mining of beautiful Blue John fluorite

*Eldon Hole, near Castleton (SK117808) – fearsome pothole slit in the limestone

*Stone walls between Lydgate Farm (SK140721) and High Dale (SK155717) – packed with Carboniferous fossils of productid brachiopods and corals

*Priestcliffe Lees Nature Reserve, above Litton Mill, Millers Dale – beautiful limestone flora

*Litton Mill railway cutting (SK153729 approx.), Monsal Trail, Millers Dale – in the Carboniferous limestone, a dark tongue of volcanic rock shows where a flow of lava terminated and shattered as it contacted cold lagoon water

*Old limekilns just east of Millers Dale Viaduct, Monsal Trail (SK142731) – massive kiln faces, drawing tunnels and hearths

*Parkhouse Hill (SK080669) and Chrome Hill (SK071673), near Earl Sterndale – striking twin limestone reef knolls like stegosaurus humps. See also High Wheeldon (SK101660) and Pilsbury Castle hills (SK114638) near Crowdecote

*Dovedale from Beresford Dale (SK128590) and Wolfscote Dale, south to dale mouth below Thorpe Cloud (SK148510) – very spectacular eroded limestone scenery: caves, pinnacles, tors

*Weaver Hills (SK095463), near Ellastone – viewpoint back north to Carboniferous limestone of the White Peak, south to Triassic sandstone and mudstone of the Midlands

*National Forest sites: Albert Village clay pit, now a lake (SK306175); Moira Furnace Museum (SK315151); Donisthorpe Colliery Woodland Park (SK317144); quarry face at Forterra Brickworks, Desford (SK463064), seen from Bagworth Heath Woods; Calke Limeyards, Calke Abbey (SK362238); Church of St Mary and St Hardulph, Breedon on the Hill, at quarry lip (SK406233)

*Charnwood Forest sites: unconformities between Precambrian and Triassic rocks spectacularly displayed at Morley Quarry near Shepshed (SK477178); Bardon Hill Quarry (view from trig pillar in Bardon Hill Wood, SK 461131)

and Old Cliffe Quarry near Markfield (viewpoint at Cliffe Lane/Stoney
    Lane roundabout, SK477101, just south of M1 Junction 22)
*Volcanic outcrop of striped uptilted Precambrian rocks at summit of Beacon
    Hill Country Park LE12 8SP (SK510148)
Limestone Way Walkers Guide (derbyshiredales.gov.uk/limestoneway, 2018)
Walking the Limestone Way by Ron and Elizabeth Haydock, and Bill and
    Dorothy Allen (Scarthin Books, 1997)
National Forest Way (e-book) by Bob Allan (2015). See nationalforest.org
Rock Trails Peak District: A Hillwalker's Guide to the Geology & Scenery by
    Paul Gannon (Pesda Press, 2011)

14. Bands of Bright Colours: Leicestershire Wolds to the Chiltern Hills

*Deserted medieval villages of the Leicestershire Wolds: Baggrave
    (SK698087), Quenby (SK701062), Cold Newton (SK716066) and Ingarsby
    (SK686050)
*Leicestershire building stone – mudstone and ironstone churches of the
    Wreake Valley; ironstone and limestone at St Peter's Church, Tilton on the
    Hill; Northampton Sand 'ferruginous sandstone' at Hallaton and
    Medbourne
*Northamptonshire building stone – sandstone stable buildings at Harlestone
    Park; marl and ironstone at the Church of St Michael and All Angels,
    Bugbrooke; Blisworth limestone at the Church of St John the Baptist at
    Blisworth; polychrome banding (Northampton sandstone, Blisworth
    limestone) in cottages at Blisworth; limestone in houses and Solomon's
    Bridge at Cosgrove
*Quaternary gravel pits along the Ouse Valley at Milton Keynes
*Chalk escarpment of the Chiltern Hills, seen from the Grand Union Canal
    south of Cheddington
Walking the Midshires Way by Ron Haydock and Bill Allen (Sigma Press, 2003)
Grand Union Canal (South): Towpath Guide by Nick Corble (thehistorypress.
    co.uk, 2005) – Milton Keynes to River Thames
Walking the Grand Union Canal from London to Birmingham by John Merrill
    (John Merrill Foundation, 2014)

15. Through the Chalk: Chiltern Hills to the River Thames

*Tring Cutting through the chalk (SP941126)

*Bradenham, Bucks (SP829970) – sarsen stones on the village green; flint, sarsen, Portland stone in St Botolph's Church and graveyard

*Whiteleaf Hill, near Princes Risborough (SP823039) – wide views of the Vale of Aylesbury, effects of the Ice Age ice sheets and meltwaters

16. Concrete and Clay: River Thames to Canvey Island

*Chafford Gorges Nature Discovery Park, Essex RM16 6RW – 'Reading the rocks at Chafford Gorges': four downloadable guides (essexwt.org.uk/reading-the-rocks)

*Thameside Nature Discovery Park, Mucking Wharf Road, Stanford-le-Hope SS17 0RN (essexwt.org.uk/nature-reserves/thurrock-thameside) – landscaped landfill site, now a nature reserve

Thames Path in London by Phoebe Clapham (Aurum Press, 2018)

17. Breaking the Wall: Wallasea Island

*Wallasea Island, Rochford, Essex SS4 2HD – managed coastal realignment; London Clay spoil from Crossrail excavations used to create islands, walls, lagoons; a magnificent RSPB reserve and a bold experiment in working with, not against, nature

Downloadable trail guide at rspb.org.uk/reserves-and-events/reserves-a-z/wallasea-island

Roach Valley Way – downloadable guide at essexhighways.org/uploads/files/roach_valley_walk.pdf

Glossary

Words in italics are explained elsewhere in the Glossary

A

alluvium clay, gravel, sand or *silt* deposited by a river on a flat plain

andesite a fine-grained *igneous* rock formed from a thick flow of *lava*

Anglian Stage a phase of *glaciation* in Britain (480,000–423,000 years ago) when the ice reached further south than before or since

Anthropocene disputed, ongoing epoch of geological time, the latest stage of the current Holocene era, during which human activity has started to have a significant impact on the planet's climate and ecosystems

anticline a fold in rocks that slopes downwards from the crest

apron reef knoll a knoll of *limestone* laid down at the rim of a *limestone* platform

B

barytes mineral form of barium sulphate, often found as 'waste' material in lead-mine spoil

basalt *magma* that has cooled and solidified as it flows on the Earth's surface

basement rocks rocks that lie above the Earth's *mantle* and below all other rocks

bedding plane meeting place of two *sedimentary* deposits

bing Scottish term for a mine's spoil heap

biosphere all the parts of the Earth where life exists

bituminous containing bitumen, a dark inflammable material, e.g. tar

blanket bog bog that is fed by rainwater and moisture in the air

Blue Lias a sequence of *shale* and *limestone*, well known for its fossils such as ammonites, nautili and bivalves, laid down in southern Britain in late Triassic/early Jurassic times (around 200 *mya*–195 *mya*)

boulder clay a.k.a. glacial *till* – a thick mix of clay, gravel, sand and boulders spread by the movement of *glaciers*

brachiopod marine creature with hinged upper and lower shells

breccia coarse-grained rock made up of angular fragments of other rocks

brick-earth mixture of sand and clay, excellent for brick-making

C

calcareous containing *calcium carbonate*, lime or *chalk*

calcite a carbonate mineral, a component of *limestone*

calcium carbonate a chemical compound found in e.g. *limestone*, *chalk* and *marble*, as well as shellfish skeletons and seashells

caldera a large volcanic *crater*, usually formed when the ground subsides above the roof of a *magma chamber*

Carboniferous geological period (358.9 *mya*–298.9 *mya*) when swamp forests thrived, their dead plants and trees being compressed to form coal in many places

cementstone clayey *limestone* with a composition like hardened cement

chalk fine-grained *limestone*, softish and grey-white, made up of uncountable trillions of *coccoliths*, tiny single-celled creatures concealed inside protective spheres of *calcium carbonate* called *coccolithophores*

chert fine-grained rock nodules often found in *chalk* or *limestone*, rich in *silica* from the skeletons of sponges or plankton

clints cobbles of *limestone* separated by *grykes*, a feature of *limestone pavements*

coccolith, coccolithophore see *chalk*

cock's tails fossil traces (zoophycos) of undersea worms burrowing and feeding in the sand of a seabed

conglomerate coarse-grained *sedimentary* rock with fragments larger than 2 mm

crag and tail rock shaped by a *glacier*, with a hard ice-resistant crag on the side where the *glacier* initially met the rock and on the other side a long tail sculpted by the ice as it pushed past and away, e.g. Castle Rock in Edinburgh, Inchgarvie in the Firth of Forth, and North Berwick Law

crater bowl-shaped depression around the *vent* of a *volcano*

craton huge conglomeration of tectonically inactive land masses

cross-bedding slanted layers of *gritstone*, formed where sediment was deposited on the sloping side of a bed

crust the layer of the Earth's surface that lies above the *mantle*

crystallizing process whereby liquid *magma* forms into solid rock

D

deposition laying down successive layers of *sedimentary* rocks and soils

diorite an *igneous* rock formed when *magma* cools slowly underground and forces itself into or between older rocks

displacement when the forces caused by a *fault* shift rocks from one place to another

dolerite an *intrusive igneous basalt*, fast-cooling and dark in colour, e.g. the Whin Sill

downthrow a downward displacement of rock along a *fault*

drumlin rounded hillock of gravel, rubble, etc. deposited by glacial ice

Dunbar marble *limestone* full of pale coral and shellfish fossils, well displayed on the southeast Scottish coast at Barns Ness

dyke rapidly cooled tongue of *basalt* cutting through another type of rock

dyke swarm a cluster of *dykes* heading in the same direction

E

edge cliff of *gritstone*

ejecta debris thrown out when e.g. a meteorite impacts with the Earth and forms a *crater*

erosion the process of e.g. wind or water wearing away rock and transporting it

erratic a piece of rock transported from one place to another, usually by a *glacier*

extinction event an event that causes mass deaths and the extinction of many species, e.g. an asteroid strike or massive volcanic eruption

F

fault a fracture in a mass of rock leading to *displacement*

feldspar minerals rich in *silica* and oxygen that make up over half the Earth's *crust*

felsic *igneous* rock such as *granite* that's rich in lighter-coloured elements, e.g. sodium, potassium, silicon and oxygen

ferruginous very rich in iron

fireclay a *sedimentary mudstone*, also known as *seat-earth*, found just below coal deposits, useful in making ceramics

flagstone *sedimentary* rock that splits into even layers along its *bedding planes*

flint rock of tiny crystals made of *silica* that filled cavities and burrows in *chalk*

fluorite mineral deposited in veins associated with metallic ores

fluorspar name for *fluorite* when it is processed for use, e.g. in steel-making to help prevent the cooling metal from cracking

G

gabbro a coarse crystalline cousin of *basalt* that cooled deep in the Earth's *crust*

gangue worthless material extracted during mining

Gault Clay sequence of *mudstones*, *siltstones* and clay deposited in the Cretaceous period *c.* 112–100 *mya*

geothermal fluids highly mineralized fluids at great temperatures

glacial valley valley sculpted into a U shape by glacial action, e.g. High Cup near Dufton, Cumbria

glaciation movement of an ice sheet across a land surface

glacier large body of ice or snow that forms on land and moves slowly downhill

glauconite iron potassium mineral that imparts a green colour to rock, e.g. *greensand*

gneiss *metamorphic* rock, often very ancient, formed by the action of high temperature and great pressure on *igneous* or sometimes *sedimentary* rock

granite *igneous*, *intrusive* rock, coarse-grained and full of *quartz* and *feldspar*

greenhouse gases gases in the atmosphere that trap heat, e.g. carbon dioxide, *methane*, nitrous oxide, and also man-made gases such as hydrofluorocarbons (HFCs)

greensand soft *chalk*, sandy rock deposited in the mid-Cretaceous period *c.* 125–100 *mya*

greenstone ancient volcanic rock coloured green by minerals, e.g. actinolite and chlorite

greywacke a hard, dark-coloured *sandstone* formed when submarine landslides carried grains from shallow into deeper water

gritstone see *millstone grit*

gryke a gully between *clints* in *limestone pavements*

H

hornfels rock baked by proximity to nearby *magma* at *c.* 700–800 °C is metamorphosed into hornfels

horst slab of rock raised into an upland

hush gash in a hillside made by releasing dammed-up water to scour away vegetation and expose suspected veins of minerals, e.g. lead

I

ice sheet an expanse of ice that has spread across land

igneous rock formed when molten rock, e.g. *magma*, cools and solidifies

ignimbrite rock mixture of volcanic ash and glass

interglacial period between Ice Ages

intrusion occurs when *magma* pushes between other rocks, then cools and solidifies

intrusive see *intrusion*

ironstone *sedimentary* rock rich in iron ore

K

karst cracked-looking landscape of *limestone* rocks dissolved by rainwater, e.g. the *limestone pavements* around Ingleborough, North Yorkshire

Kimmeridge Clay dark *mudstone* which breaks thinly along its *bedding planes* and is full of fossils

L

lamprophyre dark *igneous intrusion* full of magnesium and iron

lapilli droplets of molten rock that cool and harden as they fall from the sky after a volcanic eruption

lava molten rock or *magma* that flows across the surface of the Earth

limestone *sedimentary* rock mainly composed of *calcium carbonate*, usually full of tiny fossils – small shells and fragments of marine creatures

limestone pavement flat expanse of *limestone* weathered into *clints* and *grykes*, e.g. at Malham Cove

lithic made of tiny fragments of other rocks

London Clay mix of silty clay and sandy clay, deposited *c.* 56–49 *mya*

lycopods primitive plants, vascular (with circulatory systems), first appearing on land *c.* 425 *mya*, the ancestors of trees. Stumps displayed in Glasgow's Fossil Grove

M

mafic *igneous* rocks such as *basalt* that are rich in dark minerals, e.g. magnesium and iron

magma molten rock from beneath the Earth's *crust*

magma chamber a pool of *magma* beneath the Earth's *crust*

managed realignment man-made reshaping of a land margin, e.g. a coastline, to accommodate changes in sea levels brought about by climate change

mantle semi-solid material that lies between the Earth's hot, dense core and its outer *crust*

marble *limestone* metamorphosed by heat and pressure

marl earth-like material rich in clay and *silt*, formed in both salt and fresh water

matrix the material, e.g. clay or *silt*, that binds together elements, e.g. grains of *quartz*, that form a rock, e.g. *sandstone*

Mesolithic the Middle Stone Age, *c.* 11,600–6,000 years ago

metallophyte plants that can grow in ground impregnated with heavy metals

metamorphic rock that is transformed by great heat, great pressure or exposure to mineral fluids

methane a *greenhouse gas*, often produced by geological processes such as volcanic eruptions

mica minerals in thin layers that form part of many *metamorphic* rocks, e.g. *granite*, *gneiss* and *schist*

microdiorite a dark-coloured *igneous* rock produced when *magma* cools rapidly on contact with cold water

millstone grit coarse abrasive rock, perfect for use in the manufacture of millstones, laid down in the Carboniferous period *c.* 320 *mya* when tropical rivers spread layers of sand between layers of mud

moraine rock and/or soil left behind when a *glacier* moves on

mudstone clay or mud compressed into fine-grained *sedimentary* rock

mya million years ago

mylonite a *metamorphic* rock ground up and compressed into thin sheets, formed between two strata when the upper one is moving against the lower

N

Neolithic the New Stone Age, *c.* 6,300–4,000 years ago

O

Old Red Sandstone a thick sequence of red-coloured rock formed on land during the Devonian period, *c.* 416–359 *mya*

olivine a green-coloured mineral found in *igneous* rocks

orogeny mountain-building episode

outcrop where rocks are exposed at the surface of the Earth

Oxford Clay fine-grained *sedimentary* formation deposited under the sea in the Jurassic period *c.* 166.1–157.3 *mya*

P

Palaeolithic the Old Stone Age, *c.* 900,000–11,600 years ago

peat the compressed remains of plants that have not been able to decompose

pelite fine-grained rock metamorphosed from *mudstone* and *siltstone*

playa lake a temporary lake that forms in wet periods in a dry, flat area at the lowest part of a desert basin

pluton an *intrusive* tongue of *igneous* rock

Portland stone fine-grained white-grey *limestone* formed from naturally occurring *calcite* in seawater, and particles (e.g. shell fragments) of sea creatures

potash potassium-rich minerals

pothole round or oval hole formed when rotating particles in a river hollow out the *limestone* bed

Precambrian the earliest period of Earth's history, from the first formation of rocks down to *c.* 541 *mya*

productid large fossil *brachiopod*

psammite coarse-grained rock metamorphosed from *sandstone*

puddingstone *conglomerate* rock with rounded pebbles in a *sedimentary matrix*

pumice *igneous* rock, instantly cooled into solidified froth during explosive volcanic activity

pyroclastic flow mixture of rock, ash and gas at *c.* 800–1,500°C that races across the ground away from an explosive volcanic eruption

Q

quartz a mineral occurring in most acid rocks, e.g. *granites*, *shales* and *sandstones*

quartzite a rock formed when *quartz* in *sandstone* is *metamorphosed* by great heat or pressure

Quaternary the youngest period of Earth's history, *c.* 2.6 *mya* to the present

R

reef basin a depression in a coral reef

reef knoll pile of *calcareous* material that accumulated under the sea, now exposed on land

rhyolite a fine-grained, light-coloured volcanic rock

ring fault a circular *fault* marking the site of a collapsed *caldera*

roof pendant remnant of rock that was penetrated by an *igneous intrusion*, e.g. a *pluton*

S

sandstone *sedimentary* formation of mineral or rock grains bound together by a *silt* or clay *matrix*

sarsen folk name, probably derived from 'Saracen', for boulders of sand/gravel cemented together by *silcrete* and transported from their place of origin by glacial ice or by gravity

schist *metamorphic* rock composed of thin plates of e.g. *mudstone* or volcanic *tuff*, usually rich in *mica*

seat-earth the fossilized soil of an ancient wetland, often found just below seams of coal

sedimentary rocks formed of fragments of organisms, e.g. shellfish, or of older rocks, usually laid down in distinctive horizontal layers

semi-pelite *pelite* rock with a crystalline texture that contains *mica* and *quartz* minerals

serpentine *metamorphic* rock with snaky patterns of green and red, formed at folds in the Earth's *crust*

shake hole depression in the ground where the rock has collapsed underground

shale fine-grained rock composed of compressed clay, *silt*, mud and organic material

silcrete a glue-like compound of gravelly *sandstone* and *silica* that sets as hard as concrete

silica a pale-coloured element, SiO_2, found mostly in sand, *quartz*, *chert*, *flint*, etc.

sill a surge of *igneous* matter, e.g. *basalt*, that has pushed between layers of other rocks, then cooled to form an *intrusive* wedge, e.g. the Whin Sill

silt a *sedimentary* deposit composed of tiny particles of rock and minerals

siltstone a form of *mudstone* with a low clay content

sinkhole a.k.a. swallow hole, a circular depression into which water drains, as with the plughole in a kitchen sink

slate a fine-grained *metamorphic* rock, easily split into thin, strong layers

spreading centre where the Earth's *crust* thins and splits at the point where two *tectonic plates* are moving apart

stromatolite a layered deposit of *limestone* formed by clusters of primitive, single-celled organisms

subduction where one *tectonic plate* slides under another and is reabsorbed into the Earth's *mantle*

sugar limestone *limestone* metamorphosed by contact with *magma* into crumbly *marble*-like rock, e.g. in Upper Teesdale

sulphur dioxide an 'indirect' *greenhouse gas*, product of volcanic activity, which contributes to warming the atmosphere

supercontinent a conglomeration of the Earth's *cratons* to form one huge land mass

superimposed drainage a process in which a river switches to a new course that mirrors an older one lying below

swallow hole see *sinkhole*

syncline a fold in rocks that slopes upwards from the lowest point

T

tectonic plates large segments of the Earth's *crust* that move around the surface of the planet, occasionally colliding

terrane fragment of the Earth's *crust* separated from its neighbours by *faults*

tetrapod four-legged amphibian pioneers on land

thrust force emanating from a *fault* that can remould, distort and displace rocks

till see *boulder clay*

tuff solidified volcanic ash

turbidite slurry-like sediment deposited by a turbid current

U

unconformity a gap or missing link between geological layers, e.g. at Siccar Point where *Old Red Sandstone* (*c.* 370 *mya*) lies directly on top of *greywacke* (*c.* 435 *mya*)

V

vent gap in the Earth's *crust* through which *magma* and gases erupt

volcanic plug roughly cylindrical masses of solidified *magma* that remain where the *vent* of a *volcano* has been

volcano cone of volcanic material built up around a *vent*

W

whinstone hard dark *igneous* rock, e.g. *dolerite* and *basalt*

Y

Yoredale Group 'sandwiches' of *sandstone, limestone* and *mudstone* deposited during the Carboniferous period – see the North Yorkshire Dales, and the mini-mountains of Ingleborough, Whernside and Penyghent

Z

zircon crystals silicate crystals in *igneous* and *metamorphic* rock, so durable and long-lasting that rocks can be dated from their formation

Credits

Extract on p. xii from 'Evening Alone at Bunyah' by Les Murray, from *New Collected Poems* by Les Murray (Carcanet Press, 2003).

Extract on p. 233 from 'Northern Stone', from *No Time for Cowards* by Phoebe Hesketh, copyright © Phoebe Hesketh 1952. Extract reproduced by permission of The Random House Group Ltd.

Lyrics on p. 237 from 'Wuthering Heights', from *The Kick Inside* (EMI Records, 1978), written by Kate Bush.

Maps by LovellJohns Ltd.

Timeline on p. ix by Global Blended Learning Ltd.

Map on p.1 of picture section copyright © Philip's, a division of Octopus Publishing Group.

All photographs are from the author's collection.

Index

Christopher Somerville is the walking correspondent of *The Times*. He is one of Britain's most respected and prolific travel writers, with forty-two books, hundreds of newspaper articles and many TV and radio appearances to his name. He lives in Bristol.